Numerical Heat Transfer and Fluid Flow 2021

Numerical Heat Transfer and Fluid Flow 2021

Editor

Artur Bartosik

MDPI • Basel • Beijing • Wuhan • Barcelona • Belgrade • Manchester • Tokyo • Cluj • Tianjin

Editor
Artur Bartosik
Kielce University of
Technology,
Poland

Editorial Office
MDPI
St. Alban-Anlage 66
4052 Basel, Switzerland

This is a reprint of articles from the Special Issue published online in the open access journal *Energies* (ISSN 1996-1073) (available at: https://www.mdpi.com/journal/energies/special_issues/numerical_heat_transfer_and_fluid_flow).

For citation purposes, cite each article independently as indicated on the article page online and as indicated below:

LastName, A.A.; LastName, B.B.; LastName, C.C. Article Title. *Journal Name* **Year**, *Volume Number*, Page Range.

ISBN 978-3-0365-4091-7 (Hbk)
ISBN 978-3-0365-4092-4 (PDF)

Contents

About the Editor

Artur Bartosik, Dr. Hab. Eng. Prof. KUT, is an appointed researcher and academic teacher in the Department of Production Engineering at the Kielce University of Technology. Together with his team, he works to establish the structure for university-wide multidisciplinary collaborations in fluid mechanics, rheology, and renewable energy. His scientific interests are focused on experiments, modellings, and simulations of single- and two-phase flows, including convective heat transfer. As a researcher, Artur spent 2 years at the University of Saskatchewan in Canada and also spent some time at Ilmeanu Technical University in Germany. He is coeditor of *WSEAS Transactions on Fluid Mechanics* and Ass. Editor of the *Journal of Hydrology and Hydromechanics*. He gives lectures on fluid mechanics; heat transfer; fluid flow machinery; hydrotransport; renewable energy sources; transnational technology transfer; and applied fluid mechanics.

Editorial

Numerical Heat Transfer and Fluid Flow: A Review of Contributions to the Special Issue

Artur S. Bartosik

Department of Production Engineering, Faculty of Management and Computer Modelling, Kielce University of Technology, Al. Tysiaclecia P.P. 7, 25-314 Kielce, Poland; artur.bartosik@tu.kielce.pl

Citation: Bartosik, A.S. Numerical Heat Transfer and Fluid Flow: A Review of Contributions to the Special Issue. *Energies* **2022**, *15*, 2922. https://doi.org/10.3390/en15082922

Received: 11 April 2022
Accepted: 14 April 2022
Published: 15 April 2022

Publisher's Note: MDPI stays neutral with regard to jurisdictional claims in published maps and institutional affiliations.

1. Introduction

The paper contains a summary of successful invited papers addressed to the Special Issue on 'Numerical Heat Transfer and Fluid Flow', which were published in 2021 in the scientific journal '*Energies*'. Invitations were addressed to specialists from all over the world who deal with mathematical modeling, simulations, and experiments on heat and/or fluid flow. The submitted papers regarded the solution of problems of scientific and industrial relevance in a specific field of heat transfer and fluid transportation, including natural resources, technical devices, industrial processes, etc. Papers addressed to the Special Issue not only solved specific engineering problems, but served as a catalyst on future directions and priorities in numerical heat transfer and fluid flow. Most papers dealt with heat transfer in single-phase flow of air, in particular technical devices, while part of them regarded liquid and solid–liquid flows. Reliable predictions require reliable measurements; therefore, the majority of the papers presented experimental data and validation of mathematical models.

The importance of heat and fluid flow is still growing in all aspects of our lives, starting from nature and ending with industrial processes. In the era of digital transformation, which includes converting any process into a quantified format suitable for analysis, there is an increasing demand for modeling, simulations, and experiments on heat exchange in fluid flow for a variety of single and multiphase flows, and also boundary conditions [1]. Thanks to computational fluid dynamics and its commercial packages, especially Ansys, we can design and optimize various industrial processes. The increasing understanding of heat and mass transfer phenomena has contributed significantly to the development of new methods and techniques for solving and effectively managing many engineering processes.

Formulating any problem of prediction of heat transfer and/or fluid flow requires developing a physical model first. The next step is developing a mathematical model which fulfills the assumptions stated in the physical model and defines boundary and initial conditions. The mathematical model should be based on general governing equations, like continuity, Navier-Stokes, and energy equations. The equation set can be solved analytically—which is complicated and impractical—or numerically. If numerical methods are considered, we can use approaches like direct numerical simulation (DNS), for instance. Such a method is time-consuming, expensive, and not practical for many engineering applications. Other methods like, for instance, modelling of turbulence, which uses random averaged Navier-Stokes equations (RANS), or large eddy simulation (LES) were proved for a variety of engineering applications and are less time consuming and less expensive; however, the set of equations require closure. The problem of closure requires additional equation or equations, like those proposed by turbulence models. Requirements for turbulence models, formulated by the honorable founder of computational fluid dynamics (CFD), that is, Dudley Brian Spalding, are the following: universality, economy, extensionality, and reality [2,3]. The ability to simulate heat transfer and/or fluid flow, which includes velocity, pressure, and temperature distributions, for engineering purposes, remains one of the main challenges in CFD.

Considering the heat exchange between a transported fluid and the surrounding, we recognize methods and techniques focused on the enhancement of heat transfer, named passive or active. Passive methods, such as increasing heat transfer area and/or temperature difference, shaping an insert with dedicated perforation, or mechanically deformed pipes, have been studied for several years and have become commercial solutions [4–9]. Active methods, such as air injection, bubble or vortex generation, or proper pulsation, can lead to increased heat transfer coefficient, and, finally, can produce increased heat transfer process [10–13]. Some of such methods have been demonstrated by contributors to the Special Issue.

2. Review of Contribution to the Special Issue

The improvement of heat transfer by the use of ribs has been studied for several decades. It is known that the pitch and shape of the rib, the angle of attack of the rib, and the coefficient of channel blockage have a significant effect on heat transfer [14]. Joon Ahn et al. [15] made the LES simulation of the dependence of conjugate heat transfer in a ribbed channel on the thermal conductivity of the channel wall. The authors noted that for such a case, RANS underpredicts heat transfer and does not accurately predict local peaks, while the same predictions using LES give more accurate results. The authors considered a channel wall with a thickness three times the height of the rib. Starting points for their mathematical model were continuity, Navier-Stokes, and energy equations in 3D form, which were transformed to dimensionless governing equations. The authors assumed that the flow is fully developed with periodic boundary conditions in the stream direction. All simulations were performed for 10,000 time steps to reach a steady state. The simulations were positively validated with the available data available in the literature. On the basis of the simulations, the authors analyzed the conjugate heat transfer characteristics in a ribbed channel. The authors made simulations in a wide range of conductivity ratios between the gas turbine blade material and air. Taking into account the results of simulations of time-averaged temperature and heat transfer distribution, they analyzed whether the mechanism responsible for promoting heat transfer under pure convection conditions is valid for a variety of conductivity ratios. The authors concluded that if the conductivity ratio between the solid wall and the fluid exceeded 100, the heat transfer characteristics were similar to those under isothermal conditions, and the vortices at the corners of the ribs strongly influenced the convective heat transfer. For a conductivity ratio below 100, vortices located in the corners played an important role in heat transfer [15]. The authors concluded that the 'thermal resistance of the solid wall of the channel to convective heat transfer was observed in the turbulent flow regime' [15].

Heat pipes have been extensively developed for some decades. The main advantages are no moving parts, high reliability, and fair efficiency. Working fluid transport exists naturally, does not require energy input, and can transfer heat for significant distances. Recently, heat pipes heat exchangers have gained popularity in heat recovery applications such as air conditioning, dehumidifiers in air conditioning systems, technological processes, etc. [16]. Górecki et al. [17] conducted experimental and numerical studies on heat pipe heat exchangers with individually finned heat pipes. The authors conducted a study on modeling, design, and experiments on a heat pipe heat exchanger comprised of individually finned heat pipes utilized as a recuperator in small air conditioning systems with airflow between 300 m^3/h and 500 m^3/h. Using the available algebraic correlations, the authors developed a thermal heat pipe heat exchanger model. Based on their previous research, they used R404A refrigerant as a working fluid. The mathematical model consists of a set of algebraic equations and semi-empirical functions. On the basis of parametric studies, the authors concluded that 20 rows of finned heat pipes in the staggered arrangement guarantee stable heat exchanger efficiency equal about 60%. The authors emphasized that the designed heat pipe heat exchanger made of an individually finned heat pipe bundle is a competitive choice to the continuous plate-finely alternative. On the basis of the parametric studies, the authors designed and then constructed the heat exchanger.

Measurements made on the experimental rig were referred to the predictions made by the mathematical model, and they noted that the relative difference was about 10%. In accordance with the authors statement, the method can improve a heat pipe heat exchanger efficiency significantly. Future research should focus on intensifying heat transfer by using turbulators in an individually finned tube bundle located between heat pipes [17].

Recent interest in the liquefaction of natural gas, which occupies only 1/600 of its volume, is due to the fact that it is easier and more economical to transport and store [18–20]. Some researchers stressed that 'the development of small-scale energy in the coming years is expected to be associated with the widespread use of liquefied natural gas, which is recognized as one of the most promising types of energy carriers' [21]. By taking into account the real thermophysical properties of the gas, Avramenko et al. [21] analysed the method for solving steady-state natural convection of van Der Waals gas near a vertical heated plate. The authors proposed a novel simplified form of equations in analytical form for real gases, enabling the estimation of the effects of the dimensionless van der Waals parameters on the normalized heat transfer coefficient and Nusselt number. Using the integral method for the momentum and energy equations, they obtained an approximate analytical solution. The authors demonstrated changes in heat transfer intensity, which are due to effects that are considered by the van der Waals equation of state, but not by the ideal gas equation [21].

Some research is developing a fluidic oscillator. The fluid oscillator usually has two feedback channels and does not possess moving parts and can improve aerodynamics, such as lift forces, mixing processes, and heat transfer, for instance [22]. Kim and Kim [23] studied numerically a fluid oscillator with a bent outlet nozzle. The authors analyzed the influence of various mounting conditions, such as the arrangement and installation angles of the fluidic oscillator in the range from 0° to 40°, on the characteristics of the oscillator with and without external flow. The authors performed analyses for air which has been treated as an ideal gas. They considered 3D unsteady RANS equations with a shear stress transport turbulence model. The authors used an Ansys commercial CFD software. They concluded that the pitch angle of fluidic oscillators had the most sensitive effect on the flow control [23].

The construction of tube heat exchangers is complex. Cross-thin metal sheets, tubes, and fin pitches, with different sizes and numbers of rows, cause the complexity of fluid flow [24]. Usually, each row operates in a different way, which has an effect on the level of turbulence. For these reasons, the characteristics of tube heat exchangers are determined mainly experimentally. Marcinkowski et al. [25] presented a method to determine individual correlations for the air side Nusselt numbers on each row of tubes for a four-row finned heat exchanger with continuous flat fins and round tubes in a staggered tube layout. The authors' method was formulated using CFD modelling; however, the description of the model is limited. The authors concluded that their approach enables the selection of the optimum number of tube rows for a given heat output of the heat exchanger. The authors stated that the approach allows the reduction of investment costs of constructing heat exchangers by decreasing the tube row number. In addition, operating costs can be reduced as a result of reducing air pressure losses [25].

The increase in efficiency of flow machines dedicated to the transport of gases is mainly associated with blade impeller and labyrinth seals. This is especially important if high-power generating machines are considered. Liang et al. [26] proved that a 1% increase in seal-tooth clearance height causes a significant decrease in multistage axial compressor performance and efficiency. Some researchers are using CFDs to design seals of higher leak tightness. Joachimiak [27] proposed a method of aerodynamic sealing to reduce leakage by matching the seal geometry to the flow. The authors considered a staggered labyrinth seal for steady flow conditions. The method contains CFD predictions assuming that the air is an ideal gas. The author used the RANS method for calculations with the k-w SST turbulence model for 2D axisymmetric geometry. His calculation domain consisted of 354,000 elements. The author considered two approaches, that is, changed and unchanged seal height. The author concluded that the newly designed geometry reveals an almost stable relative reduction in the leakage rate, irrespective of the pressure ratio upstream and downstream of the seal [27]. His conclusions require validation. However, such measurements are complex and difficult to perform.

High battery costs, which contribute more than 40% of the total price of electrical vehicles, and high charging and discharging rates, variable frequency of charging, and finally battery packing, a proper cooling system and safety, make research in this field very desirable [28]. Widyantara et al. [29] made an approach to optimize the design of the lithium-ion battery pack for electric vehicles to meet the optimal operating temperature using an air-cooling system by modifying the number of cooling fans and the inlet air temperature. A 3D numerical model of the packing of the 74 V and 2.31 kWh batteries and CFD simulations based on the lattice Boltzmann method have been applied. Furthermore, the authors developed a battery thermal management system based on consideration of temperature distribution and power consumption. The authors concluded that three cooling fans with 25 °C inlet air temperature gave the best performance, with low power required. The authors' findings could be helpful in developing a standardized battery packing module and in designing low-cost battery packing for electric vehicles. The authors emphasized that in future work, an investigation is recommended on the effect of employing variable speed for cooling fans and a study on structural strength and water protection for air conditioning systems [29].

In recent years, we have observed interest in developing new oil-cooled motors with high power density, low vibration and noise, strong overload capacity, and high efficiency. Such oil-cooled motors require modern heat exchangers with an enhanced heat transfer process. The most typical liquid used in oil-cooled motors is oil, which is relatively cheap and emphasizes long service life. Some researchers are using vortex generators in radiators to improve overall heat transfer performance [30]. Junjie Zhao et al. [31] performed a numerical study on the influence of the vortex generator arrangement on the enhancement of heat transfer in oil-cooled motors. The authors carefully formulated a physical model, and then developed a mathematical model that constituted continuity, momentum, and energy equations. The closure problem was solved using the standard k-ε turbulence model. The set of equations for the 2D case was solved using the commercial software Fluent. They performed grid-independent tests concluding that the grid number equal to 6,850,000 is sufficient for simulations. Validation of the mathematical model, which includes frictional coefficient and Nusselt number, is limited, which is due to the fact that the numerical results were compared with other numerical predictions available in the literature. To increase the heat transfer coefficient, the authors studied the influence of an attack angle of the vortex generator. The authors concluded that for a single pair of rectangular vortex generators, the attack angle equal to 45° gives the best results in enhancing heat transfer [31].

The water hammer phenomenon can lead to pipeline system failures; however, such a phenomenon depends on the material of the pipeline [32]. Steel pipes are better recognized in the literature, while viscoelastic pipelines are less. Kubrak et al. [33] made experiments and predictions on hydraulic transients in a viscoelastic pipeline system with sudden cross-sectional changes. The authors focused on high-density polyethylene pipelines. The authors investigated the influence of sudden cross-sectional changes in a high-density polyethylene pipeline system on pressure oscillations during the water hammer phenomenon. The authors recorded pressure changes downstream of the pipeline system during a valve-induced water hammer. They formulated a mathematical model which constitutes the continuity and momentum equations for non-steady and one-dimensional flow. A set of transient flow equations for a polymeric pipe was numerically solved using the MacCormack explicit method. The authors concluded that for high mass flow rates, the jet frequency increases with the bending angle; however, at a bending angle equal to 40°, the oscillation of the jet disappears. The authors also observed that the effect of external flow on the frequency increases with the increase of the bending angle. The predicted numerical pressure fluctuations showed satisfactory agreement with their measurements in terms of both phase and amplitude. The authors noted that more work needs to be done to study the use of the water hammer in variable property pipeline systems. Future studies should also focus on simulating the water hammer in serially connected pipes with various inner diameters and made of different materials [33].

Saving energy is a main goal of engineers. This includes, for instance, fluid transportation and control of hydraulic systems. Controlling and adjustment of hydraulic elements, such as valves, pumps, receivers, etc., is one of the possible ways which effect the efficiency of such systems [34]. In hydrostatic systems, efficiency can be improved in a number of ways, depending on whether the system features fixed or variable displacement pumps. For example, in the system with fixed displacement pumps controlled by the throttle method, the main approach is to limit the operation of the safety valve. Bury et al. [35] performed simulations and experiments with control signals of various shapes and feedback from the hydraulic system. The authors analyzed the pressure at the pump inlet and outlet, the flow rate, and the rotational speed of the hydraulic motor as functions of time. The authors built a simple mathematical model taking into account the Hagen-Poiseuille equation for laminar flow and the Bernoulli equation for turbulent flow. The model constitutes a continuity equation at particular points in the hydraulic circuit and the equilibrium equation of torque on the shaft of the hydrostatic motor. In accordance with the authors statement, their method of modelling the opening characteristics of the proportional spool valve allows us to optimize the start-up process in terms of the execution of the objective function for the parameters like, for instance, start-up time, reaction time, or energy efficiency of the system. On the basis of the analyses, the authors concluded that the maximum pressure value at the inlet port of the hydraulic motor during its start-up can be modified by adjusting the shape of the proportional spool valve (symmetrical, asymmetrical). This solution also reduces the noise caused by the transmission during start-up and reduces the load on elements of the transmission, which extends its operational life [35].

Thermal conductivity is essential when rapid changes in temperature occur in some parts of mechanical machines. This phenomenon exists in mechanical face seals, especially during startup [36]. Błasiak [37] performed an analysis of heat transfer for non-contact mechanical face seals using the variable order derivative approach. The author proposed a physical and next mathematical model for noncontact mechanical face seals for the conduction of heat from the liquid film in the gap to the rotor and the stator, followed by convection to the water surrounding them. The author used a variable-order derivative time fractional model to describe heat transfer. His equation of heat transfer depends on time and was solved analytically; however, the characteristic features of the equation were determined through numerical simulations. The objective of the study was "to compare the results of the classical equation of heat transfer with the results of the equations involving the use of the fractional order derivative". Author concluded that his method allows one to predict the heat transfer phenomena occurring in non-contacting face seals, especially during the startup. The author emphasized that his mathematical approach, which is based on the fractional differential equation, is suitable to develop more detailed mathematical models for similar phenomena [37].

It is well known that cavitation is a dangerous process that affects pipelines and flow machines [38]. Urbanowicz et al. [39] conducted research on modeling transient pipe flow in plastic pipes with a modified discrete bubble cavitation model. In the physical model of unsteady pipe flow, they assumed the importance of three phenomena: unsteady wall shear stress, vaporous cavitation, and pipe wall retarded strain. The authors considered plastic pipelines characterized by the retarded strain (RS) that occurs on the wall of these pipes. Using the convolution integral of the local derivative of pressure and the creep function that describes the viscoelastic behavior of the pipe wall material, they calculated the RS. The authors formulated the equations of a discrete bubble cavity model by using the continuity equations for the gas and liquid phases separately, and the momentum equation for two-phase vaporous cavitation. They transformed the set of partial differential equations into a set of ordinary differential equations. The authors concluded that the modified unsteady discrete bubble cavity model allows one to simulate the pressure and velocity waveforms in which vapor areas appear as a result of the cavitation phenomenon in plastic pipes. Such a model seems to be suitable for complex networks, such as water supply, oil hydraulics, heating, etc. The authors noted that the comparison of computed results with measurements showed that the novel discrete bubble cavity model predicts pressure and velocity waveforms, including cavitation and retarded strain effects, with sufficient precision. The authors noticed that the influence of unsteady friction on the damping of pressure waves was much smaller than the influence of retarded strain [39].

Slurry pipeline transport is widely used in many industrial applications. In a situation where long distances of transportation are considered, predictions of heat exchange are usually focused on friction losses, heat transfer coefficient, velocity, and temperature distributions. Solving such a problem requires a proper physical model which includes physical properties of solid and carrier liquid phases, solid concentration, regime of flow, like homogeneous or heterogeneous, laminar or turbulent, rheological model, and boundary conditions. If the solid phase is very fine, the slurry demonstrates a yield shear stress, which increases with the increase in solid concentration. In such a case, it is quite common that the damping of turbulence appears [40]. Bartosik [41] proposed a mathematical model of heat transfer in turbulent flow of fine-dispersive slurry, including a specially designed wall damping function. The mathematical model constitutes the continuity, momentum, and energy equations. The closure problem was solved by taking into account the k-ε turbulence model and the rheological model. The study was focused on developing a new correlation of the Nusselt number for turbulent flow of fine dispersive slurry that exhibits yield shear stress and damping of turbulence. The new correlation of the Nusselt number includes Reynolds and Prandtl numbers, solid volume concentration, and dimensionless yield shear stress. The mathematical model was solved numerically and validated for heat transfer in carrier liquid flow only, as there are no data available for heat exchange in such slurries. The study demonstrates substantial differences between the slurry and velocity distributions at the pipe wall. The author concluded that for the same bulk velocity of the slurry, the Nusselt number decreases with an increase in solid volume concentration, and the highest rate of decrease in the Nusselt number is for a solid concentration below 10% by volume. The author noticed that more work needs to be done on examining the new Nusselt number for solid volume concentrations below 10% and greater than 30%, for different heat fluxes and pipe diameters [41].

3. Conclusions

Analyzing the papers contributed to the Special Issue, 'Numerical Heat Transfer and Fluid Flow', one can say that all papers are applied to specific engineering problems. All papers are encountered in fluid dynamics in machines, electronic packaging, chemical processes, and other related areas of mechanical engineering.

The majority of the papers dealt with simulations. For this reason, the papers presented physical models, which include major assumptions, physical properties of the flowing medium, boundary, and initial conditions, and also mathematical models. Mathematical models were formulated using conservation laws, such as the continuity, N-S, and energy equations. Using the boundary and initial conditions, the authors formulated the set of equations and solved them numerically, taking into account the convergence criteria and ensuring a grid-independent solution. Part of the models have been validated in full, while part have been validated in the limited scale. Some papers presented their own experimental data.

Through these approaches, the readers can find a variety of physical and mathematical models and can gain a better understanding of phenomena in the specific engineering applications of heat transfer and/or fluid flow, including interpretation of computed and measured results.

Funding: This research did not receive any specific grant from funding agencies in the public, commercial, or not-for-profit sectors.

Conflicts of Interest: The author declares no conflict of interest.

References

1. Patankar, S.V. *Numerical Heat Transfer and Fluid Flow*; Minkowych, W.J., Sparrow, E.M., Eds.; Taylor and Frances Inc.: Washington, DC, USA, 1980; p. 2014.
2. Spalding, D.B. *Turbulence Models for Heat Transfer*; Report No. HTS/78/2; Imperial College of London, Department of Mechanical Engineering: London, UK, 1978.
3. Spalding, D.B. *Turbulence Models—A Lecture Course*; Report No. HTS/82/4; Imperial College of London, Department of Mechanical Engineering: London, UK, 1983.
4. Fathinia, F.; Parsazadeh, M.; Heshmati, A. Turbulent forced convection flow in a channel over periodic grooves using nanofluids. *Int. J. Mech. Mechatron. Eng.* **2012**, *6*, 12–2782. [CrossRef]
5. Faruk, O.C.; Celik, N. Numerical investigation of the effect of flow and heat transfer of a semi-cylindrical obstacle located in a channel. *Int. J. Mech. Aerosp. Ind. Mechatron. Manuf. Eng.* **2013**, *7*, 891–896.
6. Sen, D.; Ghosh, R.A. Computational study of very high turbulent flow and heat transfer characteristics in circular duct with hemispherical inline baffles. *Int. J. Mech. Aerosp. Ind. Mechatron. Manuf. Eng.* **2015**, *9*, 1046–1051.
7. Zong, Y.; Bai, D.; Zhou, M.; Zhao, L. Numerical studies on heat transfer enhancement by hollow-cross disk for cracking coils. *Chem. Eng. Process. Process Intensif.* **2019**, *135*, 82–92. [CrossRef]
8. Nakhchi, M.E.; Esfahani, J.A. Numerical investigation of heat transfer enhancement inside heat exchanger tubes fitted with perforated hollow cylinders. *Int. J. Therm. Sci.* **2020**, *147*, 106153. [CrossRef]
9. Pandey, L.; Singh, S. Numerical Analysis for Heat Transfer Augmentation in a Circular Tube Heat Exchanger Using a Triangular Perforated Y-Shaped Insert. *Fluids* **2021**, *6*, 247. [CrossRef]
10. Bartosik, A. Numerical Modelling of Fully Developed Pulsating Flow with Heat Transfer. Ph.D. Thesis, Kielce University of Technology, Kielce, Poland, 1989.
11. Hagiwara, Y. Effects of bubbles, droplets or particles on heat transfer in turbulent channel flows. *Flow Turbul. Combust.* **2011**, *86*, 343–367. [CrossRef]
12. Bayat, H.; Majidi, M.; Bolhasan, M.; Karbalaie Alilou, A.; Mirabdolah, A.; Lavasani, A. Unsteady flow and heat transfer of nanofluid from circular tube in cross-flow. *Int. Sch. Sci. Res. Innov.* **2015**, *9*, 2078–2083.
13. Hamed, H.; Mohhamed, A.; Khalefa, R.; Habeeb, O. The effect of using compound techniques (passive and active) on the double pipe heat exchanger performance. *Egypt. J. Chem.* **2021**, *64*, 2797–2802. [CrossRef]
14. Yusefi, A.; Nejat, A.; Sabour, H. Ribbed Channel Heat Transfer Enhancement of an Internally Cooled Turbine Vane Using Cooling Conjugate Heat Transfer Simulation. *Therm. Sci. Eng. Prog.* **2020**, *19*, 100641. [CrossRef]
15. Ahn, K.; Song, J.C.; Lee, J.S. Dependence of Conjugate Heat Transfer in Ribbed Channel on Thermal Conductivity of Channel Wall: An LES Study. *Energies* **2021**, *14*, 5698. [CrossRef]
16. Tian, E.; He, Y.L.; Tao, W.Q. Research on a new type waste heat recovery gravity heat pipe exchanger. *Appl. Energy* **2017**, *188*, 586–594. [CrossRef]
17. Gorecki, G.; Łecki, M.; Gutkowski, A.N.; Andrzejewski, D.; Warwas, B.; Kowalczyk, M.; Romaniak, A. Experimental and Numerical Study of Heat Pipe Heat Exchanger with Individually Finned Heat Pipes. *Energies* **2021**, *14*, 5317. [CrossRef]
18. Banaszkiewicz, T.; Chorowski, M.; Gizicki, W.; Jedrusyna, A.; Kielar, J.; Malecha, Z.; Piotrowska, A.; Polinski, J.; Rogala, Z.; Sierpowski, K. Liquefied Natural Gas in Mobile Applications—Opportunities. *Energies* **2020**, *13*, 5673. [CrossRef]
19. Osorio-Tejada, J.; Llera-Sastresa, E.; Scarpellini, S. Liquefied natural gas: Could it be a reliable option for road freight transport in the EU? Renew. *Sustain. Energy Rev.* **2017**, *71*, 785–795. [CrossRef]
20. Staffell, I.; Scamman, D.; Abad, A.V.; Balcombe, P.; Dodds, P.E.; Ekins, P.; Shah, N.; Ward, K.R. The role of hydrogen and fuel cells in the global energy system. *Energy Environ. Sci.* **2019**, *12*, 463–491. [CrossRef]
21. Avramenko, A.A.; Shevchuk, I.V.; Kovetskaya, Y.Y.; Dmitrenko, N.P. An Integral Method for Natural Convection of Van Der Waals Gases over a Vertical Plate. *Energies* **2021**, *14*, 4537. [CrossRef]
22. Gregory, J.; Tomac, M.A. Review of fluidic oscillator development and application for flow control. *AIAA Pap.* **2013**, *2013*–2474. [CrossRef]
23. Kim, N.H.; Kim, K.Y. Effects of Bent Outlet on Characteristics of a Fluidic Oscillator with and without External Flow. *Energies* **2021**, *14*, 4342. [CrossRef]
24. Thulukkanam, K. *Heat Exchanger Design Handbook*, 2nd ed.; CRC Press: Boca Raton, FL, USA, 2013.
25. Marcinkowski, M.; Taler, D.; Taler, J.; Weglarz, K. Thermal Calculations of Four-Row Plate-Fin and Tube Heat Exchanger Taking into Account Different Air-Side Correlations on Individual Rows of Tubes for Low Reynold Numbers. *Energies* **2021**, *14*, 6978. [CrossRef]
26. Liang, D.; Jin, D.; Gui, X. Investigation of seal cavity leakage flow effect on multistage axial compressor aerodynamic performance with a circumferentially averaged method. *Appl. Sci.* **2021**, *11*, 3937. [CrossRef]
27. Joachimiak, D. Novel Method of the Seal Aerodynamic Design to Reduce Leakage by Matching the Seal Geometry to Flow Conditions. *Energies* **2021**, *14*, 7880. [CrossRef]
28. Huda, M.; Koji, T.; Aziz, M. Techno Economic Analysis of Vehicle to Grid (V2G) Integration as Distributed Energy Resources in Indonesia Power System. *Energies* **2020**, *13*, 1162. [CrossRef]

29. Widyantara, R.D.; Naufal, M.A.; Sambegoro, P.L.; Nurprasetio, I.P.; Triawan, F.; Djamari, D.W.; Nandiyanto, A.B.D.; Budiman, B.A.; Aziz, M. Low-Cost Air-Cooling System Optimization on Battery Pack of Electric Vehicle. *Energies* **2021**, *14*, 7954. [CrossRef]
30. He, Y.L.; Zhang, Y.W. Advances and Outlooks of Heat Transfer Enhancement by Longitudinal Vortex Generators. *Adv. Heat Transf.* **2012**, *44*, 119–185. [CrossRef]
31. Zhao, J.; Zhang, B.; Fu, X.; Yan, S. Numerical Study on the Influence of Vortex Generator Arrangement on Heat Transfer Enhancement of Oil-Cooled Motor. *Energies* **2021**, *14*, 6870. [CrossRef]
32. Soares, A.K.; Covas, D.I.; Reis, L.F. Analysis of PVC Pipe-Wall Viscoelasticity during Water Hammer. *J. Hydraul. Eng.* **2008**, *134*, 1389–1394. [CrossRef]
33. Kubrak, M.; Malesinska, A.; Kodura, A.; Urbanowicz, K.; Stosiak, M. Hydraulic Transients in Viscoelastic Pipeline System with Sudden Cross-Section Changes. *Energies* **2021**, *14*, 4071. [CrossRef]
34. Xiong, S.; Wilfong, G.; Lumkes, J.J. Components Sizing and Performance Analysis of Hydro-Mechanical Power Split Transmission Applied to a Wheel Loader. *Energies* **2019**, *12*, 1613. [CrossRef]
35. Bury, P.; Stosiak, M.; Urbanowicz, K.; Kodura, A.; Kubrak, M.; Malesinska, A. A Case Study of Open- and Closed-Loop Control of Hydrostatic Transmission with Proportional Valve Start-Up Process. *Energies* **2021**, *14*, 1860. [CrossRef]
36. Warbhe, S.D.; Tripathi, J.J.; Deshmukh, K.C.; Verma, J. Fractional Heat Conduction in a Thin Circular Plate with Constant Temperature Distribution and Associated Thermal Stresses. *J. Heat Transf.* **2017**, *139*, 44502. [CrossRef]
37. Blasiak, S. Heat Transfer Analysis for Non-Contacting Mechanical Face Seals Using the Variable-Order Derivative Approach. *Energies* **2021**, *14*, 5512. [CrossRef]
38. Wylie, E.B.; Streeter, V.L.; Suo, L. *Fluid Transients in Systems*; Prentice Hall: Englewood Cliffs, NJ, USA, 1993.
39. Urbanowicz, K.; Bergant, A.; Kodura, A.; Kubrak, M.; Malesinska, A.; Bury, P.; Stosiak, M. Modeling Transient Pipe Flow in Plastic Pipes with Modified Discrete Bubble Cavitation Model. *Energies* **2021**, *14*, 6756. [CrossRef]
40. Wilson, K.; Thomas, A. A new analysis of the turbulent flow of non-Newtonian fluids. *Can. J. Chem. Eng.* **1985**, *63*, 539–546. [CrossRef]
41. Bartosik, A. Numerical Modelling of Heat Transfer in Fine Dispersive Slurry Flow. *Energies* **2021**, *14*, 4909. [CrossRef]

Article

Hydraulic Transients in Viscoelastic Pipeline System with Sudden Cross-Section Changes

Michał Kubrak [1], Agnieszka Malesińska [1], Apoloniusz Kodura [1,*], Kamil Urbanowicz [2] and Michał Stosiak [3]

[1] Faculty of Building Services, Hydro and Environmental Engineering, Warsaw University of Technology, 00-653 Warsaw, Poland; michal.kubrak@pw.edu.pl (M.K.); agnieszka.malesinska@pw.edu.pl (A.M.)
[2] Department of Mechanical Engineering and Mechatronics, West Pomeranian University of Technology, 70-310 Szczecin, Poland; kamil.urbanowicz@zut.edu.pl
[3] Department of Hydraulic Machines and Systems, Wroclaw University of Technology, 50-371 Wrocław, Poland; michal.stosiak@pwr.edu.pl
* Correspondence: apoloniusz.kodura@pw.edu.pl

Abstract: It is well known that the water hammer phenomenon can lead to pipeline system failures. For this reason, there is an increased need for simulation of hydraulic transients. High-density polyethylene (HDPE) pipes are commonly used in various pressurised pipeline systems. Most studies have only focused on water hammer events in a single pipe. However, typical fluid distribution networks are composed of serially connected pipes with various inner diameters. The present paper aims to investigate the influence of sudden cross-section changes in an HDPE pipeline system on pressure oscillations during the water hammer phenomenon. Numerical and experimental studies have been conducted. In order to include the viscoelastic behaviour of the HDPE pipe wall, the generalised Kelvin–Voigt model was introduced into the continuity equation. Transient equations were numerically solved using the explicit MacCormack method. A numerical model that involves assigning two values of flow velocity to the connection node was used. The aim of the conducted experiments was to record pressure changes downstream of the pipeline system during valve-induced water hammer. In order to validate the numerical model, the simulation results were compared with experimental data. A satisfactory compliance between the results of the numerical calculations and laboratory data was obtained.

Keywords: hydraulic transients; water hammer; viscoelasticity; cross-section change

Citation: Kubrak, M.; Malesińska, A.; Kodura, A.; Urbanowicz, K.; Stosiak, M. Hydraulic Transients in Viscoelastic Pipeline System with Sudden Cross-Section Changes. *Energies* **2021**, *14*, 4071. https://doi.org/10.3390/en14144071

Academic Editor: Mehdi Esmaeilpour

Received: 2 June 2021
Accepted: 2 July 2021
Published: 6 July 2021

Publisher's Note: MDPI stays neutral with regard to jurisdictional claims in published maps and institutional affiliations.

1. Introduction

High-density polyethylene (HDPE) pipes are widely used in various pressurised pipeline systems. This is due to the excellent mechanical and chemical properties of polymers. Almost without exception, polymeric materials are known to exhibit time-dependent viscoelastic mechanical behaviour [1]. This property is particularly visible in the case of hydraulic transients, as it induces major dissipation and dispersion of the pressure waves [2]. In many practical applications, accurate computational models predicting pressure oscillations in pipeline systems are required [3]. For this reason, several studies investigating pipe wall behaviour during the water hammer phenomenon have been conducted. Most of the papers on this topic refer to single-pipe systems [4–9]. However, typical fluid distribution networks are composed of serially connected pipes with various inner diameters. Transients in viscoelastic pipes with sudden contractions and expansions have seldom been addressed in the literature. To the authors' knowledge, only [10] investigated the influence of cross-section changes on pressure waves, and [11] studied transients in a series of two polymeric pipes. Recently, in order to numerically simulate water hammer in steel pipe series, an improved junction boundary condition was established [12]. It involves assigning two sets of values, which describe flow parameters, to the connection node, thus causing it to act as two separate nodes. The present paper aims to confirm this method for HDPE pipes. As part of the study, laboratory tests were conducted, designed to record pressure changes at the downstream end of a serially connected HDPE pipeline system. To numerically solve the transient flow equations, the MacCormack explicit scheme was used. The results of the numerical calculations were compared with the experimental data.

This article is organised as follows. After this introduction, the second section gives a brief overview of the theoretical aspects of the viscoelastic behaviour of the pipe structure. In the third section, a one-dimensional numerical model of the water hammer phenomenon in serially connected viscoelastic pipes is presented. The next section looks at the experimental study. Analysis of the experimental data and numerical model validation are presented in the fifth section, and some conclusions are drawn in the final section.

2. Transient Flow Equations in Viscoelastic Pipelines

The standard system describing the one-dimensional transient flow of a compressible liquid in an elastic pipe consists of the continuity Equation (1) and momentum Equation (2) [13,14]:

$$\frac{\partial h}{\partial t} + v\frac{\partial h}{\partial x} + \frac{c^2}{g}\frac{\partial v}{\partial x} = 0 \tag{1}$$

$$\frac{\partial v}{\partial t} + v\frac{\partial v}{\partial x} + g\frac{\partial h}{\partial x} + \frac{f}{2D}v|v| = 0 \tag{2}$$

where f is the friction factor, c is the pressure wave velocity, h is the piezometric head, v is the average flow velocity, g is the gravity acceleration, x is the space coordinate, t is time and D is the internal pipe diameter.

The system of Equations (1) and (2) may be applied for steel pipelines. However, this mathematical description is no longer suitable when transient flow in polymeric pipes is considered. Polymers exhibit both viscous and elastic characteristics. In practice, to include the viscoelastic behaviour of the pipe wall during transient flow, the approach of "mechanical analogues" is usually used. In this approach, the one-dimensional mechanical response of an elastic solid is represented by the mechanical analogue of a spring, and the linear viscous response is represented by a viscous dashpot [15] (Figure 1).

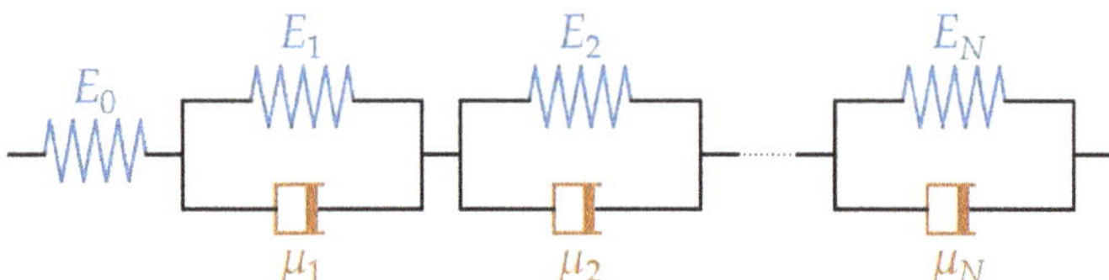

Figure 1. Generalised Kelvin–Voigt model.

Figure 1 shows the so-called generalised Kelvin–Voigt model, which consists of N elements. The total strain can be described as a sum of instantaneous (elastic) strain and retarded strain [4,5]:

$$\varepsilon = \varepsilon_0 + \varepsilon_r \tag{3}$$

where ε_0 is the instantaneous (elastic) strain and ε_r is the retarded strain. The total retarded strain results from the behaviour of each of the N elements:

$$\varepsilon_r = \sum_{k=1}^{N} \varepsilon_k \tag{4}$$

The total strain generated by the constant stress is as follows:

$$\varepsilon(t) = \frac{\sigma}{E_0} + \sum_{k=1}^{N} \frac{\sigma}{E_k}\left[1 - e^{\frac{t}{\tau_k}}\right] \tag{5}$$

where σ is the constant applied stress, E_0 is the elastic bulk modulus of the spring, E_k is the elastic bulk modulus of the k-th Kelvin–Voigt element, τ_k is the retardation time of the k-th Kelvin–Voigt element defined by $\tau_k = \mu_k/E_k$ and $f\subset_k$ is the viscosity of the k-th dashpot.

The creep compliance function corresponding to the generalised Kelvin–Voigt model is equal to the following:

$$J(t) = J_0 + \sum_{k=1}^{N} J_k \left[1 - e^{\frac{-t}{\tau_k}} \right] \tag{6}$$

where J_0 is the instantaneous creep compliance of the first spring, defined by $J_0 = 1/E_0$, and J_k is the creep compliance of the k-th Kelvin–Voigt element, defined by $J_k = 1/E_k$.

According to the Boltzmann superposition principle, the total strain is as follows:

$$\varepsilon(t) = J_0\,\sigma(t) + \int_0^t \sigma(t - \xi)\frac{\partial J(\xi)}{\partial \xi}\,d\xi \tag{7}$$

where ξ is a dummy parameter required for integration.

The hoop stress is related to pipe wall thickness and internal diameter:

$$\sigma = \frac{dp\,D\,\alpha}{2s} = \frac{(p - p_0)\,D\,\alpha}{2s} \tag{8}$$

where α is a dimensionless parameter (dependent on pipe diameter and constraints), s is the pipe wall thickness, p is the current pressure and p_0 is the steady-state pressure.

After calculating the creep function time derivative in Equation (7) and including Equation (8) in it, one obtains the following:

$$\varepsilon(t) = \frac{[p(t) - p_0]D(t)\,\alpha}{2s\,E_0} + \sum_{k=1}^{N} \frac{J_k}{\tau_k} \int_0^t \frac{[p(t - \xi) - p_0]D(t - \xi)\,\alpha}{2s}\,e^{-\frac{\xi}{\tau_k}}\,d\xi \tag{9}$$

This strain influences the cross-sectional area of the pipe, and therefore, it must be included in the continuity equation. In order to take into account the viscoelastic properties of the polymer pipe, the continuity Equation (1) has to be derived from the Reynolds transport theorem [16]. The final form of the continuity equation is as follows:

$$\frac{\partial h}{\partial t} + v\frac{\partial h}{\partial x} + \frac{c^2}{g}\frac{\partial v}{\partial x} + \frac{2c^2}{g}\frac{\partial \varepsilon_r}{\partial t} = 0 \tag{10}$$

Equations (2) and (10) constitute the mathematical description of transient flow in a polymeric pipe. A separate problem is how friction is represented in the momentum Equation (2). The way of representing friction in Equation (2) has a significant impact on the results of the numerical calculations of the water hammer equations [17]. The most common approach is to include one of the many unsteady friction models in the continuity equation [18,19]. However, the problem of unsteady friction falls outside the scope of this investigation, as energy dissipation was taken into account by applying a viscoelastic model, and the friction factor was calculated using the Darcy–Weisbach equation:

$$h = f_s \frac{v|v|}{2D} \tag{11}$$

where f_s is the Darcy friction factor.

3. Numerical Solution of Transient Flow Equations

In order to numerically solve the set of Equations (2) and (10), the MacCormack predictor–corrector finite difference explicit scheme was used [20,21]. This method is widely used in computational fluid dynamics [22], but it was only recently adapted to simulate water hammer in elastic pipes by [23] and also successfully used in [24]. In this model, the time step is no longer subjected to the spatial step, and thus, it is more convenient for meshing variable property series pipes than the classic method of characteristics is. The numerical grid was divided into sections that represent individual pipes with given properties. This allows for assigning different values of input data corresponding to each pipe (i.e., inner diameter, creep compliance, retardation time, number of Kelvin–Voigt elements). The pressure wave travels in the pipeline through the junction node connecting individual pipes. To illustrate the discretisation scheme and solving process, a numerical mesh is shown in Figure 2.

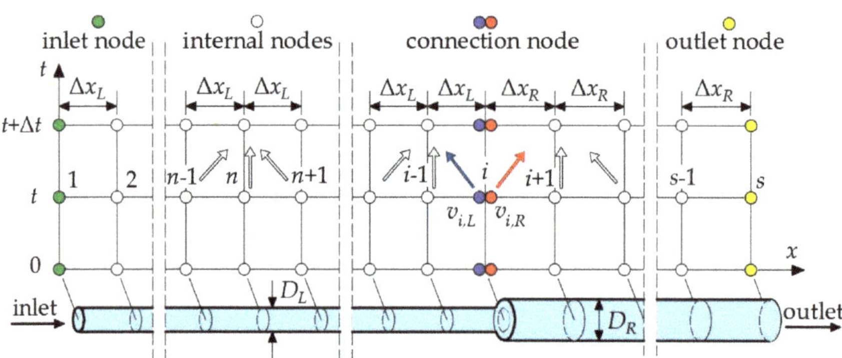

Figure 2. Numerical grid for the MacCormack explicit scheme.

Furthermore, subscripts denoting different types of numerical nodes are as in Figure 2.

- Inlet boundary condition

For a pipe connected to a constant head reservoir, the boundary condition for the first node is as follows:

$$h_1^{t+\Delta t} = h_0 \tag{12}$$

$$v_1^{t+\Delta t} = -\frac{c^2 v_2^{t+\Delta t}}{\left(h_2^{t+\Delta t} - h_1^{t+\Delta t}\right)g - c^2} \tag{13}$$

where Δt is the time step and h_0 is the initial head.

- Outlet boundary condition

For a closed valve, the outlet condition is defined as follows:

$$v_s^{t+\Delta t} = 0 \tag{14}$$

Using the backward difference, the outlet boundary condition can be calculated:

$$h_s^{t+\Delta t} = h_s^t + \Delta t \left(\frac{c^2}{g}\frac{v_{s-1}^{t+\Delta t}}{\Delta x}\right) \tag{15}$$

- Internal nodes

The derivation of the similar numerical discrete format for the internal nodes of the elastic pipeline system is in [12]. Here, Equations (1) and (10) are presented in basic time-marching format only:

$$\left(\frac{\partial h}{\partial t}\right)_n^t = -v_n\frac{h_{n+1}^t - h_n^t}{\Delta x} - \frac{c^2}{g}\frac{v_{n+1}^t - v_n^t}{\Delta x} - \frac{2c^2}{g}\frac{\partial \varepsilon_r^{t-\Delta t}}{\partial t} \tag{16}$$

$$\left(\frac{\partial v}{\partial t}\right)_n^t = -g\frac{h_{n+1}^t - h_n^t}{\Delta x} - v_n^t \frac{v_{n+1}^t - v_n^t}{\Delta x} - \frac{f v_n^t |v_n^t|}{2g} \tag{17}$$

The retarded strain in Equation (10) is calculated as a sum of each Kelvin–Voigt element:

$$\frac{\partial \varepsilon_r^{t-\Delta t}}{\partial t} = \sum_{k=1}^{N} \frac{\partial \varepsilon_{rk}^{t-\Delta t}}{\partial t} = \sum_{k=1}^{N} \left[\frac{\alpha D}{2s} \frac{J_k}{\tau_k} \rho g \left(h^{t-\Delta t} - h_0\right) - \frac{\varepsilon_{rk}^{t-\Delta t}}{\tau_k} \right] \tag{18}$$

Discrete approximation of each retarded strain is equal to the following:

$$\varepsilon_{rk}^{t-\Delta t} = J_k F^{t-\Delta t} - J_k e^{-\frac{\Delta t}{\tau_k}} F^{t-2\Delta t} - J_k \tau_k \left(1 - e^{-\frac{\Delta t}{\tau_k}}\right) \frac{F^{t-\Delta t} - F^{t-2\Delta t}}{\Delta t} + e^{-\frac{\Delta t}{\tau_k}} \varepsilon_{rk}^{t-2\Delta t} \tag{19}$$

with the function F at time $t\text{-}\Delta t$ described as follows:

$$F^{t-\Delta t} = \frac{\alpha D}{2s} \frac{J_k}{\tau_k} \rho g \left(h^{t-\Delta t} - h_0\right) \tag{20}$$

It should be noted that in this numerical model, the constant value of the pressure wave velocity describes the velocity of the transient wave travelling through the entire pipeline system.

- Connection node

In order to take into account any sudden change in the pipeline cross-section, the earlier established connection node condition was used. It involves assigning two sets of values describing the flow parameters to the connection node, which causes it to act as two separate nodes (Figure 2). The detailed derivation of the junction boundary condition is described in [12]. Here, only the final equations are given for brevity. The relation between the flow velocity on the left-hand side of the connection node and the value of the flow velocity on the right-hand side of the connection node is given by:

$$
\begin{aligned}
&\left[-\left(\frac{D_L}{D_R}\right)^2 \frac{1}{\Delta x_L} - \left(\frac{D_L}{D_R}\right)^4 \frac{1}{\Delta x_R} \right] v_{i,L}^3 + \left[\left(\frac{D_L}{D_R}\right)^2 \frac{v_{i-1}}{\Delta x_L} + \left(\frac{D_L}{D_R}\right)^2 \frac{v_{i+1}}{\Delta x_R} \right] v_{i,L}^2 \\
&+ \left[\left(\frac{D_L}{D_R}\right)^2 g\frac{h_{i-1}}{\Delta x_L} + g\frac{h_{i+1}}{\Delta x_R} - \left(\frac{D_L}{D_R}\right)^2 g\frac{1}{\Delta x_L} \frac{\left[-\frac{h_{i-1}}{\Delta x_L} + \frac{c^2}{g}\frac{1}{\Delta x_L} - \left(\frac{D_L}{D_R}\right)^2 \frac{h_{i+1}}{\Delta x_R} + \frac{c^2}{g}\left(\frac{D_L}{D_R}\right)^2 \frac{1}{\Delta x_R} \right]}{\left[-\frac{1}{\Delta x_L} - \left(\frac{D_L}{D_R}\right)^2 \frac{1}{\Delta x_R} \right]} \right. \\
&\left. -g\frac{\left[-\frac{h_{i-1}}{\Delta x_L} + \frac{c^2}{g}\frac{1}{\Delta x_L} - \left(\frac{D_L}{D_R}\right)^2 \frac{h_{i+1}}{\Delta x_R} + \frac{c^2}{g}\left(\frac{D_L}{D_R}\right)^2 \frac{1}{\Delta x_R} \right]}{\left[-\frac{1}{\Delta x_L} - \left(\frac{D_L}{D_R}\right)^2 \frac{1}{\Delta x_R} \right] \Delta x_R} \right] v_{i,L} \\
&+ \left(\frac{D_L}{D_R}\right)^2 g\frac{1}{\Delta x_L} \left[\frac{\frac{c^2}{g}\frac{v_{i-1}}{\Delta x_L} + \frac{c^2}{g}\frac{v_{i+1}}{\Delta x_R}}{-\frac{1}{\Delta x_L} - \left(\frac{D_L}{D_R}\right)^2 \frac{1}{\Delta x_R}} \right] + g\frac{1}{\Delta x_R} \left[\frac{\frac{c^2}{g}\frac{v_{i-1}}{\Delta x_L} + \frac{c^2}{g}\frac{v_{i+1}}{\Delta x_R}}{-\frac{1}{\Delta x_L} - \left(\frac{D_L}{D_R}\right)^2 \frac{1}{\Delta x_R}} \right] = 0
\end{aligned}
\tag{21}
$$

where $v_{i,L}$ is the velocity on the left-hand side of the connection node, $v_{i,R}$ is the velocity on the right-hand side of the connection node, D_L is the inner diameter of the pipe on the left-hand side of the connection node and D_R is the inner diameter of the pipe on the right-hand side of the connection node.

Equation (21) is a cubic function, so it has three roots, only one of which satisfies the physical conditions. After computing the value of the flow velocity on the left-hand side of the connection node $v_{i,L}$, the value of the flow velocity on the right-hand side of the connection node $v_{i,R}$ was calculated with the following:

$$v_{i,R} = \left(\frac{D_L}{D_R}\right)^2 v_{i,L} \tag{22}$$

4. Experimental Study

The goal of the conducted experiments was to obtain the pressure changes during the rapid water hammer phenomenon at the downstream end of a viscoelastic pipeline system with sudden contractions and expansions. Experimental data were later used to validate the numerical model. Four types of experiments were conducted (Figure 3).

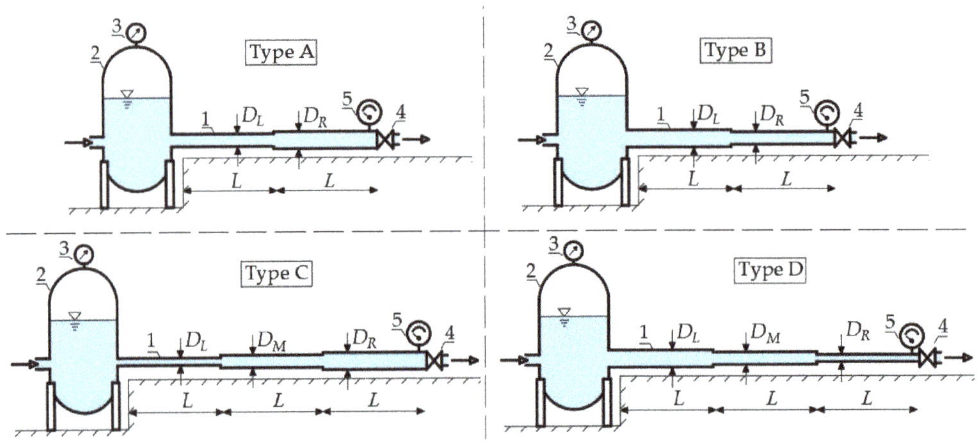

Figure 3. Experimental setup.

The main elements of the experimental setup were serially connected HDPE pipes of the same length with various inner diameters (labelled 1 in Figure 3) attached to an upstream pressure tank (labelled 2 in Figure 3) equipped with a pressure gauge (labelled 3 in Figure 3). The tested pipes were rigidly fixed to the floor. In all experiments, the absolutely sharp inlet and outlet as well as the abrupt diameter changes were considered. The water hammer positive pressure surge was induced by manual, rapid and full closure of the ball valve at the downstream end of the pipeline system (labelled 4 in Figure 3). The valve closing time was measured with an electronic gauge. In all tests, the pressure wave period was longer than the closure time, which did not exceed 0.015 s. The water pressure was measured by a high-frequency relative pressure sensor (labelled 5 in Figure 3) with a range of −0.1~1.2 MPa and a measurement uncertainty of 0.5%. Pressure samples with a frequency of 1000 Hz were converted through an analogue-to-digital card connected to a laptop.

The pipeline systems denoted as "Type A" and "Type B" consisted of two serially connected pipes, whereas the systems denoted as "Type C" and Type D" consisted of three serially connected pipes. The internal diameters of the individual pipes are denoted with "L" (as in "left"), "M" (middle) or "R" (right) subscripts. For the purpose of this investigation, 6 experiments were conducted (2 tests each for Type A and Type B experiments and 1 test each for Type C and Type D experiments). The main characteristics of the experimental installation and registered steady flow parameters before initiation of the water hammer phenomenon are reported in Table 1.

Due to the influence of temperature on the creep parameters, the measurements were carried out with care to maintain a constant water and ambient temperature. The measured pressures are presented in the next section of the paper along with the results of the numerical calculations.

Table 1. Parameters of the pipes used in experiments.

Type of Experiment	No. of Experiment	D_L (mm)	s_L (mm)	D_M (mm)	s_M (mm)	D_R (mm)	s_R (mm)	L (m)	Q (m³/h)	v_0 (m/s)
A	1	35.2	2.4	-	-	44.0	3.0	21.0	5.471	1.00
A	2	32.6	3.7	-	-	40.8	4.6	24.0	3.622	0.77
B	3	44.0	3.0	-	-	35.2	2.4	21.0	5.322	1.52
B	4	40.8	4.6	-	-	32.6	3.7	24.0	3.424	1.14
C	5	26.0	3.0	32.6	3.7	40.8	4.6	12.0	4.234	0.90
D	6	40.8	4.6	32.6	3.7	26.0	3.0	12.0	2.101	1.10

where L is the length of the individual pipe, Q is the initial steady-state volumetric flow rate and v_0 is the initial flow velocity.

5. Model Validation and Analysis of Experimental Data

Numerical calculations were performed to simulate all the conducted experiments. Calibration of the numerical model was performed by adjusting the viscoelastic parameters in order to obtain the result of calculations that match the observed pressure changes. The retardation time is a quantity that characterises one of the viscoelastic properties of the polymers. It is the duration of the retardation phenomenon that describes the moment when the stress becomes zero. Based on the constitutive linear viscoelasticity equations, it is possible to separate the creep compliance into an elastic part independent of time J_0 and a time-dependent creep function $J(t)$. Due to the fact that the numerical model takes into account a single value of pressure wave velocity for the entire pipeline system, it also includes the value of the elastic part of compliance. As mentioned earlier, only steady friction was taken into account in the momentum Equation (2). Thus, matching of the observed and calculated pressure changes was obtained by calibrating the parameters of the creep function, i.e., J and τ and the number of elements N in the Kelvin–Voigt viscoelastic model. The values of the viscoelastic parameters were chosen by trial and error to minimise the mean squared error between the calculated and observed pressure samples. In order to simplify the model calibration, if the pipeline system consisted of three pipes (experiments no. 5 and 6), the same values of viscoelastic parameters for each pipe were assumed. Moreover, if more than one Kelvin–Voight element was used in the calculations, identical viscoelastic parameters for each element were used. This simple approach, as presented later, made it possible to obtain satisfactory compliance between the calculation results and experimental data. More advanced methods of calibrating the viscoelastic parameters are presented in [25–27].

In all the conducted numerical simulations, the time step and spatial step were selected to ensure a Courant number as close as possible to 1. The parameters of the pipeline system and initial conditions listed in Table 1 were used as an input. The values of the viscoelastic parameters determined for all cases are presented in Table 2.

Table 2. Viscoelastic and transient flow parameters used as an input for the numerical model.

Type of Experiment	No. of Experiment	J_L (Pa^{-10})	τ_L (s)	N_L (-)	J_M (Pa^{-10})	τ_M (s)	N_M (-)	J_R (Pa^{-10})	τ_R (s)	N_R (-)	c (m/s)
A	1	1.85	0.040	1	-	-	-	1.85	0.040	1	336
A	2	1.20	0.030	2	-	-	-	0.90	0.060	1	420
B	3	1.15	0.030	2	-	-	-	1.35	0.030	1	349
B	4	0.40	0.065	2	-	-	-	1.30	0.030	1	410
C	5	1.00	0.025	2	1.00	0.025	2	1.00	0.025	2	420
D	6	0.75	0.024	2	0.75	0.024	2	0.75	0.024	2	397

It should be noted that the parameters related to Kelvin–Voigt elements, such as the retardation time and creep compliance, are purely mathematical. Dashpots and springs are conceptual elements with no strict attitude to the physical side of the water hammer phenomenon [17].

Furthermore, the creep and retardation of an HDPE pipe depend not only on the pipe stress history (specifically the frequency and amplitude of the load), but also on the axial and radial limitations of the pipeline system. Thus, the values of the viscoelastic parameters (Table 2) are influenced not only by a single water hammer experimental test, but also by the entire cycle of performed measurements. The calculated and observed pressure oscillations for all experiments are presented in Figures 4–9.

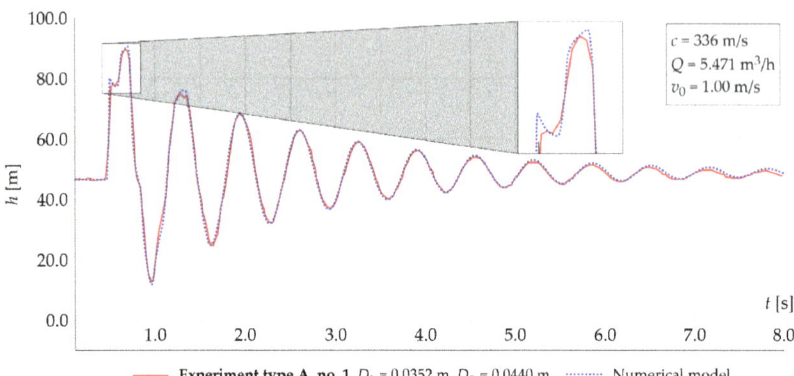

Figure 4. Calculated and observed pressure oscillations during experiment no. 1.

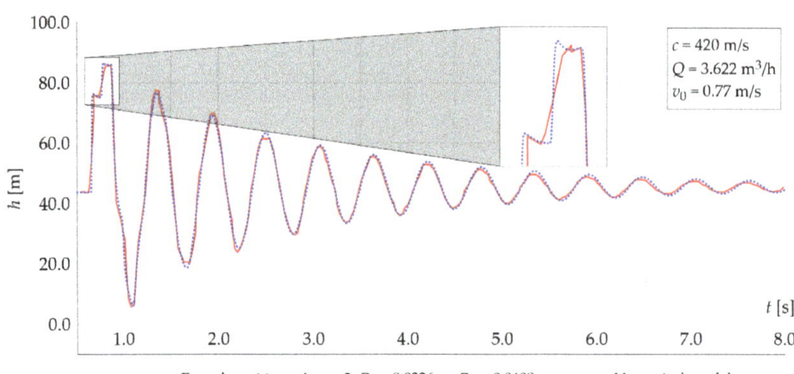

Figure 5. Calculated and observed pressure oscillations during experiment no. 2.

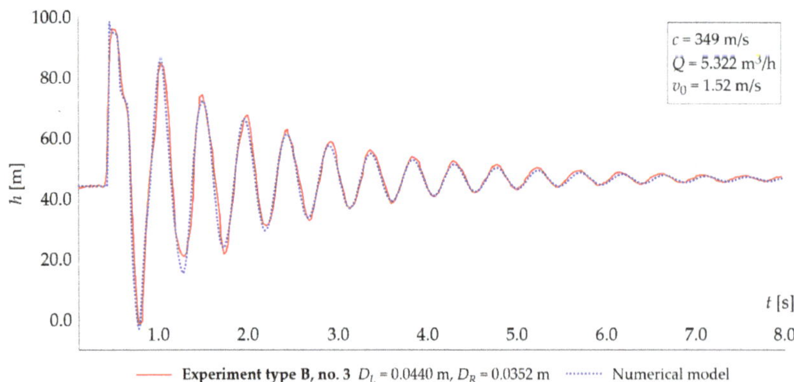

Figure 6. Calculated and observed pressure oscillations during experiment no. 3.

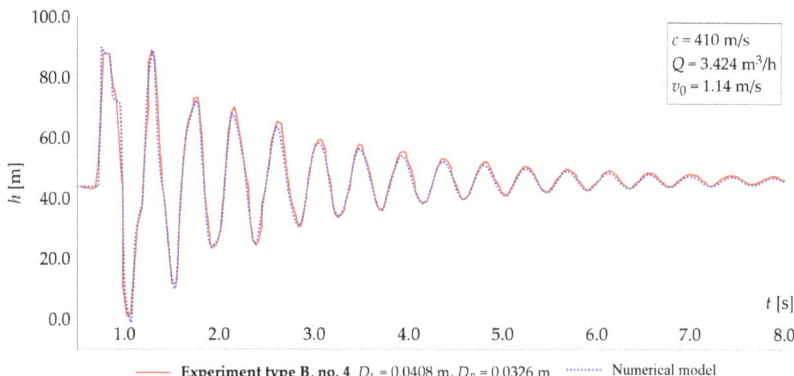

Figure 7. Calculated and observed pressure oscillations during experiment no. 4.

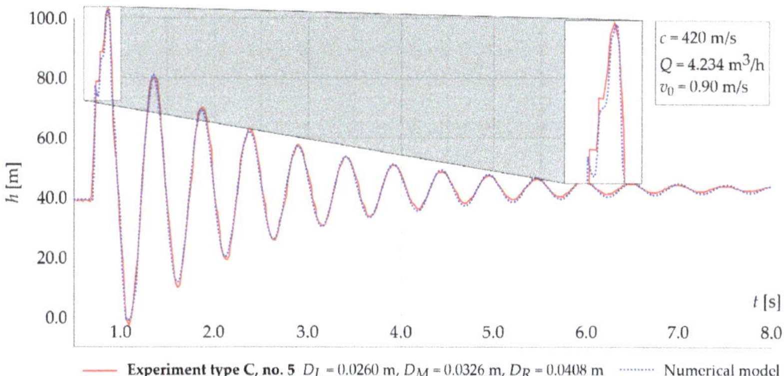

Figure 8. Calculated and observed pressure oscillations during experiment no. 5.

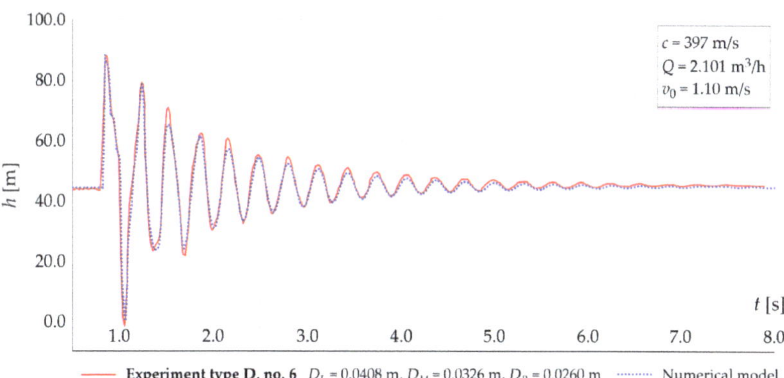

Figure 9. Calculated and observed pressure oscillations during experiment no. 6.

Figures 4–9 show that the pressure wave is strongly damped and time-dispersed. A particularly smooth $h(t)$ function was recorded during experiments no. 1 and no. 3 (Figures 4 and 6, respectively). For these tests, measurements were conducted using a pipe with a wall thickness of 2.4 mm (with an inner diameter of 35.2 mm) and a pipe with a wall thickness of 3.0 mm (with an inner diameter of 44 mm). For both of these tests (experiments no. 1 and no. 3), the ratio of the inside diameter to the wall thickness of the pipes was equal to $D/s = 15$. For other experiments (Table 1), the wall thickness to internal diameter ratio was constant and equalled $D/s = 9$. It is apparent from Figures 4 and 6 that during the water hammer experiments in the pipeline system with thinner pipe walls, a stronger damping of the pressure wave can be observed. Additionally, it can be noticed that the values of the pressure wave velocity for pipes with thinner walls are lower compared to those for the rest of the experiments (Table 2). These lower values of the pressure wave velocity confirm a faster attenuation of disturbances in viscoelastic pipes with thinner walls.

In the hydraulic transient data (Figures 4–9), additional disturbances typical of series-connected pipes are visible for the first pressure increase. The use of the MacCormack time-marching scheme along with the presented connection node equations satisfactorily reflects the pressure disturbances recorded during the measurements (zoom window in Figures 4 and 5). It can be noted that the connection sequence of the pipes with various inner diameters has an influence on the pressure disturbance during the first pressure increase. In the case of the combination of tank–smaller-diameter-pipe–larger-diameter-pipe–outlet-valve (Figures 4 and 5), the first pressure disturbance is not the maximum increase. This effect can also be observed for the water hammer test with three pipes in series (experiment no. 5). In the reverse configuration, the first recorded disturbance is also the maximum pressure increase (Figures 6, 7 and 9).

6. Conclusions

In the presented paper, the MacCormack explicit method was used to numerically solve transient flow equations in polymeric pipes. In order to take into account a sudden change in the pipeline diameter, the connection section boundary was used, which satisfies the head condition but takes into account two different values of the flow velocity on the left-hand side and the right-hand side of the connection node.

Including a term of retarded deformation in the continuity equation made it possible to take into account the viscoelastic behaviour of the pipe walls. The generalised one- and two-element Kelvin–Voigt model was used for numerical simulations. Despite including only a steady friction factor in the momentum equation, the calculated pressure fluctuations showed a satisfactory agreement with the laboratory data in terms of both phase and amplitude.

The use of the connection boundary condition made it possible to reproduce the pressure disturbances in the initial pressure increase related to the change in the diameter of the series-connected pipes.

Furthermore, a constant value of pressure wave velocity for the entire pipeline system was assumed. Further work needs to be performed to study water hammer in variable-property pipeline systems. Future studies will concentrate on simulating water hammer in serially connected pipes with various inner diameters and made of different materials. More broadly, research is also needed to address local losses during hydraulic transients due to diameter changes in pipeline systems.

Author Contributions: Conceptualisation, M.K. and A.M.; methodology, M.K., A.M. and A.K.; software, M.K.; validation, A.M., A.K., K.U. and M.S.; formal analysis, M.K., A.M. and K.U.; investigation, A.M. and A.K.; resources, A.M. and A.K.; data curation, M.K. and A.M.; writing—original draft preparation, M.K. and A.M.; writing—review and editing, A.K., K.U. and M.S.; visualisation, M.K.; supervision, M.K and A.M.; project administration, K.U. and M.S.; funding acquisition, M.K. and A.K. All authors have read and agreed to the published version of the manuscript.

Funding: This research received no external funding.

Institutional Review Board Statement: Not applicable.

Informed Consent Statement: Not applicable.

Data Availability Statement: The code generated during the study and experimental data are available from the corresponding author by request.

Conflicts of Interest: The authors declare no conflict of interest.

Nomenclature

c	pressure wave velocity (m/s)
D	pipe internal diameter (m)
E_0	elastic bulk modulus of the spring (Pa)
E_k	elastic bulk modulus of the k-th Kelvin–Voigt element (Pa)
f	friction factor (-)
f_s	Darcy-Weisbach friction factor (-)
g	gravity acceleration (m/s^2)
h	piezometric head (m)
J_0	instantaneous creep compliance (Pa^{-1})
J_k	creep compliance of the k-th Kelvin–Voigt element (Pa^{-1})
L	length of the individual pipe (m)
N	total number of Kelvin–Voigt elements (-)
Q	initial steady-state volumetric flow rate (m^3/s)
p	pressure (Pa)
s	pipe wall thickness (m)
t	time (s)
v	average flow velocity (m/s)
v_0	initial average flow velocity (m/s)
$v_{i,L}$	velocity on the left-hand side of the connection node (m/s)
$v_{i,R}$	velocity on the right-hand side of the connection node (m/s)
x	space coordinate (m)
Δt	time step (s)
Δx	spatial step (m)
ε_0	instantaneous (elastic) strain (-)
ε_r	retarded strain (-)
η_k	viscosity of the k-th dashpot (kg/sm)
σ	stress (Pa)
τ_k	retardation time of the k-th Kelvin–Voigt element (s)
Acronyms	
HDPE	high-density polyethylene

References

1. Elleuch, R.; Taktak, W. Viscoelastic Behavior of HDPE Polymer Using Tensile and Compressive Loading. *J. Mater. Eng. Perform.* **2006**, *15*, 111–116. [CrossRef]
2. Soares, A.K.; Covas, D.I.; Reis, L.F. Analysis of PVC Pipe-Wall Viscoelasticity during Water Hammer. *J. Hydraul. Eng.* **2008**, *134*, 1389–1394. [CrossRef]
3. Bertaglia, G.; Ioriatti, M.; Valiani, A.; Dumbser, M.; Caleffi, V. Numerical Methods for Hydraulic Transients in Visco-Elastic Pipes. *J. Fluids Struct.* **2018**, *81*, 230–254. [CrossRef]
4. Covas, D.; Stoianov, I.; Ramos, H.; Graham, N.; Maksimovic, C. The Dynamic Effect of Pipe-Wall Viscoelasticity in Hydraulic Transients. Part I—Experimental Analysis and Creep Characterization. *J. Hydraul. Res.* **2004**, *42*, 517–532. [CrossRef]
5. Covas, D.; Stoianov, I.; Mano, J.F.; Ramos, H.; Graham, N.; Maksimovic, C. The Dynamic Effect of Pipe-Wall Viscoelasticity in Hydraulic Transients. Part II—Model Development, Calibration and Verification. *J. Hydraul. Res.* **2005**, *43*, 56–70. [CrossRef]
6. Ferrante, M.; Massari, C.; Brunone, B.; Meniconi, S. Experimental Evidence of Hysteresis in the Head-Discharge Relationship for a Leak in a Polyethylene Pipe. *J. Hydraul. Eng.* **2011**, *137*, 775–780. [CrossRef]
7. Duan, H.-F.; Ghidaoui, M.; Lee, P.J.; Tung, Y.-K. Unsteady Friction and Visco-Elasticity in Pipe Fluid Transients. *J. Hydraul. Res.* **2010**, *48*, 354–362. [CrossRef]
8. Apollonio, C.; Covas, D.I.C.; de Marinis, G.; Leopardi, A.; Ramos, H.M. Creep Functions for Transients in HDPE Pipes. *Urban Water J.* **2014**, *11*, 160–166. [CrossRef]
9. Pezzinga, G.; Brunone, B.; Cannizzaro, D.; Ferrante, M.; Meniconi, S.; Berni, A. Two-Dimensional Features of Viscoelastic Models of Pipe Transients. *J. Hydraul. Eng.* **2014**, *140*, 04014036. [CrossRef]
10. Meniconi, S.; Brunone, B.; Ferrante, M. Water-Hammer Pressure Waves Interaction at Cross-Section Changes in Series in Viscoelastic Pipes. *J. Fluids Struct.* **2012**, *33*, 44–58. [CrossRef]

11. Ferrante, M. Transients in a Series of Two Polymeric Pipes of Different Materials. *J. Hydraul. Res.* **2021**, 1–10. [CrossRef]
12. Malesinska, A.; Kubrak, M.; Rogulski, M.; Puntorieri, P.; Fiamma, V.; Barbaro, G. Water Hammer Simulation in a Steel Pipeline System with a Sudden Cross-Section Change. *J. Fluids Eng.* **2021**. [CrossRef]
13. Wylie, E.B.; Streeter, V.L.; Streeter, V.L. *Fluid Transients*; McGraw-Hill International Book Co.: New York, NY, USA, 1978; ISBN 978-0-07-072187-6.
14. Chaudhry, M.H. *Applied Hydraulic Transients*, 3rd ed.; Springer: New York, NY, USA, 2014; ISBN 978-1-4614-8538-4.
15. Keramat, A.; Tijsseling, A.S.; Hou, Q.; Ahmadi, A. Fluid–Structure Interaction with Pipe-Wall Viscoelasticity during Water Hammer. *J. Fluids Struct.* **2012**, *28*, 434–455. [CrossRef]
16. Covas, D.; Stoianov, I.; Ramos, H.; Graham, N.; Maksimović, Č.; Butler, D. Water Hammer in Pressurized Polyethylene Pipes: Conceptual Model and Experimental Analysis. *Urban Water J.* **2004**, *1*, 177–197. [CrossRef]
17. Weinerowska-Bords, K. Viscoelastic Model of Waterhammer in Single Pipeline—Problems and Questions. *Arch. Hydro-Eng. Environ. Mech.* **2006**, *53*, 331–351.
18. Adamkowski, A.; Lewandowski, M. Experimental Examination of Unsteady Friction Models for Transient Pipe Flow Simulation. *J. Fluids Eng.* **2006**, *128*, 1351–1363. [CrossRef]
19. Bergant, A.; Ross Simpson, A.; Vìtkovsky, J. Developments in Unsteady Pipe Flow Friction Modelling. *J. Hydraul. Res.* **2001**, *39*, 249–257. [CrossRef]
20. MacCormack, R. The Effect of Viscosity in Hypervelocity Impact Cratering. In Proceedings of the 4th Aerodynamic Testing Conference, Cincinnati, OH, USA, 28 April 1969.
21. MacCormack, R. A Numerical Method for Solving the Equations of Compressible Viscous Flow. In Proceedings of the 19th Aerospace Sciences Meeting, American Institute of Aeronautics and Astronautics, St. Louis, MO, USA, 12 January 1981.
22. Ngondiep, E.; Kerdid, N.; Abdulaziz Mohammed Abaoud, M.; Abdulaziz Ibrahim Aldayel, I. A Three-level Time-split MacCormack Method for Two-dimensional Nonlinear Reaction-diffusion Equations. *Int. J. Numer. Methods Fluids* **2020**, *92*, 1681–1706. [CrossRef]
23. Wan, W.; Huang, W. Water Hammer Simulation of a Series Pipe System Using the MacCormack Time Marching Scheme. *Acta Mech.* **2018**, *229*, 3143–3160. [CrossRef]
24. Wan, W.; Zhang, B.; Chen, X.; Lian, J. Water Hammer Control Analysis of an Intelligent Surge Tank with Spring Self-Adaptive Auxiliary Control System. *Energies* **2019**, *12*, 2527. [CrossRef]
25. Weinerowska-Bords, K. Alternative Approach to Convolution Term of Viscoelasticity in Equations of Unsteady Pipe Flow. *J. Fluids Eng.* **2015**, *137*, 054501. [CrossRef]
26. Ferrante, M.; Capponi, C. Calibration of Viscoelastic Parameters by Means of Transients in a Branched Water Pipeline System. *Urban Water J.* **2018**, *15*, 9–15. [CrossRef]
27. Choura, O.; Capponi, C.; Meniconi, S.; Elaoud, S.; Brunone, B. A Nelder–Mead Algorithm-Based Inverse Transient Analysis for Leak Detection and Sizing in a Single Pipe. *Water Supply* **2021**, ws2021030. [CrossRef]

MDPI

Article

Effects of Bent Outlet on Characteristics of a Fluidic Oscillator with and without External Flow

Nam-Hun Kim and Kwang-Yong Kim *

Department of Mechanical Engineering, Inha University, 100 Inha-ro, Michuhol-gu, Incheon 22212, Korea; knh308@inha.ac.kr
* Correspondence: kykim@inha.ac.kr; Tel.: +82-32-872-3096

Abstract: A fluidic oscillator with a bent outlet nozzle was investigated to find the effects of the bending angle on the characteristics of the oscillator with and without external flow. Unsteady aerodynamic analyses were performed on the internal flow of the oscillator with two feedback channels and the interaction between oscillator jets and external flow on a NACA0015 airfoil. The analyses were performed using three-dimensional unsteady Reynolds-averaged Navier-Stokes equations with a shear stress transport turbulence model. The bending angle was tested in a range of 0–40°. The results suggest that the jet frequency increases with the bending angle for high mass flow rates, but at a bending angle of 40°, the oscillation of the jet disappears. The pressure drop through the oscillator increases with the bending angle for positive bending angles. The external flow generally suppresses the jet oscillation, and the effect of external flow on the frequency increases as the bending angle increases. The effect of external flow on the peak velocity ratio at the exit is dominant in the cases where the jet oscillation disappears.

Keywords: fluidic oscillator; bending angle; frequency; pressure drop; peak velocity ratio; aerodynamic analyses; unsteady Reynolds-averaged Navier-Stokes equations

Citation: Kim, N.-H.; Kim, K.-Y. Effects of Bent Outlet on Characteristics of a Fluidic Oscillator with and without External Flow. *Energies* **2021**, *14*, 4342. https://doi.org/10.3390/en14144342

Academic Editor: Artur Bartosik

Received: 23 June 2021
Accepted: 16 July 2021
Published: 19 July 2021

Publisher's Note: MDPI stays neutral with regard to jurisdictional claims in published maps and institutional affiliations.

1. Introduction

A fluidic oscillator has no moving parts and has the feature of creating a vibrating jet when a certain pressure is applied to the inlet. This characteristic is due to the specific geometry of the fluidic oscillator. The flow introduced into a fluidic oscillator flows along one wall of the mixing chamber due to the Coanda effect, forming a main stream. Part of this main stream goes to the outlet, and the other part flows into the feedback channels and returns to the inlet, changing the direction of the main stream to create a vibrating jet. A fluidic oscillator may or may not have one or two feedback channels, but in most cases, it has two feedback channels [1,2]. The jet frequency is determined according to the pressure applied to the oscillator inlet [3].

Fluidic oscillators have been applied and studied in various fields due to their advantages, such as robustness and simplicity. Raman [4] evaluated the effectiveness for cavity tone suppression when using a fluidic oscillator in fluid machinery through experiments. The fluidic oscillator was also effective in aeroacoustic control. The effect of a fluidic oscillator on turbine cooling is also recognized and actively studied. Hossain et al. [5] proved fluidic oscillators to be effective for film cooling. Wu et al. [6] evaluated the heat transfer using fluidic oscillators with large eddy simulation (LES) and unsteady Reynolds-averaged Navier-Stokes (URANS) analysis. They showed that the oscillating jets improved heat removal. In addition, various studies have been conducted to apply fluidic oscillators to wind turbines, flowmeters, commercial airplanes, etc. [7–10].

Recently, many studies have been conducted on the flow separation that occurs on airfoil [11–13]. In particular, it has been found to be effective to install an array of fluidic oscillators for controlling the flow separation on the wing of an aircraft or blade of a fluid machine [14,15]. Koklu and Owens [16] experimentally compared various flow control methods for flow on a ramp and confirmed that the fluidic oscillators were most effective in reducing the flow separation. Jones et al. [17] studied the flow separation control on a high-lift wing. Under the same conditions, fluidic oscillators showed similar lift performance with only 54% less mass flow rate compared with steady jet actuators. Melton et al. [18] evaluated the aerodynamic performance of a NACA0015 airfoil equipped with a simple-hinged flap and fluidic oscillators for flap angles in a range of 20° to 60°. Seele et al. [19] applied fluidic oscillators to the trailing edge of a vertical stabilizer of a commercial aircraft, improving its efficiency by about 50%.

In order to maximize the aerodynamic performance of airfoils using fluidic oscillators, many studies have been conducted on the effect of the mounting conditions of the oscillators on the performance. Koklu [20] experimentally examined two installation positions upstream of the flow separation point in an adverse-pressure-gradient ramp model. Fluidic oscillators installed closer to the separation point exhibited higher pressure recovery. Kim and Kim [21] experimentally studied various mounting conditions (arrangement, installation angles, etc.) of fluidic oscillators on a hump surface and showed that the mounting pitch angle of fluidic oscillators had the most sensitive effect on the flow control. Further, through a numerical analysis, Kim and Kim [22] showed that when fluidic oscillators were installed downstream of the separation point on an airfoil, their effect of increasing lift was excellent. The mounting position with the greatest lift depended on the angle of attack. Drag was reduced the most when the oscillators were mounted close to the leading edge for all angles of attack.

There has also been much research on the influence of the internal shape of fluidic oscillators on the flow control performance. Melton and Koklu [23] evaluated two fluidic oscillators with different exit-orifice sizes (1 mm × 2 mm and 2 mm × 4 mm). The larger oscillator size required larger mass flow rate but less power. Additionally, the separation control performance was better in the case of the larger model. Melton et al. [24] proposed a novel shape of fluidic oscillators by evaluating the performance of three types of fluidic oscillators with the same orifice size. Ostermann et al. [25] measured the pressure required to inject the same mass flow rate when the edges of fluidic oscillators were straight and curved. When the edges were curved, the same mass flow was provided with 20% lower pressure.

In addition to experiments on fluidic oscillators, numerical studies were also widely performed. Jeong and Kim [26] evaluated the peak outlet velocity ratio and pressure drop of a fluidic oscillator using URANS analysis in an investigation of the effect of the distance between the inlet nozzle and splitter on the performance. They also performed a multi-objective optimization of the oscillator. Pandey and Kim [27,28] conducted a numerical analysis of the flow in a fluidic oscillator using LES and URANS. They suggested that URANS analysis using an SST model more accurately predicted the internal flow of the fluidic oscillator than LES using the WALE model. Further, based on their analysis results, it was reported that the chamber width of the fluidic oscillator had a great influence on the flow velocity in the feedback channels.

Kim and Kim [29] investigated the flow control performance of fluidic oscillators on a NACA 0015 airfoil with a flap. They tested the effect of the pitch angle of oscillators installed just upstream of the flap on the aerodynamic performance for different flap angles. A pitch angle closer to the flap angle showed a better lift coefficient. However, due to the geometric limitations, pitch angles close to the flap angle could be achieved only by bending the outlets of the oscillators. They showed that the fluidic oscillators with a bent outlet nozzle successfully enhanced the aerodynamic performance of the airfoil, but they have not explained how the characteristics and performance (such as the oscillating jet frequency, peak velocity ratio, etc.) of the fluidic oscillator were changed by bending the outlet nozzle.

In practical applications of fluidic oscillators, especially flow control on airfoils, there may be a need for bending of the outlet nozzle to overcome geometric limitations, as in the work of Kim and Kim [29]. However, there have not been any investigations on the fluidic oscillator with a bent outlet nozzle. Thus, in the present work, the effects of a bent outlet nozzle on the performance of a fluidic oscillator with two feedback channels were investigated using URANS analysis. First, the effects of the bending angle of the outlet nozzle on the performance, such as oscillating jet frequency, peak velocity ratio, and pressure drop, were evaluated at different mass flow rates for a single fluidic oscillator without external flow. The bent fluidic oscillator was also tested for external flow over a NACA0015 airfoil with a simple hinge flap. The goal was to find how the characteristics of the fluidic oscillator change in the application to the flow control over the airfoil compared to the case without external flow.

2. Fluidic Oscillator Model and Computational Domain

The fluidic oscillator model used in this study was proposed by Melton et al. [24] and has two feedback channels. They reported that this fluidic oscillator model exhibited excellent flow separation control when mounted on a NACA0015 Airfoil. The geometrical parameters of this model are summarized in Table 1.

Table 1. Geometrical parameters of the fluidic oscillator tested.

Parameter	Value
Fluidic oscillator width, w (mm)	14.3
Inlet nozzle width, w_i (mm)	2.0
Inlet chamber width, w_c (mm)	2.03
Outlet throttle width, w_t (mm)	2.0
Fluidic oscillator height, h (mm)	1.0
Inlet nozzle feedback channel width, h_{ft} (mm)	2.41
Diffuser angle of outlet nozzle, θ_{diff} (°)	107
Distance between the inlet of mixing chamber and the throat, S(mm)	15.6

In the present work, the effect of a bent outlet nozzle on the fluidic oscillator performance was investigated with and without external flow, and two different cases were considered: the internal flow of a single fluidic oscillator and the interaction between the internal and external flows of fluidic oscillators mounted on a NACA0015 airfoil with a simple hinge flap. Figure 1 shows the computational domain for the internal flow of the fluidic oscillator. Figure 1a,b show the domains for the standard and bent oscillators, respectively. In the case of the bent oscillator, the bending occurs at the throat of the outlet nozzle with a bending angle (β), as shown in Figure 1b. The bending angle (β) varies in the range of 0–40°.

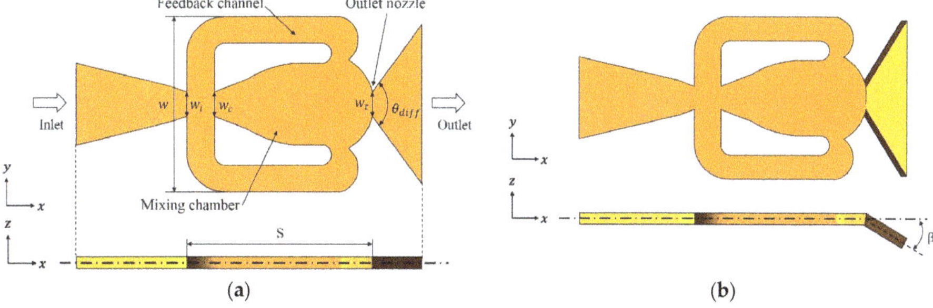

Figure 1. Flow configuration and fluidic oscillator model: (**a**) Reference model, (**b**) Model with bent outlet nozzle.

Figure 2 shows the computational domain for the interaction between the internal and external flows of the fluidic oscillators on the airfoil with a flap. This computational domain consists of the internal and external domains of the fluidic oscillators. Using the periodic conditions, only three oscillators are included in the domain. The reason for selecting this oscillator number was presented in the previous work of Kim and Kim [29]. Melton's experimental work [30] was referenced for the external flow configuration. The angle of attack (α) is fixed at 8°, and the flap deflection angle (δ_f) is 40°. When $\delta_f = 0°$, the chord length (c) is 305 mm, and the span (S) is 99 mm. The flap is located at 70% chord from the leading edge of the airfoil, and fluidic oscillators are installed just in front of the flap hinge.

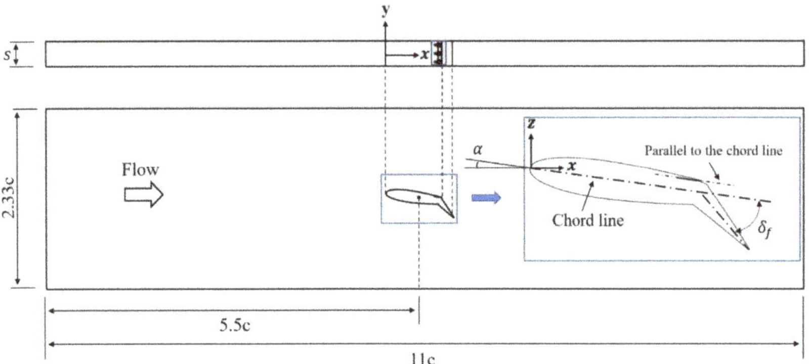

Figure 2. Computational domain for the external flow over NACA0015 airfoil with a flap and fluidic oscillators [29].

For external flow, the fluidic oscillator's mounting conditions are shown in Figure 3, which are the same as in the study of Kim and Kim [29]. The main body of the fluidic oscillator was fixed so that it was always parallel to the airfoil's cord line. Since the pitch and bending angles of the fluidic oscillator coincide, both are expressed as β, and the test range of this angle was determined as $\beta = 0$–40° considering the results of the previous work [29]. In the case of $\beta = 0°$ (reference model), the exit surface of the fluidic oscillator is mounted close to point A′ on plane A-A′, and this plane is perpendicular to the airfoil chord line. However, as β increased, the exit plane was inevitably moved to plane B-B′ to avoid interference with the flap surface. The B-B′ plane is translated 7 mm downstream from the A-A′ plane and rotated by the angle β around point B.

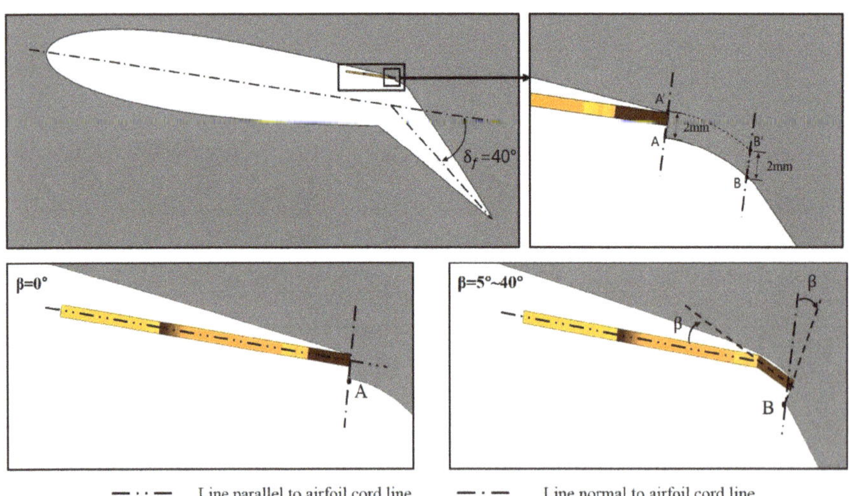

Figure 3. Installation conditions of fluidic oscillators.

3. Performance Parameters

Two performance parameters were defined to evaluate the performance of the fluidic oscillator depending on the bending angle. The first one is the peak velocity ratio of the oscillating jet at the exit of the fluidic oscillator (F_{VR}), which is defined as follows:

$$F_{VR} = \frac{U_{peak}}{U_{ref}} \tag{1}$$

where reference velocity U_{ref} is the velocity at the outlet throat.

$$U_{ref} = \frac{\dot{m}_{inlet}}{A_{ref}\rho} \tag{2}$$

U_{peak} is the peak value of the time-averaged jet velocity at the oscillator outlet, $\dot{m}_{inlet}$ is the mass flow rate at the inlet, A_{ref} is the area of the outlet throat, and ρ is the air density at 25 °C.

The second performance parameter is the dimensionless pressure drop (F_f), which directly affects the pumping power:

$$F_f = \frac{\Delta p D_h}{2\rho U_{ref}^2 S} \tag{3}$$

where Δp is the pressure drop through the fluidic oscillator, D_h is the hydraulic diameter of the outlet throat, and S is the distance between the inlet and the outlet throat shown in Figure 1.

On the other hand, to evaluate the aerodynamic performance of the airfoil with fluidic oscillators, the lift and drag coefficients are defined as follows:

$$C_L = \frac{L}{\frac{1}{2}\,\rho_\infty\,U_\infty^2\,c\,s} \tag{4}$$

$$C_D = \frac{D}{\frac{1}{2}\,\rho_\infty\,U_\infty^2\,c\,s} \tag{5}$$

where $L, D, \rho, U, c,$ and s indicate the lift force, drag force, fluid density, velocity, airfoil cord length, and width of the computational domain, respectively. The subscript ∞ indicates free-stream values.

4. Numerical Analysis

In the present work, the commercial CFD software ANSYS CFX 15.0® [31] was used for the flow analysis. In both the analyses of the internal flow of the fluidic oscillator and the external flow on the airfoil, three-dimensional URANS equations with the shear stress transport (SST) turbulence model and continuity equation were calculated numerically. Pandey and Kim [28] reported that URANS analysis with the SST model predicted the internal flow of a fluidic oscillator better than LES with the WALE model. The SST model is also known to predict the flow separation well under adverse pressure gradients [32,33].

For the computational domain inside the fluidic oscillator (Figure 1), the following boundary conditions were used. Uniform velocity was assigned at the oscillator inlet, and no-slip conditions were used at the walls. For the external flow domain shown in Figure 4, a uniform velocity of 25 m/s was assigned at the inlet of the domain, which corresponds to a Reynolds number of 5.0×10^5 based on the inlet velocity and chord length. Constant pressure is assigned at the outlet of the domain, and no-slip boundary conditions are used at the wall. In both the analyses, the working fluid is air at 25 °C, which is assumed to be an ideal gas.

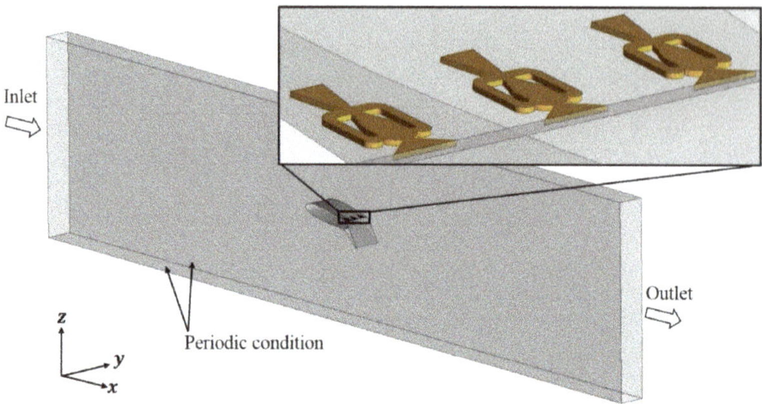

Figure 4. Computational domain using periodic boundary conditions [29].

For the external flow, Melton et al. [18] measured the jet frequency of the fluidic oscillator in a range of mass flow rates, $\dot{m} = 0 - 1.3$ g/s, and evaluated the aerodynamic performance of the airfoil in a range of momentum coefficients (C_μ), 0–6.63%, for different flap angles. They found that as the mass flow rate increased, the frequency increased but converged to a value. The performance of flow control generally improved as C_μ increased, and a larger value of C_μ is required for a larger flap angle to achieve effective flow control. The momentum coefficient is defined as follows:

$$C_\mu = \frac{n\,\rho_{jet}U_{jet}\,A_{nozzle}\,U_{jet}}{\frac{1}{2}\,\rho_\infty\,U_\infty^2\,c\,n\,l} = 2\frac{A_{nozzle}}{c\,l}\left(\frac{U_{jet}}{U_\infty}\right)^2 \tag{6}$$

$$U_{jet} = \frac{\dot{m}}{\rho_{jet}A_{nozzle}} = \frac{\dot{m}}{\rho_\infty A_{nozzle}} \tag{7}$$

where U_{jet} is the velocity at the oscillator outlet, $\dot{m}$ is the mass flow rate through the oscillator, and A_{nozzle} is the area of the oscillator outlet. l and n are the space between adjacent oscillators and the number of oscillators. The fluid density in the fluidic oscillator (ρ_{jet}) was assumed to be the same as the density in the external flow (ρ_∞).

In the present study, the bent oscillator was tested in a range of mass flow rates, $\dot{m}$ =0.19–0.72 g/s, and the aerodynamic performance of the airfoil was evaluated in a C_μ range of 0.41–5.94%.

Grid structures in the internal and external domains are shown in Figures 5 and 6, respectively. In both grids, prism meshes were constructed near a wall, and unstructured tetrahedral meshes were used in the other regions. To adopt the SST model developed for low Reynolds numbers, the first grid points near the wall were located at $y^+ < 2$. In both the unsteady analyses of the internal and external flows, the time step was 5×10^{-6} s. As a convergence criterion, the root-mean-square of the relative residuals was kept less than 1.0×10^{-4}. The total time interval calculated was 0.03 s in each case. The results of the steady RANS analysis were used as the initial assumption for the URANS analysis. The simulation was carried out using a supercomputer employing an Intel Xeon Phi 7250/1.4 GHz processor with 68 CPU cores. The internal flow analysis of the fluidic oscillator took about 8 h, and the analysis of the external flow and three internal flows in the computational domain took about 48 h.

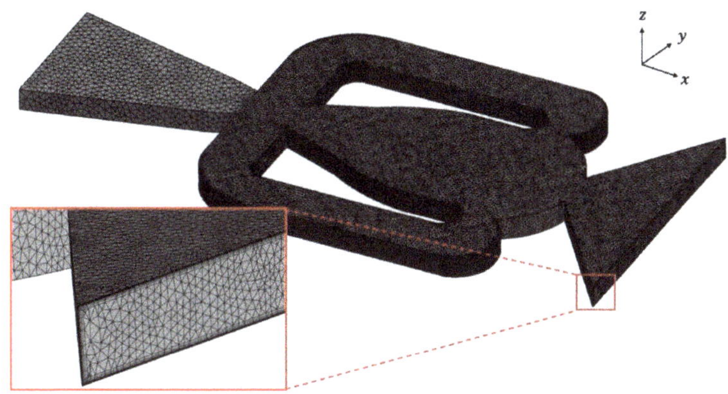

Figure 5. Grid system in the fluidic oscillator.

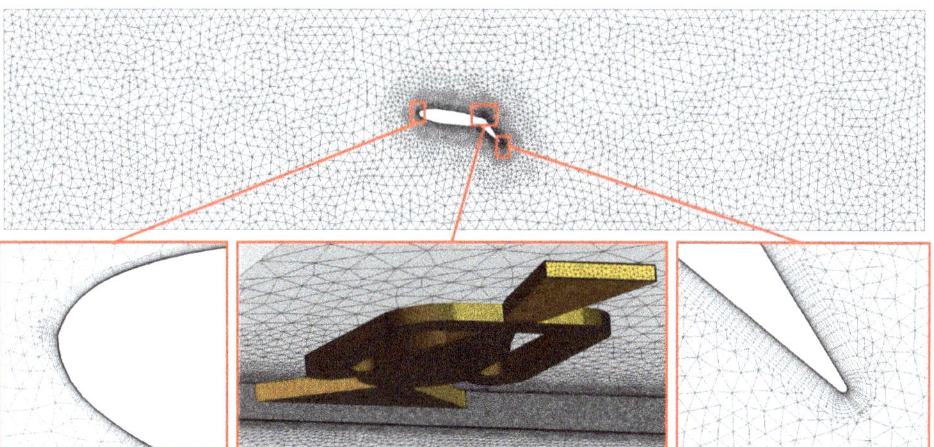

Figure 6. Grid system in internal and external domains ($\alpha = 8°$ and $\delta_f = 40°$).

5. Results and Discussion

5.1. Grid Conver

Grid-convergence index (GCI) analyses based on Richardson extrapolation [34] were performed to evaluate the grid dependency of the numerical results. Table 2 shows the results of the GCI analysis for the jet frequency in the internal domain of the fluidic oscillator. The relative discretization error $\left(\text{GCI}_{\text{fine}}^{21}\right)$ of the finally selected grid with 4.7×10^5 cells (N_1) was 0.957%. The results of the GCI analysis for the lift coefficient in the external domain are shown in Table 3. The grid with 3.8×10^6 cells was selected in this domain, including internal domains of the fluidic oscillators with a $\text{GCI}_{\text{fine}}^{21}$ of 0.072%. The grid selected in Table 3 was also used in each oscillator included in the domain shown in Figure 2.

Table 2. Grid-convergence index (GCI) analysis for the internal domain of a fluidic oscillator.

Parameter		Value
Number of cells	$N_1/N_2/N_3$	$4.7 \times 10^5/3 \times 10^5/2.1 \times 10^5$
Grid refinement factor	r	1.3
Computed jet frequencies corresponding to N_1, N_2, and N_3	f_1 f_2 f_3	1298.7 1272.4 1176.5
Apparent order	p	4.93
Extrapolated values	ϕ_{ext}^{21}	1308.6
Approximate relative error	e_a^{21}	2.025%
Extrapolated relative error	e_{ext}^{21}	0.759%
Grid-convergence index	GCI_{fine}^{21}	0.957%

Table 3. Grid-convergence index(GCI) analysis for the external domain of airfoil with fluidic oscillator.

Parameter		Value
Number of cells	$N_1/N_2/N_3$	$3.8 \times 10^6/3.1 \times 10^6/2.6 \times 10^6$
Grid-refinement factor	r	1.3
Computed lift coefficients (C_L) corresponding to N_1, N_2, and N_3	C_{L1} C_{L2} C_{L3}	2.135 2.147 2.198
Apparent order	p	5.51
Extrapolated values	ϕ_{ext}^{21}	2.131
Approximate relative error	e_a^{21}	0.562%
Extrapolated relative error	e_{ext}^{21}	0.188%
Grid-convergence index	GCI_{fine}^{21}	0.072%

5.2. Validation of Numerical Results

The numerical results for the internal and external flow domains were validated using the experimental data of Melton et al. [18]. Figure 7 shows the comparison between predicted and measured jet frequencies for the internal flow of the fluidic oscillator. The relative error is reduced as the mass flow rate increases and reaches about 2% at mass flow rate larger than 0.7 g/s.

In previous work [29], numerical results were obtained using the same numerical methods used in the present work and validated for the interaction between internal and external flows of the fluidic oscillators. The results were compared with experimental data [18] for the pressure distribution and lift coefficient of a NACA0015 airfoil with a flap angle of 40°, as shown in Figure 8 and Table 4, respectively. In Figure 8, the distribution of the pressure coefficient shows some deviations near the leading edge on the lower surface in both cases with and without oscillators. However, it shows good agreement on the upper surface, which shows lower pressure. In the case of the lift coefficient, the difference between numerical and experimental [18] results decreases rapidly as the angle of attack increases, as shown in Table 4. Several factors such as flap angle, oscillator location, and flow rate seem to be involved in the error.

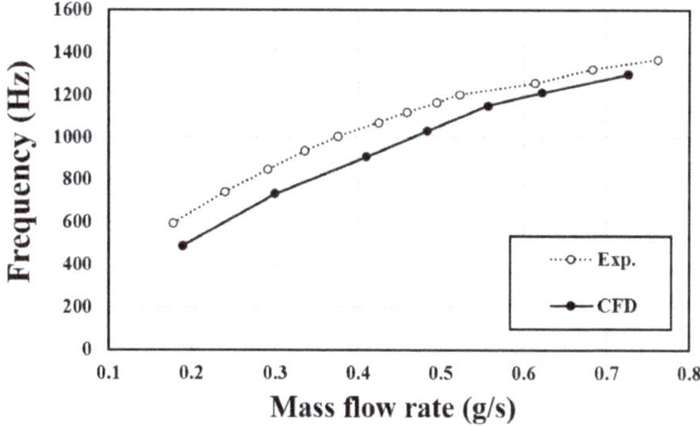

Figure 7. Validation of numerical results for frequency of fluidic oscillator using experimental data (Melton et al. [18]).

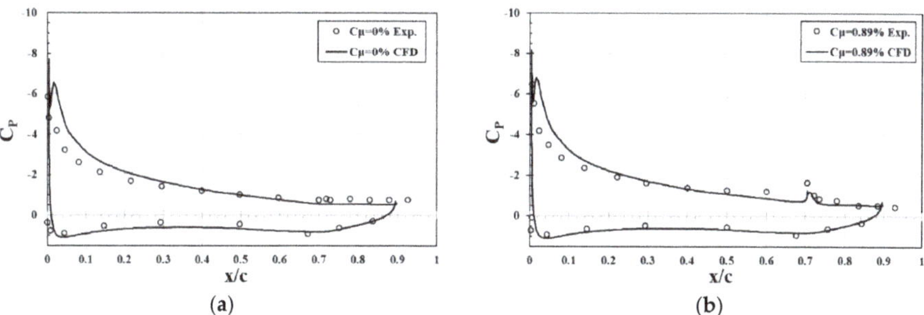

(**a**)　　　　　　　　　　　(**b**)

Figure 8. Validation of numerical results for pressure coefficient using experimental data (Melton et al. [18]) for the NACA 0015 airfoil with a flap and fluidic oscillators ($x_0/c = 0.7$, $\alpha = 8°$, $\delta_f = 40°$ and $y/s = 0.5$) performed by Kim and Kim [29]. (**a**) Case I and (**b**) Case II.

Table 4. Validation of numerical results for lift coefficient using experimental data (Melton et al. [18]) for the NACA 0015 airfoil with a flap and fluidic oscillators ($x_0/c = 0.7$, $C_\mu = 2.15\%$, $\alpha = 8°$, $\delta_f = 40°$).

Angle of Attack (α)	Lift Coefficient		Relative Error (%)
	Experiment	CFD	
0	1.87	1.45	28.9
4	2.13	1.79	18.9
8	2.34	2.13	9.8
10	2.43	2.35	3.4
11	2.44	2.41	1.2

5.3. Single Fluidic Oscillator without External Flow

The effect of a bent outlet on the oscillator performance was first examined for the single fluidic oscillator shown in Figure 1. Figure 9 shows the variations of jet frequency with the mass flow rate at different bending angles for the single fluidic oscillator without external flow (Figure 1). The range of mass flow rate is $\dot{m} = 0.19 - 0.72$ g/s. The frequency generally increases as the mass flow rate increases. It also increases with the bending angle (β) for the mass flow rates larger than 0.41 g/s. However, $\beta = 20°$ and $25°$ show almost the same values of frequency throughout the whole mass flow range.

All the tested bending angles show larger frequencies than the reference model with $\beta = 0$ (Figure 1a), regardless of the mass flow rate. In the low mass flow range of 0.2–0.4, the frequency shows similar variations at $\beta = 5$–$15°$ and $\beta = 30$–$35°$. However, beyond the mass flow rate of 0.4, the frequency varies differently according to the bending angle. At $\beta = 40°$, the frequency largely increases for low mass flow rates ($\dot{m} = 0.19$ and 0.30 g/s), but for the mass flow rates larger than 0.3 g/s, the oscillation of the jet disappears, and the jet becomes steady.

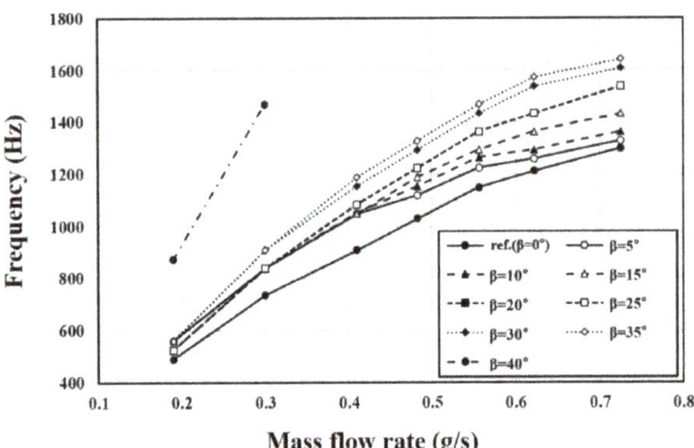

Figure 9. Frequency variations with mass flow rate for various bending angles (β).

Figures 10 and 11 show the variations of peak velocity ratio at the outlet (F_{VR}) and the friction coefficient (F_f) with the mass flow rate at different bending angles, respectively. As shown in Figure 10, the peak velocity ratio generally increases with the mass flow rate for positive bending angles. However, in the case of the reference model ($\beta = 0$), F_{VR} has a maximum value of about 0.9 at around $\dot{m} = 0.4$ g/s and shows the lowest values among the tested bending angles for the mass flow rates larger than 0.4 g/s. At $\dot{m} > 0.5$ g/s, F_{VR} increases with β in a range of $\beta = 0$–15°, but it decreases with β at $\beta = 15$–25°, and $\beta = 15°$ shows the highest peak velocity ratios among the tested bending angles. In the other range, the variation of F_{VR} with the bending angle (β) is quite complicated.

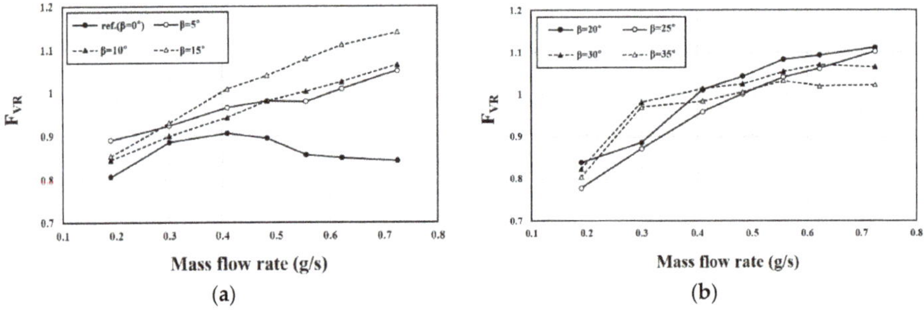

Figure 10. Variations of peak velocity ratio with mass flow rate for various bending angles. (**a**) in the case of the reference model ($\beta = 0$) and (**b**) in the case of $\beta = 20°$, 25°, 30° and 35°.

However, in the case of the friction coefficient, the variations with the bending angle and mass flow rate are relatively simple, as shown in Figure 1. Except for the case of the reference model ($\beta = 0$), F_f shows maxima for all the tested bending angles and it increases almost uniformly with the bending angle for $\beta > 0°$ throughout the whole mass flow range. The maximum F_f occurs around $\dot{m} = 0.4$ g/s for $\beta \leq 25°$, but it shifts to around $\dot{m} = 0.5$ g/s for $\beta = 30°$ and 35°. The reference model with $\beta = 0$ shows values of F_f between those of $\beta = 5°$ and 10° for mass flow rate less than $\dot{m} = 0.5$ g/s, but it shows values similar to or less than those of $\beta = 10°$ for $\dot{m} > 0.5$ g/s.

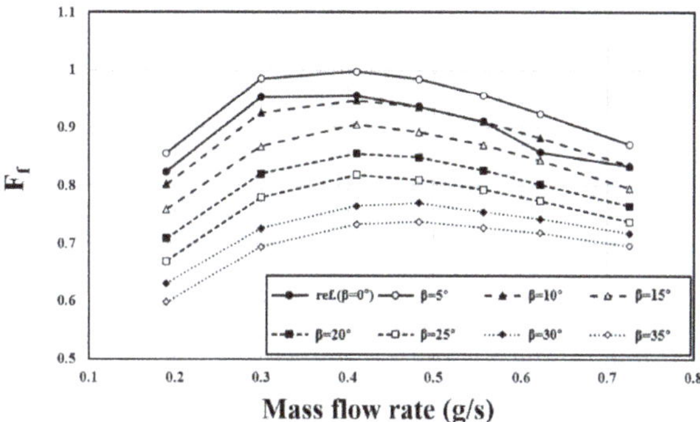

Figure 11. Variations of dimensionless pressure drop with mass flow rate for various bending angles.

Figure 12 shows the velocity fields in the fluidic oscillator at $\dot{m}$ = 0.299 g/s and 0.622 g/s for four bending angles (β = 0°, 15°, 35° and 40°). Two different phases (Φ = 90° and 270°) of the oscillation are shown for each case. It is observed that the angle of jet oscillation at the outlet is reduced slightly as the mass flow rate increases, especially at β = 35°. As the bending angle (β) increases in a range of β less than 40°, the angle of jet oscillation and the jet width increase at both mass flow rates. This is related to the phenomena where the main flow shifts to the wall of the bent outlet opposite to the direction of bending, as shown in Figure 12. This flow shift causes an increase in the peak velocity inside the outlet nozzle and a shift in its location, which seem related to the increase in the jet frequency with bending angle shown in Figure 9. Additionally, the increase in the peak velocity causes a decrease in the pressure, which becomes the reason for the decrease in the pressure drop with bending angle shown in Figure 11. It is also found that the size of the main vortex in the mixing chamber is reduced as β increases. However, the vortex in the feedback channel near the inlet increases with β until β = 35°, especially at the higher mass flow rate. However, at β = 40°, the oscillation disappears as discussed above, even though a slight oscillation remains inside the chamber at the higher mass flow rate.

5.4. Effects of External Flow on the Characteristics of the Fluidic Oscillator

The effect of the bending angle on the characteristics of the fluidic oscillators was also evaluated for external flow over a NACA0015 airfoil with a simple hinge flap. With the installation conditions shown in Figure 3, four bending angles (i.e., pitch angles) of β = 0°, 20°, 35°, and 40° were tested. The values of C_μ used for the mass flow rates are presented in Table 5.

Table 5. C_μ values according to mass flow rates.

$\dot{m}$ (g/s)	C_μ (%)
0.300	1.02
0.410	1.90
0.483	2.64
0.622	4.38

Figure 13 shows the comparison between frequencies of the fluidic oscillator with and without the external flow in a range of mass flow rates, $\dot{m}$ = 0.30–0.62 g/s. Generally, the external flow reduces the frequency except at β = 40°. The average relative difference in the frequency increases with the bending angle β. For β < 40°, the largest relative difference of 7.62% is found at β = 35° and $\dot{m}$ = 0.30 g/s.

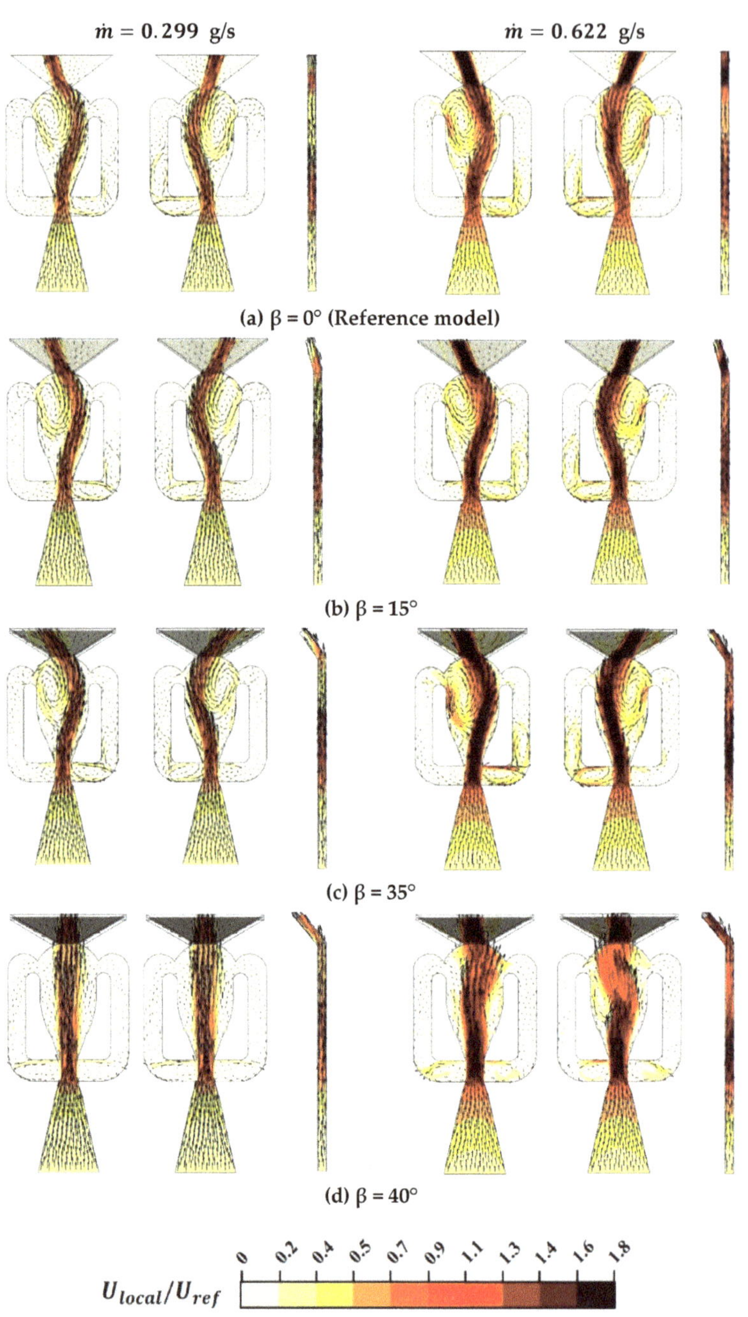

Figure 12. Velocity contours and vectors for different bending angles and flow rates (Left: $\Phi = 90°$, Center: $\Phi = 270°$, Right: x-z cross section).

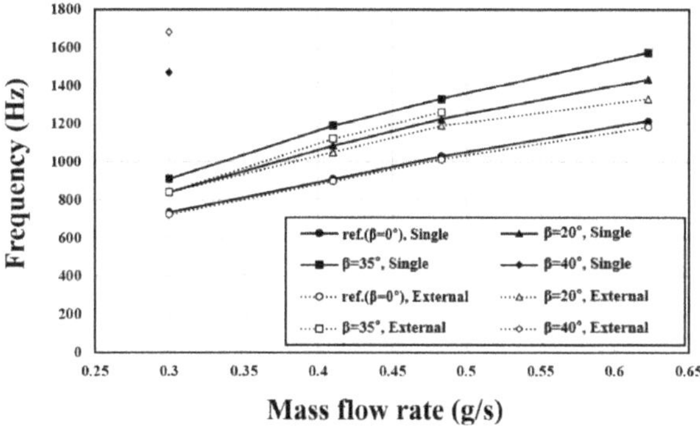

Figure 13. Comparison of jet frequency variation with mass flow rate between the cases with and without external flow for various bending angles.

With the external flow, the jet from the fluidic oscillator becomes steady at $\beta = 40°$ beyond $\dot{m} = 0.30$ g/s, as in the case without external flow (Figure 9). At this bending angle, the external flow increases the frequency for $\dot{m} = 0.30$ g/s, unlike the other cases, and the relative difference in the frequency increases up to 14.4%. In the case with the external flow, oscillation of the jet also disappears at $\beta = 35°$ for the highest mass flow rate of $\dot{m} = 0.62$ g/s, unlike the case without external flow. Therefore, the external flow acts to suppress the oscillation earlier.

The effects of the external flow on the peak velocity ratio of the fluidic oscillator for different bending angles are shown in Figure 14. Except at $\beta = 40°$, where oscillation of the jet disappears for high mass flow rates, the external flow increases the peak velocity ratio, regardless of the mass flow rate. However, at $\beta = 40°$, the peak velocity ratio shows almost uniform variation with the mass flow rate, and the external flow largely reduces the peak velocity ratio, even for $\dot{m} = 0.30$ g/s (relative difference of 24.3%), where the jet still oscillates. At $\beta = 35°$, the peak velocity ratio increases rapidly for $\dot{m} = 0.62$ g/s, where the jet oscillation disappears, showing the largest relative difference of 33.8%. Except for the cases where the jet oscillation disappears, the effect of bending angle on the relative difference in the peak velocity ratio is not remarkable, and the range of the relative difference is 2.33–8.50%.

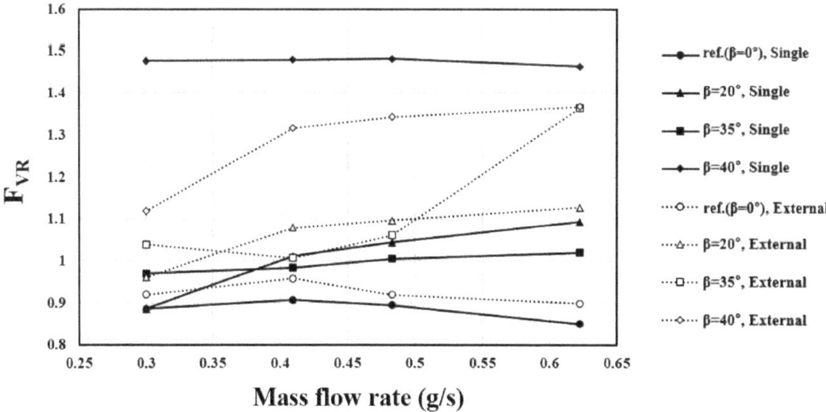

Figure 14. Comparison of peak velocity ratio variation with mass flow rate between the cases with and without external flow for various bending angles.

Figure 15 shows the effects of the external flow on the pressure drop in the fluidic oscillator. The external flow increases the pressure drop by 3–14% at all the tested bending angles, regardless of mass flow rate. The external flow reduces the pressure at the outlet of the oscillator by increasing the velocity there, and this becomes the reason for the increase in the pressure drop through the oscillator with the external flow. The relative difference in F_f does not vary largely with the mass flow rate. $\beta = 20°$ shows the minimum relative differences, and $\beta = 35°$ and $40°$ show similar F_f variations. Thus, there are similar relative differences in the tested range of mass flow rate. The existence of the jet oscillation does not seem to affect the pressure drop.

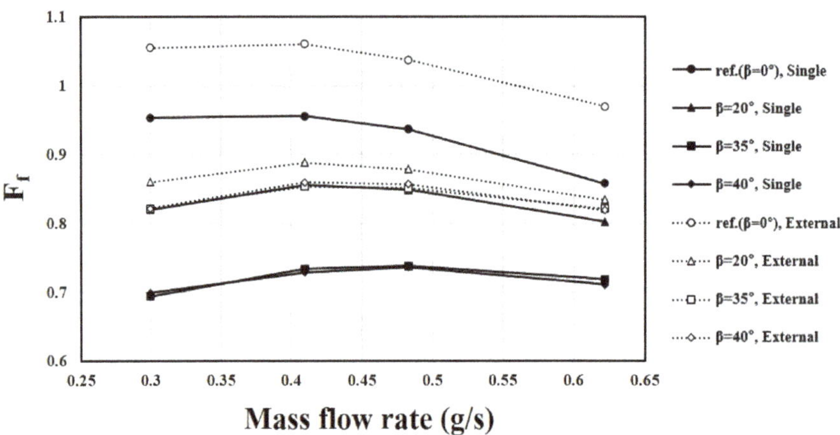

Figure 15. Comparison of F_f variation with mass flow rate between the cases with and without external flow for various bending angles.

Figure 16 shows the variations of the lift coefficient (C_L) with C_μ for different bending angles. The lift coefficient generally increases with C_μ. The relationship between C_μ and the mass flow rate in the oscillator is shown in Table 5. Except for the steady jets at $\beta = 40°$, the lift coefficient increases as the bending angle increases for all C_μ values. Therefore, $\beta = 35°$ shows the highest lift coefficients among the tested bending angles, but there is no further increase with C_μ for $C_\mu = 4.38$ (i.e., $\dot{m} = 0.62$ g/s), where the jet oscillation disappears. The steady jets at $\beta = 40°$ show C_L values similar to those at $\beta = 20°$. This reflects the combined effects of the increase in the pitch angle and the disappearance of the jet oscillation on the lift coefficient. Variations of C_L in the tested C_μ range for non-zero bending angles are much larger than that of the reference model.

Figure 17 shows the effects of bending angle on the drag coefficient (C_D). The variation of C_D with C_μ is generally not large except at $\beta = 0°$. Similar to the case of the lift coefficient shown in Figure 16, except for the case of $\beta = 40°$, the drag coefficient decreases as the bending angle increases for all C_μ values. However, for the lowest value of $C_\mu = 1.02$, $\beta = 0°$ and $20°$ show similar drag coefficients. $\beta = 40°$ shows much larger drag coefficients than those at $\beta = 35°$ except for $C_\mu = 1.02$, where the jet oscillation still exists. However, these values with the steady jets are still lower than those of $\beta = 20°$, which was probably due to the larger pitch angle.

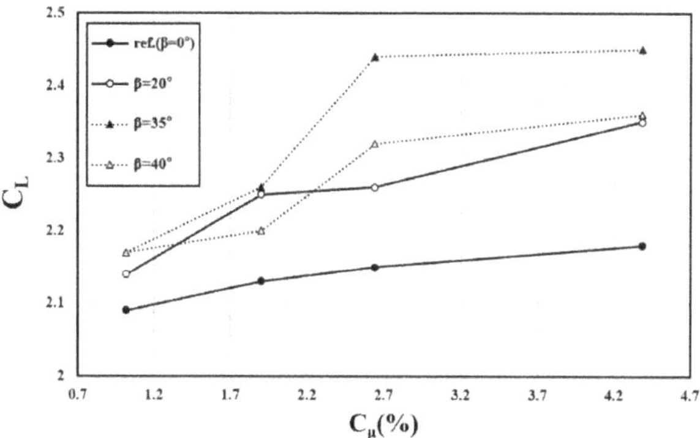

Figure 16. Variations of lift coefficient with momentum coefficient for various bending angles ($x_0/c = 0.7$, $\alpha = 8°$, $\delta_f = 40°$).

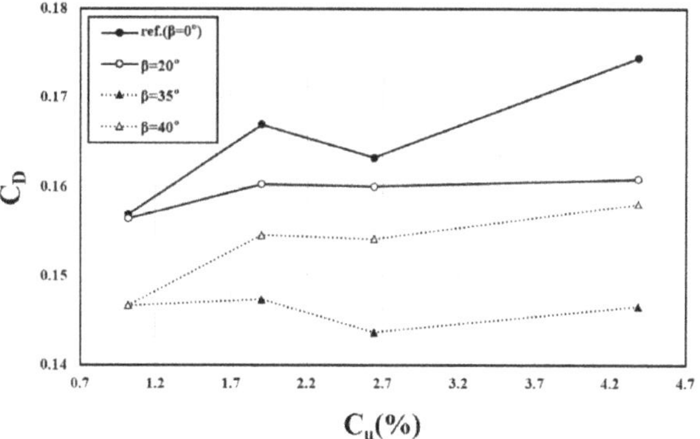

Figure 17. Variations of drag coefficient with momentum coefficient for various bending angles ($x_0/c = 0.7$, $\alpha = 8°$, $\delta_f = 40°$).

6. Conclusions

The effects of bending outlet nozzles on the characteristics of a fluidic oscillator were investigated using URANS analysis with and without external flow in a range of bending angles (β) of 0–40°. In the case without external flow, the frequency increased with the mass flow rate and also with the bending angle in a range of mass flow rates larger than 0.41 g/s. The reference model ($\beta = 0°$) showed the lowest frequencies for all tested mass flow rates. The largest frequencies were shown at $\beta = 40°$ for low mass flow rates, but the jet stopped oscillating for the mass flow rates larger than 0.30 g/s.

The peak velocity ratio (F_{VR}) also increased with the mass flow rate except for the reference model, which showed the lowest F_{VR} values for the mass flow rates larger than 0.4 g/s. For mass flow rates larger than 0.5 g/s, F_{VR} increased with β in a range of $\beta = 0$–15°, but it decreased thereafter until $\beta = 25°$. The pressure drop through the oscillator (F_f) increased almost uniformly with β throughout the mass flow range for $\beta > 0°$. F_f showed maxima around $\dot{m} = 0.4$ g/s for $\beta \leq 25°$, which shifted to $\dot{m} = 0.5$ g/s for $\beta = 30°$ and 35°. The reference model showed an F_f level similar to that of $\beta = 10°$.

The external flow was found to reduce the jet frequency, except at $\beta = 40°$. The average relative difference in the frequency between the cases with and without external flow increased as β increased. For bending angles less than 40°, the largest relative difference was 7.62% at $\beta = 35°$ and $\dot{m} = 0.30$ g/s. As in the case without external flow, the jet oscillation disappeared at $\beta = 40°$ for $\dot{m} > 0.30$ g/s. At this bending angle, the external flow increased the frequency by 14.4% for $\dot{m} = 0.30$ g/s. With the external flow, the jet also became steady at $\beta = 35°$ for the highest mass flow rate $\dot{m} = 0.62$ g/s, unlike the case without external flow. Therefore, it seems that the external flow generally suppresses the oscillation.

Except at $\beta = 40°$, the external flow increased the peak velocity ratio for all mass flow rates. At $\beta = 40°$, however, the external flow largely reduced F_{VR}, especially for $\dot{m} = 0.30$ g/s, where the jet oscillation still existed. The effect of external flow on F_{VR} was dominant in the cases where the jet oscillation disappeared, and the effect of β on the relative difference in F_{VR} was not remarkable in the other cases, where the range of the relative difference was $2.33-8.50\%$. The external flow increased F_f by 3–14% in the tested β range, regardless of mass flow rate. The lowest relative differences in F_f were found at $\beta = 20°$. The existence of the jet oscillation did not affect F_f.

As for the lift coefficient of the airfoil, $\beta = 40°$ did not show the lowest values, but the values were similar to those at $\beta = 20°$. This reflects both the positive effect of the increase in the pitch angle and the negative effect of the disappearance of the jet oscillation on the lift coefficient. The drag coefficient generally decreased as the bending angle increased, except at $\beta = 40°$, where the drag coefficients were much larger than those at $\beta = 35°$, except for $C_\mu = 1.02$, where the jet still oscillated. The results obtained in this study provide information how the characteristics of a fluidic oscillator change with the bending angle of the outlet nozzle, which may be necessary in some practical applications. Improvement in the aerodynamic performance of airfoils using fluidic oscillators would contribute to energy savings in the operation of aircraft. Further research is required to find the difference in the aerodynamic performance of an airfoil between the cases using straight and bending fluidic oscillators.

Author Contributions: N.-H.K. performed numerical analysis and provided original draft manuscript K.-Y.K. revised and finalized the manuscript. Both authors have read and agreed to the published version of the manuscript.

Funding: This work was supported by the National Research Foundation of Korea (NRF) grant funded by the Korean government (MSIT) (No. 2019R1A2C1007657).

Institutional Review Board Statement: Not applicable.

Informed Consent Statement: Not applicable.

Data Availability Statement: Not applicable.

Conflicts of Interest: The authors declare no conflict of interest.

References

1. Tesar, V.; Zhong, S.; Rasheed, F. New fluidic-oscillator concept for flow-seperation control. *AIAA J.* **2013**, *51*, 397–405. [CrossRef]
2. Gregory, J.; Tomac, M. A review of fluidic oscillator development and application for flow control. *AIAA Pap.* **2013**, 2013–2474. [CrossRef]
3. Cattafesta, L.N.; Sheplak, M. Actuators for active flow control. *Annu. Rev. Fluid Mech.* **2011**, *43*, 247–272. [CrossRef]
4. Raman, G.; Raghu, S. Cavity resonance suppression using miniature fluidic oscillators. *AIAA J.* **2004**, *42*, 2608–2612. [CrossRef]
5. Hossain, M.A.; Prenter, R.; Lundgreen, R.K.; Ameri, A.; Gregory, J.W.; Bons, J.P. Experimental and Numerical Investigation of Sweeping Jet Film Cooling. *ASME J. Turbomach.* **2018**, *140*. [CrossRef]
6. Wu, Y.; Yu, S.; Zuo, L. Large eddy simulation analysis of the heat transfer enhancement using self-oscillating fluidic oscillators. *Int. J. Heat Mass Transf.* **2019**, *131*, 463–471. [CrossRef]
7. Cerretelli, C.; Wuerz, W.; Gharaibah, E. Unsteady separation control on wind turbine blades using fluidic oscillators. *AIAA J.* **2010**, *48*, 1302–1311. [CrossRef]
8. Feikema, D.; Culley, D. Computational fluid dynamic modeling of a fluidic actuator for flow control. *AIAA Aerosp. Sci. Meet. Exhib.* **2008**, 1–13. [CrossRef]
9. Seifert, A.; Greenblatt, D.; Wygnanski, I.J. Active separation control: An overview of Reynolds and Mach numbers effects. *Aerosp. Sci. Technol.* **2004**, *8*, 569–582. [CrossRef]
10. Wen, X.; Liu, J.; Li, Z.; Peng, D.; Zhou, W.; Kim, K.C.; Liu, Y. Jet impingement using an adjustable spreading-angle sweeping jet. *Aerosp. Sci. Technol.* **2020**, *105*, 105956. [CrossRef]

11. Al-Battal, N.H.; Cleaver, D.J.; Gursul, I. Unsteady actuation of counter-flowing wall jets for gust load attenuation. *Aerosp. Sci. Technol.* **2019**, *89*, 175–191. [CrossRef]
12. Lei, J.; Zhang, J.; Niu, J. Effect of active oscillation of local surface on the performance of low Reynolds number airfoil. *Aerosp. Sci. Technol.* **2020**, *99*, 105774. [CrossRef]
13. Zhu, H.; Hao, W.; Li, C.; Ding, Q.; Wu, B. Application of flow control strategy of blowing, synthetic and plasma jet actuators in vertical axis wind turbines. *Aerosp. Sci. Technol.* **2019**, *88*, 468–480. [CrossRef]
14. Lee, K.Y.; Chung, H.S.; Cho, D.H.; Sohn, M.H. Flow Separation Control Effects of Blowing Jet on an Airfoil. *J. Korean Soc. Aeronaut. Space Sci.* **2007**, *35*, 1059–1066. [CrossRef]
15. Nagib, H.; Kiedaisch, J.; Reinhard, P.; Demanett, B. Control Techniques for Flows with Large Separated Regions: A New Look at Scaling Parameters. *AIAA Pap.* **2006**, 2006–2857. [CrossRef]
16. Koklu, M.; Owens, L.R. Flow Separation Control over a Ramp using Sweeping Jet Actuators. In Proceedings of the 7th AIAA Flow Control Conference, Atlanta, GA, USA, 16–20 June 2014. [CrossRef]
17. Jones, G.S.; Milholen, W.E.; Chan, D.T.; Melton, L.; Goodliff, S.L.; Cagle, C.M. A Sweeping Jet Application on a High Reynolds Number Semispan Supercritical Wing Configuration. In Proceedings of the 35th AIAA Applied Aerodynamics Conference, Denver, CO, USA, 5–9 June 2017. [CrossRef]
18. Melton, L.P.; Koklu, M.; Andino, M.; Lin, J.C. Active Flow Control via Discrete Sweeping and Steady Jets on a Simple-Hinged Flap. *AIAA J.* **2018**, *56*. [CrossRef]
19. Seele, R.; Graff, E.; Lin, J.; Wygnanski, I. Performance enhancement of a vertical tail model with sweeping jet actuators. In Proceedings of the 51st AIAA Aerospace Sciences Meeting Including the New Horizons Forum and Aerospace Exposition, Grapevine, TX, USA, 7–10 January 2013.
20. Koklu, M. The Effects of Sweeping Jet Actuator Parameters on Flow Separation Control. *AIAA J.* **2018**, *56*, 100–110. [CrossRef] [PubMed]
21. Kim, S.-H.; Kim, K.-Y. Effects of Installation Conditions of Fluidic Oscillators on Control of Flow Separation. *AIAA J.* **2019**, *57*, 5208–5219. [CrossRef]
22. Kim, S.-H.; Kim, K.-Y. Effects of installation location of fluidic oscillators on aerodynamic performance of an airfoil. *Aerosp. Sci. Technol.* **2020**, *99*, 105735. [CrossRef]
23. Melton, L.P.; Koklu, M. Active Flow Control using Sweeping Jet Actuators on A Semi-Sapn Wing Model. In Proceedings of the 54th AIAA Aerospace Sciences Meeting, San Diego, CA, USA, 4–8 January 2016. [CrossRef]
24. Melton, L.P.; Koklu, M.; Andino, M.; Lin, J.C.; Edelman, L. Sweeping Jet Optimization Studies. In Proceedings of the 8th AIAA Flow Control Conference, Washington, DC, USA, 13–17 June 2016. [CrossRef]
25. Ostermann, F.; Woszidlo, R.; Nayeri, C.; Paschereit, C.O. Experimental Comparison between the Flow Field of Two Common Fluidic Oscillator Designs. In Proceedings of the 53 rd AIAA Aerospace Sciences Meeting, Kissimmee, FL, USA, 5–9 January 2015; Volume 10, p. 6. [CrossRef]
26. Jeong, H.S.; Kim, K.Y. Shape Optimization of a Feedback-Channel Fluidic Oscillator. *Eng. Appl. Comput. Fluid Mech.* **2017**, *12*, 169–181. [CrossRef]
27. Pandey, R.J.; Kim, K.Y. Numerical Modeling of Internal Flow in A Fluidic Oscillator. *J. Mech. Sci. Technol.* **2018**, *32*, 1041–1048. [CrossRef]
28. Pandey, R.J.; Kim, K.Y. Comparative Analysis of Flow in a Fluidic Oscillator Using Large Eddy Simulation and Unsteady Reynolds-Averaged Navier-Stokes Analysis. *Fluid Dyn. Res.* **2018**, *50*, 065515. [CrossRef]
29. Kim, N.H.; Kim, K.Y. Flow control using fluidic oscillators on an airfoil with a flap. *Eng. Appl. Comput. Fluid Mech.* **2021**, *15*, 377–390. [CrossRef]
30. Melton, L.P. Active Flow Separation Control on a NACA 0015 Wing using Fluidic Actuators. In Proceedings of the 7th AIAA Journal Flow Control Conference, Atlanta, GA, USA, 16–20 June 2014. [CrossRef]
31. ANSYS. *ANSYS CFX-Solver Theory Guide-Release 15.0*; ANSYS Inc.: Canonsburg, PA, USA, 2014.
32. Menter, F.R. Two-Equation Eddy-Viscosity Turbulence Models for Engineering Applications. *AIAA J.* **1994**, *32*, 1598–1605. [CrossRef]
33. Bardina, J.E.; Huang, P.G.; Coakley, T.J. Turbulence Modeling Validation. *AIAA Pap.* **1997**, 1997–2121. [CrossRef]
34. Celik, I.B.; Ghia, U.; Roache, P.J.; Freitas, C.J.; Coleman, H.; Raad, P.E. Procedure for Estimation and Reporting of Uncertainty due to Discretization in CFD Applications. *J. Fluids Eng.* **2008**, *130*, 078001. [CrossRef]

Article

An Integral Method for Natural Convection of Van Der Waals Gases over a Vertical Plate

A. A. Avramenko [1], I. V. Shevchuk [2,*], Yu. Yu. Kovetskaya [1] and N. P. Dmitrenko [1]

[1] Institute of Engineering Thermophysics, National Academy of Sciences, 03057 Kiev, Ukraine; aaav1@i.com.ua (A.A.A.); kovetskajulia@ukr.net (Y.Y.K.); NatDmitrenko@i.ua (N.P.D.)

[2] Faculty of Computer Science and Engineering Science, TH Köln–University of Applied Sciences, 51643 Gummersbach, Germany

* Correspondence: igor_v.shevchuk@th-koeln.de

Abstract: This paper focuses on a study of natural convection in a van der Waals gas over a vertical heated plate. In this paper, for the first time, an approximate analytical solution of the problem was obtained using an integral method for momentum and energy equations. A novel simplified form of the van der Waals equation for real gases enabled estimating the effects of the dimensionless van der Waals parameters on the normalized heat transfer coefficients and Nusselt numbers in an analytical form. Trends in the variation of the Nusselt number depending on the nature of the interaction between gas molecules and the wall were analyzed. The results of computations for a van der Waals gas were compared with the results for an ideal gas.

Keywords: natural convection; van der Waals gas; analytical solution

Citation: Avramenko, A.A.; Shevchuk, I.V.; Kovetskaya, Y..Y..; Dmitrenko, N.P. An Integral Method for Natural Convection of Van Der Waals Gases over a Vertical Plate. *Energies* **2021**, *14*, 4537. https://doi.org/10.3390/en14154537

Academic Editor: Artur Bartosik

Received: 23 June 2021
Accepted: 24 July 2021
Published: 27 July 2021

Publisher's Note: MDPI stays neutral with regard to jurisdictional claims in published maps and institutional affiliations.

1. Introduction

Nowadays, the need for knowledge of the physical and chemical properties of gases, the features of their behavior in real production conditions, and during their transportation and storage has significantly increased. The rational and efficient design of various technological processes involving gases can significantly reduce the cost of industrial production.

In technical thermodynamics, it is customary to refer to real gases as the gaseous state of any substance in the entire range of its existence, that is, at any pressures and temperatures. Under appropriate conditions, real gas can be liquefied or converted into a solid state. It is expected that the development of small-scale energy in the coming years will be associated with the widespread use of liquefied natural gas, which is recognized as one of the most promising types of energy carriers. The liquefaction of gases is important when storing and transporting this type of fuel. In its liquefied form, natural gas occupies only about 1/600 of its gaseous volume; therefore, it is easier and more economical to transport it [1–3].

The vapors of various substances, such as water, ammonia, methyl chloride, sulfur dioxide, and others are widely used in technological applications. The most widely used is water vapor, which is the main working medium of steam engines, heating, and other devices [4–7].

It is known that the ideal gas model allows a satisfactory description of the state of real gases only in a relatively small region of the variation of the state parameters. To design and determine the optimal operating conditions for heat exchange equipment, calculations must take into account the real thermophysical properties of the gas. The use of the ideal gas equation can lead to a significant error in determining the parameters of the gas state, up to 100%. The authors in [8] emphasized that for simple, mono-, and diatomic gases, such as air, helium, nitrogen, etc., under certain conditions, the model for ideal gases is well-suited. For gases with a more complex organic structure, the effects of real gases are more significant.

Taking air as an example, as a mixture of mono- and diatomic gases, the limits of the applicability of the ideal gas model for air are low densities (large specific volumes v), low pressures (<10 bar), and moderate/high temperatures (up to 600 °C) [9,10].

From the point of view of the molecular theory of the structure of matter, a real gas is a gas, the properties of which depend on the interaction and size of the molecules. To date, more than 150 equations of state for real gases are known. One of the most widely known approximate equations of state for real gases is the van der Waals equation [9,10]. Some of the other equations of state for real gases are refinements of the van der Waals equation, while others were obtained experimentally.

Three forms of the van der Waals equation are used by different authors [9–13]:

$$\left(p + n^2 \frac{a_*}{V^2}\right)(V - nb_*) = nR_m T,\tag{1}$$

$$\left(p + \frac{a_*}{V_m^2}\right)(V_m - b_*) = R_m T\tag{2}$$

$$\left(p + \frac{a}{v^2}\right)(v - b) = RT,\tag{3}$$

where $V_m = V/m = M_i/\rho$ is molar volume, $n = m/M_i$ is the number of moles, whereas $R = R_m/M_i$ and $R_m = 8314 \text{ J}/(\text{mol·K})$.

Because the first and the second derivatives of p with respect to v at the critical point must be zero, one can obtain the following relations, which link the empirical constants a and b (or a_* and b_*) in the Equations (1)–(3) with the parameters in the critical point [11–13]

$$p_R = \frac{a_*}{27b_*^2}, \ T_R = \frac{8a_*}{27R_m b_*},\tag{4}$$

$$a = 3p_{cr}v_{cr}^2, \ b = \frac{v_{cr}}{3}, \ Z_{cr} = \frac{p_{cr}v_{cr}}{RT_{cr}} = \frac{3}{8},\tag{5}$$

$$a = \frac{a_*}{M_i^2}, \ b = \frac{b_*}{M_i}.\tag{6}$$

Here the compressibility factor Z accounts for the deviation of a real gas from ideal-gas behavior at a given temperature and pressure.

$$Z = \frac{p}{RT\rho} = \frac{pv}{RT\rho} = \frac{pV}{mRT}.\tag{7}$$

The experimentally measured values of the compressibility factor Z_{cr} for real gases vary over the range $Z_{cr} = 0.2 \ldots 0.3$. To make the van der Waals equation of state more accurate for each particular real gas, the constants a and b are determined via comparisons with precise experiments over a wider range of the parameters instead of from a single point [11–13].

For ideal gases, $Z = 1$ by definition, which results in the known ideal-gas equation [9,10]:

$$pV = m\frac{R_m}{M_i}T, \ pV = nR_m T, \ pv = \frac{R_m}{M_i}T = RT.\tag{8}$$

where $n = m/M_i$ is the number of moles.

For example, in recent works [14–16] on battery design, the van der Waals equation was used as a real gas model versus an ideal gas model. In these studies, various trends in the behavior of a real gas during its compression and expansion are presented and analyzed in both isothermal and adiabatic processes.

In the studies [16,17], several equations of state of a real gas are used, taking into account the interaction of water vapor molecules. The actual physical properties of humid air have been determined, and their influence on the conjugate heat and mass transfer for various conditions has been assessed. This provided an increase in the accuracy of predictions of heat and mass transfer processes when designing contact heat exchangers, convective drying plants, hygroscopic desalination plants, compressors with the injection of water or steam, as well as combustion chambers where flue gases are mixed with steam.

Heat transfer during natural and mixed convection has been studied extensively. In particular, some works [18–24] investigated the cases of natural and mixed convection in different geometries for the different ranges of physical parameters in the frames of the ideal gas model. The cases of real gases were considered in the studies [25–27].

Avramenko et al. [28] analytically solved the problem of natural convection in van der Waals gases near a heated vertical plate. Based on the use of the simplified van der Waals equation, analytical solutions were obtained for the profiles of velocity, temperature, and normalized Nusselt numbers. The limits of the applicability of the simplified van der Waals equation were determined. The data obtained were compared with the ideal gas model.

The objective of this work is to further study the influence of the thermophysical properties of a real gas in the framework of the above-mentioned simplified van der Waals model on heat transfer during natural convection near a heated vertical plate. A novel solution to the problem will be obtained for the first time using the integral method and compared with the previously obtained analytical solution. To authors' knowledge, such a solution has not been published in the literature yet.

2. Mathematical Model

We will here solve a problem of the steady-state natural convection over a vertical heated plate with the temperature T_w located in a non-moving gas, whose temperature is also the constant T_∞. We assume that $T_w > T_\infty$, however, the solution and its results obtained below will be valid for the case of $T_\infty > T_w$ as well. As a result of heating, a boundary layer of heated gas with a thickness of δ is formed near the plate, is formed near the plate with a vertical lifting movement affecting it. In the coordinate system we have chosen, the origin is at the lower edge of the plate, the x-axis is directed longitudinally upward, and the y-axis is directed perpendicular to the plate (Figure 1). The problem is solved in a two-dimensional statement under the assumption that the plate is infinite in the z-direction.

In the present paper, we investigate the influence of the thermophysical properties of a gas within the framework of the van der Waals equation of state on the characteristics of natural convection in comparison with the case of an ideal gas. Therefore, as a basis for comparison, we take the results for an ideal gas, which are also obtained based on the integral approach.

Within the framework of the adopted model, we assume that the physical properties of the gas, except for the density, are constant. In this regard, we only consider the buoyancy arising from the dependence of density on temperature. The energy dissipation is not considered, since the flow rate with free convection is small. The considered process of free convection is stationary.

As a result, the temperature and velocity fields can be described by the following differential equations in the boundary layer approximation [29]:

$$\frac{\partial u}{\partial x} + \frac{\partial v}{\partial y} = 0, \tag{9}$$

$$u\frac{\partial u}{\partial x} + v\frac{\partial u}{\partial y} = \nu\frac{\partial^2 u}{\partial y^2} + g\left(1 - \frac{\rho}{\rho_\infty}\right), \tag{10}$$

$$u\frac{\partial T}{\partial x} + v\frac{\partial T}{\partial y} = \alpha\frac{\partial^2 T}{\partial y^2}, \tag{11}$$

where x and y are the Cartesian coordinates, u and v are the streamwise (along the x-coordinate) and normal (along the y-coordinate) velocity components, respectively, and T is the local temperature, ν is the kinematic viscosity, g is the gravitational acceleration, ρ is the density, α is the thermal diffusivity, and the subscript "∞" refers to the parameters outside of the boundary layer.

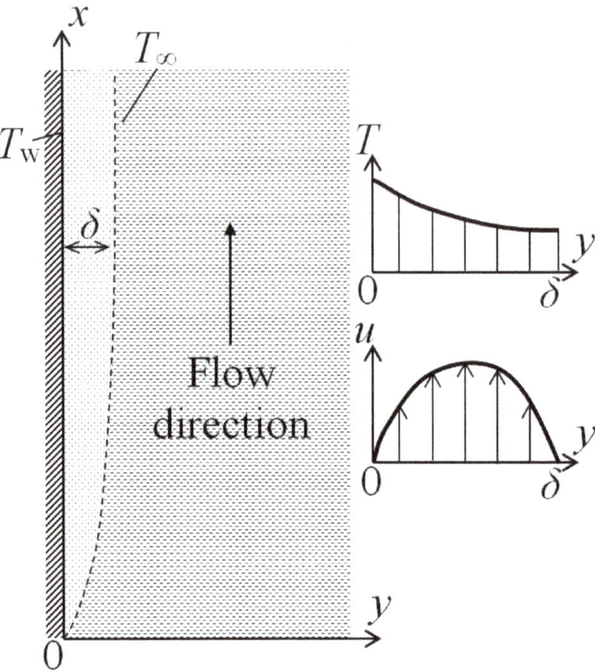

Figure 1. Schematic representation of two-dimensional natural convection boundary layer near a heated vertical plate.

As one can see from Equations (9) to (11), this model considers inertial terms and convective heat transfer. The lifting force is taken into account based on the last term of Equation (10).

In contrast to the simplified problem statement in the existing work [28], the problem statement in the present study considers the complete system of the equations of continuity, momentum, and energy, in which the inertial and convective components are explicitly taken into consideration.

The boundary conditions for system (9)–(11) are set as follows:

$$u = 0, \ T = T_w, \ \text{at} \ y = 0, \tag{12}$$

$$u = 0, \ T = T_\infty, \ \text{at} \ y = \delta, \tag{13}$$

where δ is the boundary layer thickness, and the subscript "w" refers to the parameters at the wall. The boundary conditions for the x-coordinate are described below in Section 3.

This problem statement must be completed with the van der Waals equation of state (3), rewritten using the gas density ρ instead of the specific volume υ:

$$\left(p + a\rho^2\right)\left(\frac{1}{\rho} - b\right) = RT, \tag{14}$$

where p is pressure, R is the individual (specific) gas constant, and a and b are the van der Waals constants.

The last term on the right-hand side of the equation of motion (10) can be transformed using the van der Waals equation of state (14). To do this, we solve Equation (14) with respect to density. The result is a cubic equation that has two complex conjugate roots and one real root.

$$\rho = \frac{1}{6b}\left(2 + \frac{\sqrt[3]{16}(a-3b(bp+RT))}{\sqrt[3]{2a^3+9a^2b(2bp-RT)-\sqrt{a^3\left(a(2a+9b(2bp-RT))^2-4(a-3b(bp+RT))^3\right)}}} + \frac{\sqrt[3]{2a^3+9a^2b(2bp-RT)-\sqrt{a^3\left(a(2a+9b(2bp-RT))^2-4(a-3b(bp+RT))^3\right)}}}{a\sqrt{8}}\right), \tag{15}$$

The solution to this equation is rather cumbersome, so it is difficult to use it when integrating the system of Equations (9)–(11). Avramenko et al. [28] showed that in the approximation of small values of the constants a and b, Equation (15) can be expanded in a Maclaurin series, which, in what follows, considers only the first three terms:

$$\rho = \frac{p}{RT} - \frac{bp^2}{R^2T^2} + \frac{ap^2}{R^3T^3}, \tag{16}$$

This equation can be represented in the dimensionless form [28]:

$$Z(1 + Wa_a - Wa_b) = 1, \tag{17}$$

where

$$Wa_a = \frac{ap}{R^2T^2}, \quad Wa_b = \frac{bp}{RT}, \tag{18}$$

are van der Waals numbers.

Avramenko et al. [28] validated the simplified van der Waals Equation (16) for ethylene. It was shown that Equation (16) is more accurate than the ideal gas Equation (8), and agrees well with the full van der Waals Equation (8) up to the pressure of $p = 40$ bar (or $p/p_{cr} = 0.4$) and $T = T_{cr}$. In this case, the relative error of Equation (16) is about 8%, whereas the relative error of the ideal gas Equation (8) is 22.7%.

As it was demonstrated by Avramenko et al. [28], for these pressures and temperatures, the values of both van der Waals numbers for ethylene are $Wa_a = 2.78 \cdot 10^{-6}$ and $Wa_b = 9.03 \cdot 10^{-7}$. This completely justifies the assumption of the small values of the parameters Wa_a and Wa_b that lie in the background of the simplified van der Waals Equation (16) in the series form.

The density ratio ρ/ρ_∞ in Equation (10) can be expressed using Equations (16)–(18). As a result, we get:

$$u\frac{\partial u}{\partial x} + v\frac{\partial u}{\partial y} = v\frac{\partial^2 u}{\partial y^2} + g\left(1 - \frac{Z}{1+\beta\Delta T\theta}\left(1 - \frac{Wa_b}{1+\beta\Delta T\theta} + \frac{Wa_a}{(1+\beta\Delta T\theta)^2}\right)\right), \tag{19}$$

Here, the compressibility factor Z and van der Waals numbers Wa_a and Wa_b are defined using the parameters outside of the boundary layer (subscript "∞").

The dimensionless local temperature in the boundary layer is defined as:

$$\theta = \frac{T - T_\infty}{T_w - T_\infty} = \frac{T - T_\infty}{\Delta T}, \quad \Delta T = T_w - T_\infty, \tag{20}$$

whereas

$$\beta = \frac{1}{T_\infty}, \tag{21}$$

is the volume expansion coefficient [9–11].

3. Integral Forms of Equations

The solution of the stated problem will be performed using the integral method. The essence of the integral method is the reduction in the original system of differential transport equations in partial derivatives into a system of ordinary differential equations, in which the marching coordinate x remains the only independent variable [29,30]. To do this, one should assume proper approximating functions for the profiles of the longitudinal velocity component u and the local temperature T in the boundary layer, and integrate transport Equations (10) and (11) within the limits from the wall to the outer boundary δ of the boundary layer. In this case, the continuity equation is incorporated into the integral equation of momentum transfer (to exclude the transverse velocity component v). The unknown quantities in the so-called integral equations of the boundary layer (which, in fact, after integration across the boundary layer, become ordinary differential equations) are: (a) the boundary layer thickness δ, which for gases is the same for the velocity and temperature profiles; and (b) the scaling (maximum) velocity in the boundary layer U, which increases with the development of the boundary layer and is itself unknown, and which must be found from the solution of the problem. Therefore, the only boundary conditions along the marching coordinate x, which are needed for solving the integral equations of the boundary layer, are the equality to zero of the thickness of the boundary layer δ and the scaling velocity U at the leading edge of the plate at $x = 0$. These boundary conditions will be implicitly incorporated into the mathematical form of the functions $\delta(x)$ and $U(x)$ in the subsequent solution of the problem.

Integral equations for the boundary layer in the case of the free convection of an ideal gas are given in the classical works [30,31]. Let us derive integral equations for the case of a real gas considered in this paper and described by the van der Waals equation. To obtain the integral equation of motion, we multiply the continuity Equation (9) by the velocity component and add it to the left side of Equation (19). As a result, we have:

$$\frac{\partial u^2}{\partial x} + \frac{\partial uv}{\partial y} = \nu \frac{\partial^2 u}{\partial y^2} + g\left(1 - \frac{Z}{1 + \beta\Delta T\theta}\left(1 - \frac{Wa_b}{1 + \beta\Delta T\theta} + \frac{Wa_a}{(1 + \beta\Delta T\theta)^2}\right)\right), \qquad (22)$$

Let us integrate Equation (22) over the thickness of the boundary layer, taking into account the boundary conditions (12) and (13). This gives:

$$\frac{d}{dx}\int_0^\delta u^2 dy = -\nu\left(\frac{\partial u}{\partial y}\right)_{y=0} + g\int_0^\delta\left(1 - \frac{Z}{1 + \beta\Delta T\theta}\left(1 - \frac{Wa_b}{1 + \beta\Delta T\theta} + \frac{Wa_a}{(1 + \beta\Delta T\theta)^2}\right)\right)dy, \qquad (23)$$

Let us further rewrite this equation using dimensionless variables and functions:

$$\frac{d(\delta U^2)}{dx}\int_0^1 w^2 d\eta = -\nu\left(\frac{\partial w}{\partial \eta}\right)_{\eta=0}\frac{U}{\delta} + g\delta\int_0^1\left(1 - \frac{Z}{1 + \beta\Delta T\theta}\left(1 - \frac{Wa_b}{1 + \beta\Delta T\theta} + \frac{Wa_a}{(1 + \beta\Delta T\theta)^2}\right)\right)d\eta, \qquad (24)$$

where U is a scaling velocity that depends on the streamwise coordinate (see below), whereas:

$$w = \frac{u}{U}, \qquad\qquad \eta = \frac{y}{\delta}, \qquad (25)$$

In a similar way, one can integrate the energy Equation (11). This brings:

$$\frac{d(U\delta)}{dx}\int_0^1 w\theta d\eta = -\frac{\alpha}{\delta}\left(\frac{\partial\theta}{\partial\eta}\right)_{\eta=0}, \qquad (26)$$

To solve Equations (24) and (26), it is necessary to specify the velocity and temperature profiles in the boundary layer. Authors [30,31] proposed the following equations for these profiles:

$$w = \eta(1 - \eta)^2, \qquad (27)$$

$$\theta = (1 - \eta)^3, \qquad (28)$$

The substitution of profiles (27) and (28) into the integral Equation (24) gives:

$$\frac{1}{105}\frac{d(\delta U^2)}{dx} = -\nu\frac{U}{\delta} + g\delta F, \qquad (29)$$

where

$$F(\beta\Delta T, Wa_a, Wa_b) =$$
$$\frac{1}{54}\left(\begin{array}{c} 54 - \dfrac{3\left(-6Wa_b(1+\beta\Delta T)+Wa_a(8+5\beta\Delta T)\right)}{(1+Wa_a-Wa_b)(1+\beta\Delta T)^2} + \\[2ex] + \dfrac{(9+5Wa-6Wa_b)\left(6\arctan\left(\dfrac{1-2\sqrt[3]{\beta\Delta T}}{\sqrt{3}}\right)+\sqrt{3}\ln\left(1-\dfrac{3\sqrt[3]{\beta\Delta T}}{\left(1+\sqrt[3]{\beta\Delta T}\right)^2}\right)-\pi\right)}{\sqrt[3]{\beta\Delta T}\sqrt{3}(1+Wa_a-Wa_b)} \end{array}\right) \tag{30}$$

In the limiting case of $\beta\Delta T \to 0$, Equation (29) transforms into:

$$\frac{1}{105}\frac{d(\delta U^2)}{dx} = -\nu\frac{U}{\delta} + \frac{1}{4}g\delta\beta\Delta T\frac{1+3Wa_a-2Wa_b}{1+Wa_a-Wa_b}. \tag{31}$$

For an ideal gas ($Wa_a = Wa_b = 0$), Equation (31) reduces to the respective equation obtained in the classical works [30,31]. Energy Equation (26) takes the following form:

$$\frac{1}{42}\frac{d(\delta U)}{dx} = 3\frac{\alpha}{\delta}, \tag{32}$$

If Equation (28) for the temperature profile is replaced by another profile used in our study [28]:

$$\theta = (1-\eta)^2, \tag{33}$$

then the form of the function F in Equation (30) will change as follows:

$$F(\beta\Delta T, Wa_a, Wa_b) =$$
$$1 - \frac{-4Wa_b(1+\beta\Delta T)+Wa_a(5+3\beta\Delta T)}{8(1+Wa_a-Wa_b)(1+\beta\Delta T)^2} - \frac{(8+3Wa_a-4Wa_b)\arctan\left(\sqrt{\beta\Delta T}\right)}{8\sqrt{\beta\Delta T}(1+Wa_a-Wa_b)} \tag{34}$$

Then, in the limiting case of $\beta\Delta T \to 0$ used in the Boussinesq approximation [30,31], the integral Equation (29) looks as follows:

$$\frac{1}{105}\frac{d(\delta U^2)}{dx} = -\nu\frac{U}{\delta} + \frac{1}{3}g\delta\beta\Delta T\frac{1+3Wa_a-2Wa_b}{1+Wa_a-Wa_b}, \tag{35}$$

and the energy equation takes the form:

$$\frac{1}{30}\frac{d(\delta U)}{dx} = 2\frac{\alpha}{\delta}, \tag{36}$$

Summing up, we can write the system of integral equations in a general form:

$$s\frac{d(\delta U^2)}{dx} = -\nu\frac{U}{\delta} + g\delta F, \tag{37}$$

$$t\frac{d(\delta U)}{dx} = z\frac{\alpha}{\delta}, \tag{38}$$

where the function F is defined either by Equation (30) or Equation (34). Here s and t are numerical coefficients. Substituting Equation (27) into Equation (24), one can obtain $s = 1/105$; a substitution of Equation (27) into Equation (24) yields $t = 1/30$.

4. Heat Transfer

Following the authors [30,31], we will solve the system (37) and (38) using power-law equations for the scaling velocity U and the boundary layer thickness δ:

$$U(x) = \Psi x^m, \tag{39}$$

$$\delta(x) = \Phi x^n, \tag{40}$$

where m, n, Ψ, and Φ are yet unknown constants. Solutions (39) and (40) satisfy the requirement that the scaling velocity U and the boundary layer thickness δ are equal to zero at the leading edge of the plate at $x = 0$. To find these constants, we substitute Equations (39) and (40) into Equations (37) and (38). As a result, we have:

$$(2m + n)s\Phi\Psi^2 x^{2m+n-1} = -\nu\frac{\Psi^2}{\Phi}x^{m-n} + gF\Phi x^n, \tag{41}$$

$$(m + n)t\Phi\Psi x^{m+n-1} = \alpha\frac{z}{\Phi}x^{-n}. \tag{42}$$

Let us equate the exponents at the coordinate x. As a result, we have a system of equations, whose solution is:

$$m = \frac{1}{2}, \qquad n = \frac{1}{4}, \tag{43}$$

Substituting these values into Equations (41) and (42), we obtain a system of equations for the unknown constants Ψ and Φ. Eliminating one of these unknowns, we obtain a fourth-order algebraic equation with respect to the other unknown. As a result of the solution, we have four pairs of roots, of which only one has a positive real form:

$$\Psi = 2\sqrt{\frac{zgF}{5sz + 3t\mathrm{Pr}}}, \tag{44}$$

$$\Phi = \sqrt{\frac{4z}{3t}\frac{\alpha}{\Psi}}, \tag{45}$$

Having obtained relations (44) and (45), it is possible to derive a solution for the heat transfer coefficient:

$$h = \frac{k}{\delta}\left(\frac{d\theta}{d\eta}\right)_{\eta=0} = z\frac{k}{\delta} = k\sqrt{\frac{3zt}{4}\frac{\Psi}{\alpha\sqrt{x}}} = k\sqrt{\frac{3zt}{2}\frac{1}{\alpha\sqrt{x}}}\sqrt{\frac{zgF}{5sz + 3t\mathrm{Pr}}}, \tag{46}$$

Equation (46) can be rewritten in the form of the normalized Nusselt number:

$$\frac{\mathrm{Nu}}{\mathrm{Nu}_0} = \sqrt[4]{\frac{F}{r\beta\Delta T}}, \tag{47}$$

where

$$\mathrm{Nu} = \frac{hx}{k} \tag{48}$$

The Nusselt number is commonly interpreted as a dimensionless heat transfer coefficient, which characterizes the heat transfer rate at the boundary between the wall and the flow [29–31].

Here, the subscript "0" refers to the ideal gas ($\mathrm{Wa}_a = \mathrm{Wa}_b = 0$).

The coefficient r is determined from the equation:

$$\lim_{\beta\Delta T\to 0}\frac{F}{\beta\Delta T} = r\frac{1 + 3\mathrm{Wa}_a - 2\mathrm{Wa}_b}{1 + \mathrm{Wa}_a - \mathrm{Wa}_b}, \tag{49}$$

For example, $r = 1/4$ for Equation (30), whereas $r = 1/3$ for Equation (34).

As a result, Equation (47) can be rewritten as follows:

$$\frac{\mathrm{Nu}}{\mathrm{Nu}_0} = \sqrt[4]{\frac{1 + 3\mathrm{Wa}_a - 2\mathrm{Wa}_b}{1 + \mathrm{Wa}_a - \mathrm{Wa}_b}}, \tag{50}$$

where Nu and Nu_0 are the Nusselt numbers for the van der Waals gas and the ideal gas, respectively.

In the integral method considered here, the main sought quantity is the Nusselt number. If necessary, the final relation for the velocity profile can be easily found using the profile (27) for the dimensionless velocity w, as well as the combination of Equations (39) and (44) for the function U. The profile of the local temperature T is found from either Equation (28) or Equation (33), whereas the gas density ρ is determined by Equation (16).

5. Results and Discussion

Figures 2 and 3 elucidate the effect of the dimensionless van der Waals numbers on the normalized Nusselt number Nu/Nu_0, as predicted by Equation (50). The numerical values of Nu/Nu_0 coincide with the data obtained in our work [28] using an approximate analytical solution.

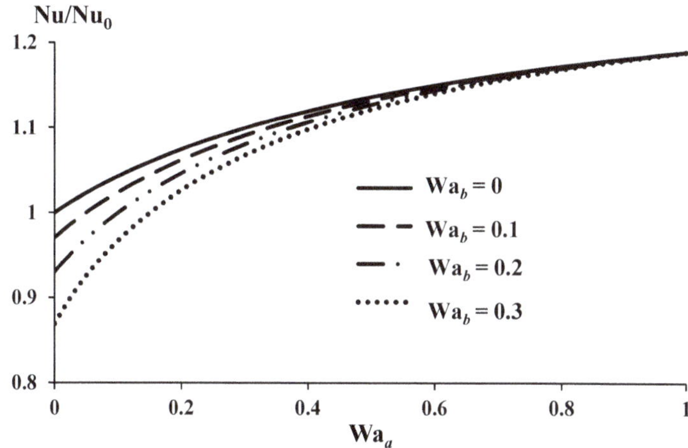

Figure 2. Effect of the van der Waals number Wa_a on the normalized Nusselt number under condition $Wa_b = \text{const}$: (1) $Wa_b = 0$; (2) $Wa_b = 0.1$; (3) $Wa_b = 0.2$; (4) $Wa_b = 0.3$.

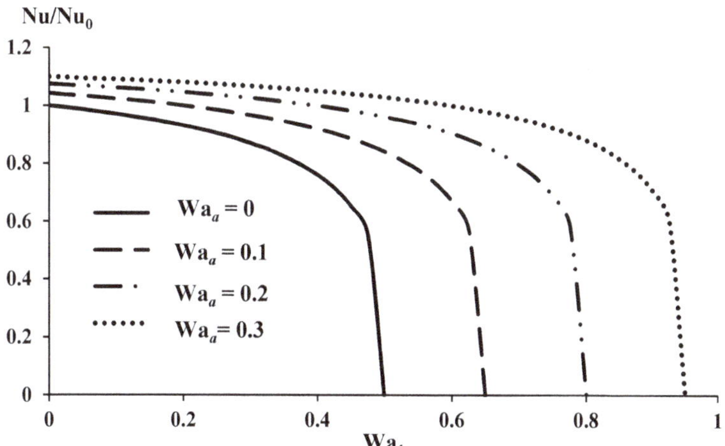

Figure 3. Effect of the van der Waals number Wa_b on the normalized Nusselt number under condition $Wa_a = \text{const}$: (1) $Wa_a = 0$; (2) $Wa_a = 0.1$; (3) $Wa_a = 0.2$; (4) $Wa_a = 0.3$.

The parameter a in the van der Waals equation characterizes the additional pressure in the near-wall layer, which increases the probability of collision of real gas molecules with the wall in comparison with the ideal one. Obviously, this leads to an increase in lifting force. As a result, the velocity in the boundary layer increases, which causes an increase in heat transfer with an increase in the Wa_a number.

As it is known, the nature of the interaction of gas molecules with the wall has a significant effect on heat transfer, which occurs due to the exchange of energy between molecules and the surface. The parameter a in the van der Waals equation characterizes the additional pressure in the near-wall layer, which increases the probability of a collision between real gas molecules and the wall in comparison with the ideal one. Obviously, this leads to an increase in the lifting force. As a result, the velocity in the boundary layer increases, which causes an increase in heat transfer, with an increase in the Wa_a number.

In turn, the parameter b describes the additional volume not filled with molecules. With its increase, the heat transfer between the molecules and the wall decreases, which causes deterioration in the conditions of interaction between them. Consequently, as can be seen from Figures 2 and 3, with an increase in the Wa_b number, the normalized Nusselt number decreases compared to ideal gas case.

Calculations have shown that with an increase in the Wa_a number, the effect of the Wa_b number noticeably weakens, and at values of $Wa_a \geq 0.6$, cannot be observed at all (Figure 2). The more pronounced effect of the Wa_a number on the flow characteristics can be explained by the fact that the constant a describes the quadratic effect on the density variation in the van der Waals equation, whereas the parameter b describes the linear effect on the density variation.

We must, however, emphasize that Equation (16) is to be used for small values of the parameters Waa and Wab for the approximate series expansion of the full van der Waals equation to remain in force. Thus, for the larger values of Waa and Wab, Figures 2 and 3 demonstrate only qualitative trends.

6. Conclusions

The study has dealt with the problem of the steady-state natural convection of a van der Waals gas near a vertical heated plate. The problem was solved analytically based on an integral method. The novel solution was obtained for the first time using a novel, simplified form of the van der Waals equation of state proposed in our work [28]. To the authors' knowledge, such a solution has not yet been published in the literature.

The effects of the dimensionless Wa_a and Wa_b numbers on the normalized Nusselt number in the real gas compared to the ideal gas were estimated. The analysis of the calculation results showed that, with an increase in the Wa_a number (which characterizes the additional pressure in the real gas), the normalized Nusselt number increases. This is due to an increase in the lifting force and velocity in the boundary layer. The effect of additional volume, which manifests itself in an increase in the Wa_b number, causes a deterioration in the conditions for interaction between gas molecules and the wall. This is accompanied by a decrease in the Archimedes force and flow rate in the boundary layer, which leads to a weakening of heat transfer when compared with an ideal gas.

It was therefore shown that the change in the heat transfer intensity is due to effects that are considered by the van der Waals equation of state, but not by the ideal gas equation.

Author Contributions: Analytical investigation, A.A.A.; conceptualization, A.A.A. and I.V.S.; methodology, A.A.A. and I.V.S.; validation and formal analysis, A.A.A., I.V.S., and N.P.D.; writing—original draft preparation, review and editing, A.A.A., I.V.S., and Y.Y.K.; writing of the final manuscript, A.A.A. and I.V.S. All authors have read and agreed to the published version of the manuscript.

Funding: The research contributions of A.A.A. and Y.Y.K. were funded via the program of research grants of the NAS of the Ukraine "Support of priority for the state scientific research and scientific and technical (experimental) developments" 2020–2021. Project 1.7.1.892: "Development of scientific and technical fundamentals of heat of mass transfer intensification in porous media for materials of building designs and the thermal engineering equipment".

Institutional Review Board Statement: Not applicable.

Informed Consent Statement: Not applicable.

Data Availability Statement: Not applicable.

Acknowledgments: Not applicable.

Conflicts of Interest: The authors declare no conflict of interest.

Abbreviations

a, b	van der Waals constants;
g	gravitational acceleration;
h	heat transfer coefficient;
k	thermal conductivity;
p	pressure;
p_{cr}	critical pressure;
R	individual (specific) gas constant;
T	local temperature;
T_{cr}	critical temperature;
u	streamwise velocity component;
v	spanwise (wall-orthogonal) velocity component;
V	volume;
υ	specific volume;
x	streamwise coordinate;
y	spanwise (wall-orthogonal) coordinate
Greek symbols	
α	thermal diffusivity;
δ	boundary layer thickness;
ν	kinematic viscosity;
ρ	density
Dimensionless values	
Nu	Nusselt number;
Pr	Prandtl number;
U	dimensionless axial (streamwise) velocity;
Θ	dimensionless temperature;
Wa_a, Wa_b	van der Waals numbers
Subscripts:	
0	ideal gas;
w	value of a parameter at the wall;
∞	value at the outer boundary of the boundary layer.

References

1. Banaszkiewicz, T.; Chorowski, M.; Gizicki, W.; Jedrusyna, A.; Kielar, J.; Malecha, Z.; Piotrowska, A.; Polinski, J.; Rogala, Z.; Sierpowski, K.; et al. Liquefied Natural Gas in Mobile Applications—Opportunities. *Energies* **2020**, *13*, 5673. [CrossRef]
2. Osorio-Tejada, J.; Llera-Sastresa, E.; Scarpellini, S. Liquefied natural gas: Could it be a reliable option for road freight transport in the EU? *Renew. Sustain. Energy Rev.* **2017**, *71*, 785–795. [CrossRef]
3. Staffell, I.; Scamman, D.; Abad, A.V.; Balcombe, P.; Dodds, P.E.; Ekins, P.; Shah, N.; Ward, K.R. The role of hydrogen and fuel cells in the global energy system. *Energy Environ. Sci.* **2019**, *12*, 463–491. [CrossRef]
4. Bystritsky, G.F. *Heat Engineering and Power Equipment of Industrial Enterprises*, 5th ed.; Yurayt: Moscow, Russia, 2018. Available online: https://urait.ru/bcode/414423 (accessed on 26 July 2021).
5. Hu, B.; Wu, D.; Wang, R.Z. Water vapor compression and its various applications. *Renew. Sustain. Energy Rev.* **2018**, *98*, 92–107. [CrossRef]
6. Perdiguer-López, R.; Berné-Valero, J.L.; Garrido-Villén, N. Application of GNSS Methodologies to Obtain Precipitable Water Vapor (PWV) and Its Comparison with Radiosonde Data. *Proceedings* **2019**, *19*, 24. [CrossRef]
7. Hanley, E.S.; Deane, P.; Gallachóir, B. The role of hydrogen in low carbon energy futures–A review of existing perspectives. *Renew. Sustain. Energy Rev.* **2018**, *82 Pt 3*, 3027–3045. [CrossRef]

8.   Taleb, A.I.; Barfuss, C.; Sapin, P.; White, A.J.; Willich, C.; Fabris, D.; Markides, C.N. The influence of real gas effects on thermally induced losses in reciprocating piston-cylinder systems. In Proceedings of the 12th International Conference on Heat Transfer, Costa de Sol, Spain, 11–13 July 2016. Available online: http://hdl.handle.net/2263/61968 (accessed on 26 July 2021).

9.   Çengel, Y.A.; Boles, M.A. *Thermodynamics: An Engineering Approach*, 5th ed.; McGraw-Hill Education: New York, NY, USA, 2004.

10. Baehr, H.-D.; Kabelac, S. *Thermodynamik—Grundlagen und Technische Anwendungen*, 15th ed.; Springer: Berlin/Heidelberg, Germany, 2012.

11. Weigand, B.; Köhler, J.; von Wolfersdorf, J. *Thermodynamik Kompakt*, 4th ed.; Springer: Berlin/Heidelberg, Germany, 2016.

12. Reid, R.C.; Prausnitz, J.M.; Poling, B.E. *The Properties of Gases and Liquids*, 4th ed.; McGraw-Hill: New York, NY, USA, 1987.

13. Eremin, V.V.; Kargov, S.I.; Kuzmenko, N.E.; Gases, R. Equations of state, thermodynamic properties, statistical description. In *Methodical Guidelines for Students of Chemical Faculties of Universities*; Poltorak, O.M., Ed.; Chemical Faculty of Moscow State University: Moscow, Russia, 1998. Available online: http://www.chem.msu.ru/rus/teaching/realgases/welcome.html (accessed on 26 July 2021).

14. Zhang, S.; Iwashita, H.; Sanada, K. Thermal performance difference of ideal gas model and van der Waals gas model in gas-loaded accumulator. *Int. J. Hydromechatronics* **2018**, *1*, 293–307. [CrossRef]

15. Miyashita, L.; Iwashita, H.; Sanada, K. A study on a mathematical model of gas in accumulator using van der Waals equation. In Proceedings of the 15th Scandinavian International Conference on Fluid Power, SICFP'17, Linköping, Sweden, 7–9 June 2017. [CrossRef]

16. Kozlova, M.V.; Sokolov, P.S.; Bannikov, A.V. Investigation of humid air physical properties influence on the heat and mass transfer processes calculation accuracy. *Vestn. IGEU* **2020**, *4*, 5–13. [CrossRef]

17. Bulygin, V.S. Heat capacity and internal energy of Van der Waals. In *Course Textbook General Physics*; MFTI: Moscow, Russia, 2012; p. 13.

18. Avramenko, A.A.; Tyrinov, A.I.; Shevchuk, I.V.; Dmitrenko, N.P.; Kravchuk, A.V.; Shevchuk, V.I. Mixed convection in a vertical circular microchannel. *Int. J. Ther. Sci.* **2017**, *121*, 1–12. [CrossRef]

19. Avramenko, A.A.; Tyrinov, A.I.; Shevchuk, I.V.; Dmitrenko, N.P.; Kravchuk, A.V.; Shevchuk, V.I. Mixed convection in a vertical flat microchannel. *Int. J. Heat Mass Transf.* **2017**, *106*, 1164–1173. [CrossRef]

20. Avramenko, A.A.; Kovetska, Y.Y.; Shevchuk, I.V.; Tyrinov, A.I.; Shevchuk, V.I. Mixed Convection in Vertical Flat and Circular Porous Microchannels. *Transp. Porous Media* **2018**, *124*, 919–941. [CrossRef]

21. Eriksson, L.; Nielsen, L. *Modelling and Control of Engines and Drivelines*, 1st ed.; John Wiley & Sons: Hoboken, NJ, USA, 2014; p. 588. [CrossRef]

22. Tilgner, A. Convection in an ideal gas at high Rayleigh numbers. *Phys. Rev.* **2011**, *84 Pt 2*, 026323. [CrossRef]

23. Kumar, R. Modelling of natural convection theory for gases neglecting effect of gravity. *AIP Conf. Proc.* **2013**, *1547*, 124. [CrossRef]

24. Heath, M.; Hall, W.; Monde, M. An experimental investigation of convection heat transfer during filling of a composite-fibre pressure vessel at low Reynolds number. *Exp. Therm. Fluid Sci.* **2014**, *54*, 151–157. [CrossRef]

25. Corcione, M.; Fontana, D.M. New dimensionless correlation-equations for laminar free convection heat transfer in real gases with high wall-fluid temperature differences. *Int. J. Therm. Sci.* **2004**, *43*, 87–94. [CrossRef]

26. Johnston, D.C. *Advances in Thermodynamics of the van der Waals Fluid*; Morgan Claypool: San Rafael, CA, USA, 2014. Available online: https://iopscience.iop.org/book/978-1-627-05532-1.pdf (accessed on 26 July 2021).

27. Vallance, C. *An Introduction to the Gas Phase*; Morgan Claypool Publishers: San Rafael, CA, USA, 2017. Available online: https://www.amazon.com/Introduction-Gas-Phase-Concise-Physics/dp/168174693 (accessed on 26 July 2021).

28. Avramenko, A.A.; Shevchuk, I.V.; Kovetskaya, M.M. An analytical investigation of natural convection of a van der Waals gas over a vertical plate. *Fluids* **2021**, *6*, 121. [CrossRef]

29. Schlichting, H. *Boundary Layer Theory*, 7th ed.; McGraw-Hill: New York, NY, USA, 1979.

30. Squire, H.B. Integral solution. In *Modern Developments in Fluid Dynamics*; Goldstein, S., Ed.; Academic Press: New York, NY, USA, 1965; Volume 2, pp. 641–643.

31. Eckert, E.R.G.; Drake, R.M. *Heat and Mass Transfer*, 2nd ed.; McGraw Hill: New York, NY, USA, 1959.

Article

Numerical Modelling of Heat Transfer in Fine Dispersive Slurry Flow

Artur Bartosik

Division of Production Engineering, Faculty of Management and Computer Modelling, Kielce University of Technology, Al. Tysiaclecia P.P. 7, 25-314 Kielce, Poland; artur.bartosik@tu.kielce.pl

Abstract: Slurry flows commonly appear in the transport of minerals from a mine to the processing site or from the deep ocean to the surface level. The process of heat transfer in solid–liquid flow is especially important for the long pipeline distance. The paper is focused on the numerical modelling and simulation of heat transfer in a fine dispersive slurry, which exhibits yield stress and damping of turbulence. The Bingham rheological model and the apparent viscosity concept were applied. The physical model was formulated and then the mathematical model, which constitutes conservative equations based on the time average approach for mass, momentum, and internal energy. The slurry flow in a pipeline is turbulent and fully developed hydrodynamically and thermally. The closure problem was solved by taking into account the Boussinesque hypothesis and a suitable turbulence model, which includes the influence of the yield shear stress on the wall damping function. The objective of the paper is to develop a new correlation of the Nusselt number for turbulent flow of fine dispersive slurry that exhibits yield stress and damping of turbulence. Simulations were performed for turbulent slurry flow, for solid volume concentrations 10%, 20%, 30%, and for water. The mathematical model for heat transfer of the carrier liquid flow has been validated. The study confirmed that the slurry velocity profiles are substantially different from those of the carrier liquid and have a significant effect on the heat transfer process. The highest rate of decrease in the Nusselt number is for low solid concentrations, while for C > 10% the decrease in the Nusselt number is gradual. A new correlation for the Nusselt number is proposed, which includes the Reynolds and Prandtl numbers, the dimensionless yield shear stress, and solid concentration. The new Nusselt number is in good agreement with the numerical predictions and the highest relative error was obtained for C = 10% and Nu = 44.3 and is equal to -12%. Results of the simulations are discussed. Conclusions and recommendations for further research are formulated.

Keywords: heat transfer in non-Newtonian slurry; damping of turbulence; Nusselt number for slurry

Citation: Bartosik, A. Numerical Modelling of Heat Transfer in Fine Dispersive Slurry Flow. *Energies* **2021**, *14*, 4909. https://doi.org/10.3390/en14164909

Academic Editor: Dmitri Eskin

Received: 21 July 2021
Accepted: 8 August 2021
Published: 11 August 2021

Publisher's Note: MDPI stays neutral with regard to jurisdictional claims in published maps and institutional affiliations.

1. Introduction

Slurry pipeline transport is widely used in the food and chemical industry and in the transfer of minerals from a mine to the processing site or from the deep ocean to the surface level [1]. Pipeline transportation is more economical and environmentally friendly than other modes of transport, such as rail or trucking, when moving minerals over long distances. They also have a high safe delivery rate [2,3]. In the literature, we can find various approaches for the analysis of solid–liquid flow and its predictions [4–14]. Analyses are mainly dedicated to a proper class of flow, such as homogeneous, heterogeneous, and with steady or mowing beds. The proposed models are mainly algebraic. If predictions are considered, we use analytical or numerical methods [15–24]. Analytical methods are impractical and can be applied to very simple flow conditions. If numerical methods are considered, we recognize Direct Numerical Simulation (DNS), which is computationally costly, Large Eddy Simulation (LES), which is less expensive, Reynolds-averaged Navier–Stokes equations (RANS), which is the computationally cheapest, or a combination of the mentioned methods, called hybrid methods. The slurry flow phenomenon is very complex; therefore, mathematical models are strictly dedicated to a certain flow situation.

The heat transfer process in slurry transportation is essential, especially for long distances. Temperature affects the viscosity and rheological properties of the slurry and thus changes the energy loss during pipeline transport, which affects the operation of the pipeline system. Changes in the temperature of the transported slurry can be caused by heat transfer between the slurry pipeline and the environment and by the conversion of part of the mechanical energy into heat. In the case of pipeline operation in the Arctic Sea, for example, it could cause permafrost melting and instability of pipeline routes. In other cases, maintaining a constant temperature of the flowing medium is imperative, while in some other cases, the risk of freezing or loss of heat appears, which requires proper pipe insulation to avoid plug formation. Therefore, the ability to predict the flow of the slurry with heat exchange is a great challenge in CFD.

To predict the heat transfer between the flowing slurry and the environment, the first point is to set up a reliable physical model and then a mathematical model. The mathematical model should be able to predict the velocity and temperature fields of the slurry. When formulating a physical model for slurry flow with heat transfer, it is necessary to define the slurry properties and flow conditions like, for instance:

- particle size distribution and its shape, elasticity, porosity, degradation, hydrophobicity, averaged solid particle diameter and density;
- carrier liquid viscosity and density;
- type of slurry: Newtonian or non-Newtonian, with or without yield stress;
- type of solid concentration: dilute suspension or highly concentrated;
- type of slurry flow: laminar, transient, turbulent, homogeneous, pseudo-homogeneous, heterogeneous, with steady or moving beds;
- geometry and flow direction: circular or noncircular conduit, one-, two-, or three-dimensional, vertical, horizontal, or inclined conduit;
- boundary conditions for velocity and temperature;

The study focuses on the modelling and simulation of convective heat transfer between a fine dispersive slurry and a pipe wall under a turbulent flow regime in a range of solid concentrations for commercial applications. The solid phase constitutes fine solid particles with an average diameter of about 20 μm, since small particles are responsible for increasing viscosity, non-Newtonian behavior, and damping of turbulence. Therefore, the relation between shear stress and shear rate is crucial to include in a mathematical model.

Solid particles are known to affect the flow structure and can enhance or suppress turbulence [25–33]. Generally, one can conclude that small particles can suppress, while large particles can increase turbulence. Of course, the phenomenon is more complex as the particle density, carrier liquid properties, and flow conditions, such as the velocity and geometry effect on the level of turbulence as well. Experimental data on frictional head loss for turbulent flow of fine dispersive slurry clearly demonstrate a lower frictional head loss than expected [6,25,34,35]. Wilson and Thomas hypothesized that in a turbulent flow of fine dispersive slurry, the viscous sublayer becomes thicker compared to a single-phase flow under the same flow conditions [25]. Therefore, the Wilson and Thomas hypothesis was taken into account to develop a mathematical model.

Considering the heat exchange between the transported slurry and the surrounding, we recognize methods and techniques focused on the enhancement of heat exchange, named passive or active. Passive methods, such as shape insert with dedicated perforation or mechanically deformed pipes, have been studied for several years and have become commercial solutions [36–41]. Active methods, such as air injection, bubble generation, or proper pulsation, can produce increases in the heat transfer process [42–45].

Heat transfer in the solid–liquid flow has been experimentally investigated by several studies [46–49]. However, experiments and studies on heat transfer in a turbulent flow with fine solid particles, especially for solid concentrations of commercial interest, above 20% by volume, are scattered in the literature. Existing experiments refer mainly to water-air or water-oil or nanofluids or slurries with low solid concentrations [50–53]. Research mainly includes solid particles made of metals, oxides, or carbides suspended in a liquid. Such experiments and predictions focus on examining whether solid particles can increase or decrease the heat transfer process between the suspension and a pipe wall. The results show that such an increase depends mainly on the properties of the particles and the solid concentration [54,55].

Wang et al. [48] experimentally examined the thermal conductivity of Al_2O_3 and CuO nanoparticles mixed with water, vacuum pump liquid, engine oil and ethylene glycol. The average particle diameter of Al_2O_3 was 28 nm and the average particle diameter of CuO was 23 nm. All particles were loosely agglomerated in the chosen liquid. Experiments have shown that the heat transfer of the mixture increases with the volume fraction of Al_2O_3 particles [48].

Rozenblit et al. [56] conducted an experimental study on the heat transfer coefficient associated with solid–liquid transport in a horizontal pipe using an electro-resistance sensor and infrared imaging. They considered a flow of acetyl-water mixture on a moving bed with solid volume concentrations of 6%, 9%, 12%, and 15% by volume. They noted that the local heat transfer coefficient changes from its lowest value at the bottom of the pipe to its highest value above the carrier liquid at the upper heterogeneous layer. These experiments indicated that the heat transfer coefficient is strongly influenced by the cross-sectional distribution of the solid phase [56].

Ku et al. [49] experimentally examined the effect of solid particles on heat transfer. They used spherical fly ash particles with a mass median particle diameter of 4 to 78 μm and a particle density of 2270 kg/m^3, and Reynolds numbers of 4000 to 11,000 in an 8 mm inner diameter pipe. The authors noted that the highest heat transfer coefficient was obtained for the dilute solution with C = 3%, and then gradually decreased with increasing solid volume concentration. They proposed a correlation for the Nusselt number. The correlation depends on Reynolds and Prandtl numbers, solid concentration, and the ratio of pipe diameter to average particle diameter. The correlation is limited to: Re = 3000–11,000; Pr = 3.8–5.0; D/d_P = 102–615; C = (1–10)%. However, it is not clear whether such suspensions possess or do not possess the yield stress. Furthermore, the apparent viscosity of the slurry in their research was 1.75 higher compared to water if C = 30%. This contrasts with current studies, as the apparent viscosity of C = 30% is at least 13 times higher compared to water and increases as the wall shear stress decreases. Additionally, the Prandtl number in the current study is much higher compared to the limitation of Ku et al. 49]. In conclusion, we can say that it is impossible to compare the numerical predictions with the experimental data if the slurry properties differ much.

Bartosik performed simulations of the influence of the solid volume concentration [57] and the yield shear stress [58] on heat transfer in kaolin slurry. The author considered two solid concentrations of kaolin equal to 7.4% and 38.3%, which correspond to yield shear stresses of 4.92 Pa and 9 Pa, respectively. The author found that both solid concentration and yield stress strongly affect the Nusselt number and the Nusselt number decreases with increasing solid concentration or yield stress.

Analysis of the literature indicates that most of the research deals with heat transfer in a single phase, liquid-air flow, nanofluids, ice slurry, laminar flow, or with low solid concentrations. Experiments and modelling of heat transfer in non-Newtonian slurries containing minerals with fine particles, yield stress, and solid volume concentration at least 20% by volume are scattered. Reliable predictions require reliable measurements. Prediction difficulties arise when the solid concentration increases and the particle-liquid and/or particle–wall interaction occurs. A better approach to understanding the slurry flow mechanism and its correlation with the heat transfer process, especially at high solid concentrations, is needed, as it is important for better optimization design and operation control. The ability to simulate turbulent solid–liquid flow to predict frictional head loss, velocity, and temperature distributions remains one of the main challenges in CFD. However, from an engineering point of view, we need simple empirical or semiempirical correlations that allow the calculation of the heat transfer coefficient for known slurry. Therefore, proper correlations for the Nusselt number are still required.

The objective of the paper is to develop a new correlation of the Nusselt number for turbulent flow of fine dispersive slurry that exhibits yield stress and damping of turbulence. For such a slurry, a new correlation of the Nusselt number is proposed. The new correlation of the Nusselt number includes Reynolds and Prandtl numbers, solid volume concentration, and dimensionless yield shear stress.

2. Physical Model

The slurry is assumed to be a mixture of fine solid particles and water, which plays the role of a carrier liquid. The assumptions made in the physical model are as follows:

- solid particles are round, smooth, rigid, non-hydrophobic, and the influence of such properties on slurry flow is not considered;
- solid particles are sufficiently fine, with a median particle diameter $d_P \cong 20$ μm, so the damping of turbulence, described by Wilson and Thomas [25], occurs;
- the slurry flows in a straight and smooth horizontal pipeline, which is rigid and its inner diameter is constant;
- the profile of the solid concentration distribution on the vertical axis is constant; therefore, the slurry flow is homogeneous, which is in accordance with the experiments of Sumner et al. [34] and Ramisetty [59];
- the rheological properties of the slurry can be approximated by the Bingham model;
- the flow of the slurry is turbulent, steady, axially symmetric (V = 0) and without circumferential eddies (W = 0);
- the slurry flow is fully developed hydrodynamically, which means that the time-averaged velocity component U does not change in the flow direction ($\partial U / \partial x = 0$).
- the wall temperature is steady;
- the heat flux is homogeneous; the heat flux acts from the slurry to the pipe wall; its value is steady, known, and sufficiently small, so the following arbitrarily chosen condition is met: $|T_b - T_w| < 7$ K;
- the apparent viscosity and specific heat are constant and are determined by the thermal condition at the pipe wall, while the slurry density and conduction coefficient are calculated from the bulk temperature;
- there is no inner heat production in a slurry, and conversion of mechanical energy into internal heat is not considered;
- heat due to radiation and conduction through a pipe wall are negligible;
- the slurry flow is fully developed thermally, which means that the temperature changes linearly in the flow direction 'ox' ($\partial T / \partial x = \text{const} \neq 0$).

For simulation purposes, the cylindrical coordinate system is used, where 'ox' is the main flow direction, while 'or' is the radial coordinate, as presented in Figure 1. It is assumed that the wall temperature is equal to $T_w = 293.15$ K, and the heat flux density $q = -500$ W/m and both quantities are steady. Simulations are performed for carrier liquid flow (C = 0%) and for slurry with a solid particle density equal to 2550 kg/m^3, solid concentrations 10%, 20%, 30% and for an inner pipe diameter D = 0.02 m.

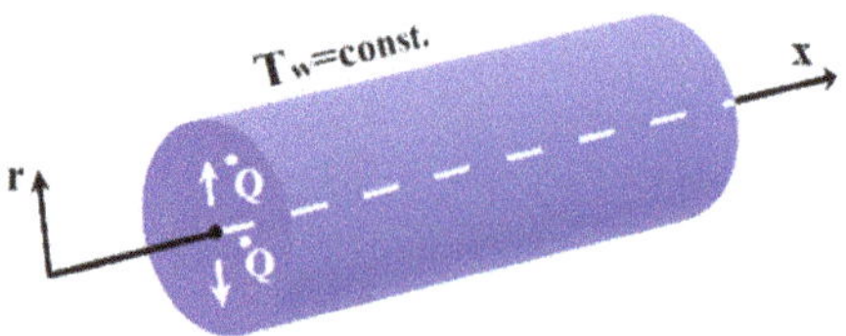

Figure 1. The cylindrical coordinate system used for numerical predictions.

3. Mathematical Model

The starting points for building a mathematical model for non-isothermal slurry flow in a pipeline are continuity, Navier–Stokes, and internal energy equations. A mathematical model is formulated for Newtonian liquid and non-Newtonian slurry.

3.1. Newtonian Liquid

The mathematical model for the Newtonian liquid regards the carrier liquid, which is pure water. Taking into account the time-averaged procedure proposed by Reynolds [60], and neglecting terms with fluctuations of density and viscosity, it is possible to obtain the conservative equations of mass, momentum, and internal energy as follows:

$$\frac{\partial}{\partial x_i}\left(\bar{\rho}\overline{U}_i\right) = 0 \tag{1}$$

$$\frac{\partial}{\partial x_j}\left(\bar{\rho}\overline{U}_i\overline{U}_j\right) = -\frac{\partial \bar{p}}{\partial x_i} + \frac{\partial}{\partial x_j}\left(\mu\frac{\partial \overline{U}_i}{\partial x_j} - \bar{\rho}\,\overline{u_i'u_j'}\right) + \bar{\rho}g_i \tag{2}$$

$$\frac{\partial}{\partial x_j}\left(\bar{\rho}\overline{U}_j\overline{T}\right) = \frac{\partial}{\partial x_j}\left(\frac{\mu}{Pr}\frac{\partial \overline{T}}{\partial x_j}\right) - \frac{\partial}{\partial x_j}\left(\bar{\rho}\,\overline{u_j't'}\right) \tag{3}$$

where the Prandtl number in Equation (3) is defined as follows:

$$Pr = \frac{\mu\,c_p}{\lambda} \tag{4}$$

The momentum Equation (2), also named the Reynolds Averaged Navier–Stokes equation, possesses an additional term $\bar{\rho}\,\overline{u_i'u_j'}$, which appears on the right-hand side, named turbulent stress tensor or Reynold's stress tensor. It is a symmetric tensor and requires additional equations to calculate its components. To solve the set of equations, the number of dependent variables must be the same as the number of equations, which is called the closure problem. Therefore, additional equations are proposed that allow one to calculate the components of the turbulent stress tensor. Equation (3) possesses an additional term $\bar{\rho}\,\overline{u_j't'}$. This term is responsible for the intensification of heat exchange caused by the presence of velocity and temperature fluctuations and requires closure, too. The additional terms in Equations (2) and (3) can be modelled by a direct approach (DNS) or an indirect approach. To calculate each component of $-\bar{\rho}\,\overline{u_i'u_j'}$ and $-\bar{\rho}\,\overline{u_j't'}$ the direct or indirect methods require additional equations, either algebraic or partial differential. Although DNS and LES methods can give a high level of accuracy, they are time-consuming and computationally costly. Indirect methods can give accurate predictions with less computational power. Moreover, we deal with time-averaged quantities of velocity and temperature, instead of random quantities, which are more useful to engineers. For this study, an indirect method is employed. The indirect method employs the Boussinesque hypothesis, which introduced a new dependent variable, called turbulent viscosity, as a measure of turbulence rather than viscosity [61]. The Boussineque hypothesis assumes that the turbulent stress tensor in Equation (2) can be expressed as follows:

$$-\bar{\rho}\,\overline{u_i'u_j'} = \mu_t\left(\frac{\partial \overline{U}_i}{\partial x_j} + \frac{\partial \overline{U}_j}{\partial x_i}\right) - \frac{2}{3}\bar{\rho}\,k\,\delta_{i,j} \tag{5}$$

and the fluctuating components of velocity and temperature in the energy Equation (3), as:

$$-\bar{\rho}\,\overline{u_j't'} = \frac{\mu_t}{Pr_t}\frac{\partial \overline{T}}{\partial x_i} \tag{6}$$

The turbulent Prandtl number (Pr_t) in Equation (6) was intensively examined by several researchers. Studies indicated that in the flow on a plate, $Pr_t \cong 0.5$, while for the boundary layer, which exists in a pipe flow, the turbulent Prandtl number is estimated to be $Pr_t = 0.9$ [62].

If an axially symmetric flow is considered, the most suitable coordinate system is cylindrical, where the dependent variables are functions of x, r, θ. The assumption that the flow is axially symmetric causes the velocity component V = 0. The lack of circumferential eddies causes that component of the velocity W = 0. The last term on the right-hand side of Equation (2) is also zero because the flow is horizontal.

Taking into account the assumptions stated in the physical model, the time-averaged Equations (1)–(3) together with (5), (6) can be formulated in the final form as follows:

$$\frac{\partial}{\partial x}\left(\bar{\rho}\bar{U}\right) = 0 \tag{7}$$

$$\frac{1}{r}\frac{\partial}{\partial r}\left[r\left(\mu + \mu_t\right)\frac{\partial \bar{U}}{\partial r}\right] = \frac{\partial \bar{p}}{\partial x} \tag{8}$$

$$\bar{\rho}\bar{U}\frac{\partial \bar{T}}{\partial x} = \frac{1}{r}\frac{\partial}{\partial r}\left[r\left(\frac{\mu}{Pr} + \frac{\mu_t}{Pr_t}\right)\frac{\partial \bar{T}}{\partial r}\right] \tag{9}$$

Taking into account the assumption that the flow is fully developed thermally, which means that the temperature changes linearly in the main flow direction 'ox', the convective term in Equation (9) will be delivered by an energy balance. An energy balance is developed for the slurry flowing in a tube with inner diameter D and unit length $L = 1$ m. Such an approach simplifies the mathematical model to one-dimensional and substantially reduces the time of computation.

Let us consider a heat flux acting between the flowing liquid and a pipe—Figure 1.

$$\dot{Q} = \alpha\, A_*\, \Delta\bar{T}_r \tag{10}$$

where

$$A_* = \pi DL \tag{11}$$

and $\Delta\bar{T}_r = T_b - T_w$ is temperature difference on radius r.

The change in the inner energy of the flowing liquid at distance L is the following:

$$\dot{Q} = \dot{m}\, c_p\, \Delta\bar{T} = \bar{\rho}_b\, \bar{U}_b\, A\, c_p\, \Delta\bar{T}_x \tag{12}$$

where the pipe cross section $A = \pi D^2/4$, and $\Delta\bar{T}_x$ is the slurry temperature difference at distance $x = L$.

Comparing (10) and (12), we can write:

$$\alpha\, \pi\, D\, L\, \Delta\bar{T}_r = \bar{\rho}_b\, \bar{U}_b\, \pi\, \frac{D^2}{4}\, c_p\, \Delta\bar{T}_x \tag{13}$$

Dividing each side of Equation (13) by the unit length of the pipe L, we can write:

$$\dot{q} = \bar{\rho}_b\, \bar{U}_b\, \pi\, \frac{D^2}{4}\, c_p\, \frac{\Delta\bar{T}_x}{L} \tag{14}$$

where the heat flux density can be written as:

$$\dot{q} = \frac{\dot{Q}}{L} = \alpha\, \pi\, D\, \Delta\bar{T}_r \tag{15}$$

The final form of the convective term can be established using Equation (14), as follows:

$$\frac{\partial \overline{T}}{\partial x} = \frac{\dot{q}}{\overline{\rho}_b \, \overline{U}_b \, \pi \, R^2 c_p} = const \tag{16}$$

The convective term (16) delivered from the energy balance replaces the convective term on the left-hand side of Equation (9). Taking into account Equation (16), one can say that the convective term $\partial T/\partial x = 0$ if q = 0. The heat flux acting on the pipeline is positive if $T_w > T_b$ or negative if $T_w < T_b$. The thermal properties of the slurry are treated as known quantities and will be presented. The influence of temperature on a carrier liquid density and a thermal conductivity was approximated using the Taylor expansion as follows:

$$\Phi = a + b\overline{T} + c\overline{T}^2 \tag{17}$$

while slurry properties, such as density, coefficient of heat conduction, and specific heat, were calculated as follows:

$$\Phi_{SL} = C\Phi_S + (1 - C)\Phi_L \tag{18}$$

Finally, the mathematical model of heat transfer in liquid flow constitutes two partial differential equations, namely, (8) and (9), since the continuity Equation (7) is included in the momentum Equation (8), and the complementary Equation (16). Equation sets possess the following dependent variables: U(r), μ_t(r), p(x), T(r). To calculate the velocity and temperature profiles across a pipe, a set of equations requires closure. Turbulent viscosity μ_t(r) will be calculated using an indirect method together with a suitable turbulence model, while the pressure gradient $\partial p/\partial x$ will be treated as a known value. For the preset value of $\partial p/\partial x$, the velocity and temperature profiles will be calculated. In other words, for known values of $\partial p/\partial x$ the time-averaged profiles of U and T will be calculated.

To calculate the turbulent viscosity μ_t(r), the Launder and Sharma turbulence model was chosen [63]. The chosen model fulfills the requirements formulated by Spalding [64,65], an honorable founder of Computational Fluid Dynamics (CFD). The Launder and Sharma model was successfully examined for comprehensive types of flow, including homogeneous slurries. The researchers emphasized that it is one of the first and most widely used models and has been shown to agree well with the experimental and DNS data for a wide range of turbulent flow problems, performing better than many other k–ε models [66–69].

$$k = \frac{\overline{u_i' u_i'}}{2} \tag{19}$$

$$\varepsilon = \nu \overline{\frac{\partial u_i'}{\partial x_k} \frac{\partial u_i'}{\partial x_k}} \tag{20}$$

The Launder and Sharma turbulence model assumes that the turbulent viscosity depends on the kinetic energy of the turbulence, defined by Equation (19), and the dissipation rate of the kinetic energy of the turbulence, defined for homogeneous turbulence by Equation (20) [70]. On the basis of dimensional analyzes, Launder and Sharma assumed that the turbulent viscosity depends on k, ε and ρ as follows:

$$\mu_t = f_\mu \frac{\overline{\rho}}{\varepsilon} k^2 \tag{21}$$

where f_μ is called the damping of the turbulence function or the wall function. The function causes a reduction in turbulent viscosity if y → 0. The wall function was developed empirically by matching the predictions with measurements and is the following:

$$f_\mu = 0.09 \, exp \left[\frac{-3.4}{\left(1 + \frac{Re_t}{50}\right)^2} \right] \tag{22}$$

where the turbulent Reynolds number (Re_t) was developed from dimensionless analysis, as follows:

$$Re_t = \frac{\rho \, k^2}{\mu \, \varepsilon}$$ (23)

Equations for the kinetic energy of turbulence and its dissipation rate are obtained from the Navier–Stokes equations, using a time-averaged procedure [63].

Taking into account the assumptions made in the physical model, the final form of the equations for the kinetic energy of turbulence and its dissipation rate, in cylindrical coordinates, are as follows:

$$\frac{1}{r}\frac{\partial}{\partial r}\left[r\left(\mu + \frac{\mu_t}{\sigma_k}\right)\frac{\partial k}{\partial r}\right] + \mu_t\left(\frac{\partial \overline{U}}{\partial r}\right)^2 = \rho\varepsilon + 2\mu\left(\frac{\partial k^{1/2}}{\partial r}\right)^2$$ (24)

$$\frac{1}{r}\frac{\partial}{\partial r}\left[r\left(\mu + \frac{\mu_t}{\sigma_\varepsilon}\right)\frac{\partial \varepsilon}{\partial r}\right] + C_1\frac{\varepsilon}{k}\mu_t\left(\frac{\partial \overline{U}}{\partial r}\right)^2 = C_2\left[1 - 0.3\,exp\left(-Re_t^2\right)\right]\frac{\rho\varepsilon^2}{k} - 2\frac{\mu}{\rho}\mu_t\left(\frac{\partial^2 \overline{U}}{\partial r^2}\right)^2$$ (25)

The set of partial differential Equations (8), (9), (24) and (25) together with the complementary Equations (16), (21)–(23) is dedicated to a Newtonian liquid.

3.2. Non-Newtonian Fine Dispersive Slurry

If non-Newtonian slurry is considered, the first step is to set up a suitable rheological model. According to the physical model, we consider a slurry with fine solid particles. It is well known that fine solid particles are responsible for increased viscosity and non-Newtonian behavior [6]. The slurry with fine solid particles can be described by several rheological models, like for instance, Bingham, Ostwalda–de Waele, Carreau, Casson, or Herschel–Bulkley. The Bingham model is very simple as it demonstrates the yield stress and constant viscosity. Two- and three-parameter models, such as Casson or Herschel-Bulkley, describe the influence of the shear rate on the shear stress more accurately. In the case of laminar flow, a rheological model plays a crucial role in determining shear stress, while in a turbulent flow the dominant role plays turbulence. Therefore, the Bingham model is used as a simpler one. It is also important that experimental data for such a model are available in the literature.

The Bingham rheological model is described by Equations (26) and (27), as follows:

$$\tau = \tau_o + \mu_{PL}\dot{\gamma} \text{ for } \tau > \tau_o$$ (26)

and

$$\dot{\gamma} = 0 \text{ for } \tau \leq \tau_o$$ (27)

Taking into account the concept of apparent viscosity [25,71,72], one can write:

$$\tau = \mu_{app}\,\dot{\gamma}$$ (28)

Taking into account the right-hand side of Equations (26) and (28), the following equation for apparent viscosity can be obtained:

$$\mu_{app} = \mu_{PL} + \frac{\tau_o}{\dot{\gamma}} = \frac{\mu_{PL}}{1 - \frac{\tau_o}{\tau_w}}$$ (29)

Of course, Equation (29) has the limitation that the yield shear stress cannot be equal to or higher than the wall shear stress. If the wall shear stress increases, the apparent viscosity decreases.

Considering the forces that act on the slurry flowing in a horizontal pipeline of length L and inner diameter D, we can write that the force responsible for the movement is:

$$\vec{F}_1 = \Delta p\,\pi\,D^2\frac{1}{4}$$ (30)

while the resistance force is the following:

$$\vec{F}_2 = \tau_w \, \pi \, D \, L \tag{31}$$

For a steady flow of slurry, both forces must be equal. Therefore, we can write:

$$\tau_w = \frac{\Delta p}{L} \frac{D}{4} = \frac{\partial p}{\partial x} \frac{D}{4} \tag{32}$$

Finally, for known plastic viscosity, yield stress and $\partial p / \partial x$, the apparent viscosity can be calculated as follows:

$$\mu_{app} = \frac{\mu_{PL}}{1 - \frac{\tau_o}{\frac{\partial p}{\partial x} \frac{D}{4}}} \tag{33}$$

Analyzing Equation (33) it can be concluded that for the Bingham model the apparent viscosity is constant across a pipe stream if $\partial p / \partial x$ = const and τ_o = const. To consider a non-Newtonian slurry, the coefficient of dynamic viscosity, which exists in Equations (8), (9) and (23)–(25), has to be replaced by the apparent viscosity, defined by Equation (33).

$$f_\mu = 0.09 \, exp \left[\frac{-3.4 \left(1 + \frac{\tau_o}{\tau_w} \right)}{\left(1 + \frac{Re_t}{50} \right)^2} \right] \tag{34}$$

The mathematical model for isothermal flow, which constitutes Equations (8), (24) and (25) together with the complementary Equations (21)–(23) and (33) fails if the predictions of fine dispersive slurry flow, defined in the physical model, are considered. Such a slurry exhibits lower head loss than expected. Therefore, taking into account the Wilson and Thomas hypothesis [25], Bartosik [73] proposed a modified the wall function, defined by Equation (34). The wall Function (34) is important for a close distance to a pipe wall. For a turbulent Reynolds number greater than 100, the wall Function (34) gives similar results to the function defined by Equation (22). Now, the Equation (34) replaces Equation (22) proposed by Launder and Sharma.

Finally, the mathematical model for heat transfer in fine dispersive slurry flow constitutes the partial differential Equations (8), (9), (24) and (25), together with the complementary Equations (16), (21), (23), (33) and (34). The constants in the k-ε turbulence model are the same as in the Launder and Sharma model and are the following: C_1 = 1.44; C_2 = 1.92; σ_k = 1.0; σ_ε = 1.3 [63].

4. Numerical Procedure

The mathematical model assumes the following boundary conditions:

$$\text{at the pipe wall for } r = R: T = T_w; q = \text{const}; U = 0; k = 0; \varepsilon = 0; \tag{35}$$

$$\text{symmetry axis for } r = 0: \partial T / \partial r = 0; \partial U / \partial r = 0, \partial k / \partial r = 0 \text{ and } \partial \varepsilon / \partial r = 0. \tag{36}$$

The boundary condition (35) indicates that there is no sleep velocity, k and ε on the pipe wall (r = R). Very near the wall, the dissipation rate is equal to 2 $\mu (\partial k^{0.5} / \partial r)^2$. This term was added to Equation (24) to allow to be set to zero on the pipe wall (y = 0). The boundary condition (36) indicates that axially symmetric conditions were applied to all dependent variables.

The set of partial differential Equations (8), (9), (24) and (25) was computed using a finite difference scheme and its own computer code. The set of equations was solved taking into account the TDMA approach, with an iteration procedure, and using the control volume method [74]. The control volume was obtained for a pipe of length equal to L = 1 m and rotating the radius of the pipe around the symmetry axis at an angle of 1 radian.

The calculations were carried out in a proper order. For the predetermined value of $\partial p / \partial x$, the following inlet conditions were used:

- $T(r) = T_w$;
- $U(r) = U_{max}[(R - r)/R]^{1/7}$, where U_{max} was arbitrarily chosen;
- k was set up on the basis of the intensity of the assigned turbulence, that is, $k(r) = \frac{1}{2} U^2(r)$.
- ε was established based of the relation: $\varepsilon(r) = 0.09^{3/4} k^{3/2}(r)/L$, where L was determined from the Nikuradse formula.

Attention was paid to the effect of the number of grid points localized across a radius of a pipe. It is well known that the number of nodal points strongly affects the accuracy of the predictions [74,75]. Therefore, the radius of the pipe was divided into 80 nodal points that were not uniformly distributed across the pipe. Most of the nodal points were located near the wall of the pipe. The number of nodal points was experimentally set to provide nodally independent predictions.

The iteration cycles were repeated until the convergence criterion, defined by Equation (37), was achieved.

$$\sum_j \left| \frac{\varnothing_j^n - \varnothing_j^{n-1}}{\varnothing_j^n} \right| \leq 0.001 \tag{37}$$

where $\varnothing_j^n$ is a general dependent variable $\varnothing = U, T, k, \varepsilon$; the jth is the nodal point after the nth iteration cycle and the $\varnothing_j^{n-1}$ is the (n − 1)th iteration cycle.

Finally, the hydrodynamically and thermally developed turbulent flow of the fine dispersive slurry was calculated and the following profiles were obtained: T(r), U(r), k(r), ε(r) and the following dependent variables were calculated: U_b, T_b, Re, Pr, Nu.

5. Validation of the Mathematical Model

The mathematical model for the isothermal flow of fine dispersive slurry has been validated in a comprehensive range of solid concentrations, yield stresses, Reynolds number and pipe diameters, providing fairly good predictions of frictional head loss and velocity profiles [20,21,73,76].

The validation of the heat transfer was performed only for the carrier liquid flow. Heat transfer validation for the fine dispersive slurry was not performed because, to the best knowledge of the author, no experimental data are available. Existing experiments and correlations of Nusselt number, made for instance, by Harada et al. [46], Ku et al. [49], Salamone and Newman [77], Ozbelge and Somer [78] do not take into account the slurry defined in the physical model.

Validation of heat transfer for carrier liquid flow includes the prediction of the Nusselt number. The results were compared with empirical data expressed by the Dittus–Boelter correlation [79]. The Dittus–Boelter correlation is valid for fully developed flow and for $Re > 10{,}000$ and $0.7 \leq Pr \leq 160$ [79]. The Dittus–Boelter correlation is reasonably consistent with the experimental data [80,81] and is expressed as follows:

$$Nu = 0.02296 \, Re^{0.8} \, Pr^{1/3} \tag{38}$$

where Reynolds and Prandtl numbers were expressed for this study by Equations (39) and (40), respectively.

$$Re = \frac{\rho \, U_b \, 2 \, R}{\mu_{app}} \tag{39}$$

$$Pr = \frac{\mu_{app} \, c_p}{\lambda} \tag{40}$$

Of course, the apparent viscosity (μ_{app}) in case of carrier liquid is equal to carrier liquid viscosity (μ).

The Dittus–Boelter correlation, expressed by Equation (38), approximates the physical situation quite well for the case of constant wall temperature and constant wall heat flux, which is just exactly as it is assumed in the physical model.

Figure 2 presents comparisons of the Nusselt number calculated using the mathematical model with the Dittus–Boelter correlation (38) for the carrier liquid in the range of Reynolds numbers of 6900 to 100,000. The results of the comparison presented in Figure 2 show good agreement; however, there are some discrepancies for low and high Reynolds numbers. The average relative error in the range of Re = 6900–100,000 is approximately −3.5%.

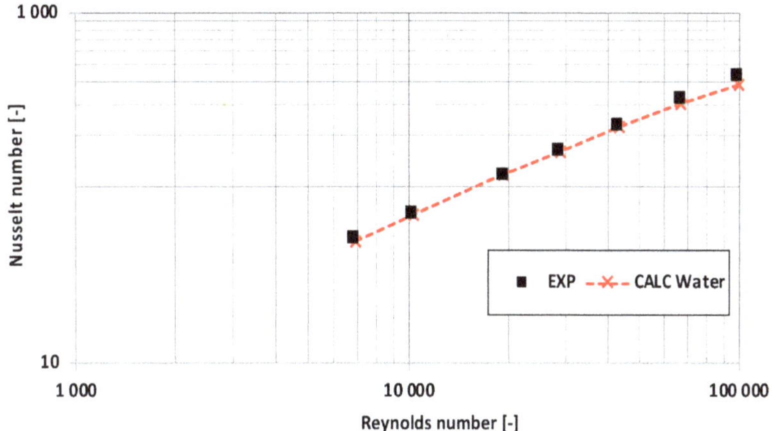

Figure 2. Validation of numerical predictions for carrier liquid flow. D = 0.02 m; T_w = 293.15 K, q = −500 W.

As mentioned above, the mathematical model for the isothermal flow of fine dispersive slurries, which exhibits turbulence damping, was successfully examined. The model is also positively verified for heat transfer in the carrier liquid flow—Figure 2. Therefore, it is assumed that the mathematical model is suitable for predicting the thermally fully developed flow of the fine dispersive slurry with constant wall heat flux and constant wall temperature.

6. Simulation Results

To perform simulations of heat transfer in a turbulent flow of a fine dispersive slurry, which is non-Newtonian and exhibits yield stress and turbulence damping, the rheological parameters must first be determined. Therefore, rheological measurements of Shook and Roco of a fine limestone slurry, with a solid density similar to that assumed in the physical model, were considered [6]. Because Shook and Roco experimental data are for a narrow range of solid concentrations, the Slatter measurements were used to extend the range of rheological parameters [82]. In the case of the Shook and Roco experiments, solid particles with diameter below 44 μm constitute 67%, while in the case of the Slatter experiment the median particle diameter was equal to 28 μm. Therefore, the assumption made in the physical model that the median particle diameter is about 20 μm is an estimation rather than a precise diameter determination.

The results of the approximation of the measurements of Shook and Roco [6] and Slatter [82] are collected in Figure 3a,b. Taking into account the approximation of the measurements, the yield stress and plastic viscosity have been estimated for solid volume concentrations equal to 10%, 20% and 30%. The rheological parameters of the slurry for the Bingham model are collected in Table 1.

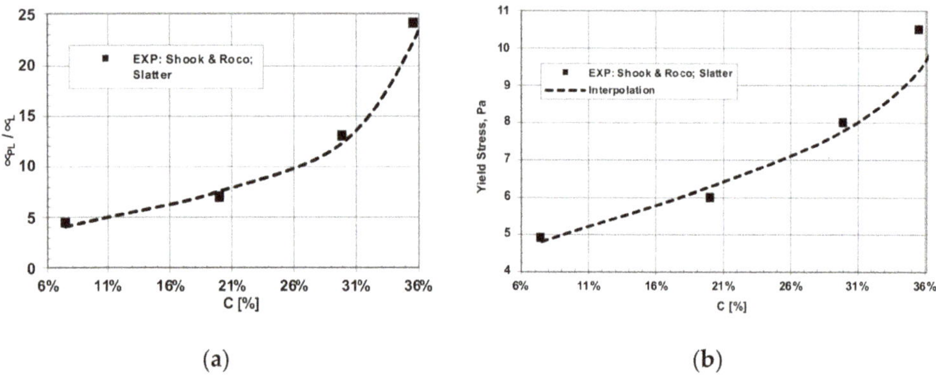

(**a**)   (**b**)

Figure 3. (**a**) Comparison of assumed relative viscosity with experimental data of fine dispersive slurry for the Bingham model (Shook and Roco [6]; Slatter [82]). (**b**) Comparison of assumed relative yield stress with experimental data of fine dispersive slurry for the Bingham model (Shook and Roco [6]; Slatter [82]).

Table 1. Parameters of the fine dispersive slurry for the Bingham model.

Solid Volume Concentration	Bingham Model	
%	τ_o, Pa	μ_{PL}, Pa s
10	5.10	0.004521
20	6.00	0.007032
30	8.00	0.013061

Numerical simulations were performed for a constant heat flux that acts from the slurry to the pipe wall. The thermal properties of the solid and liquid phases are presented in Table 2. Simulations of heat transfer in turbulent flow of fine dispersive slurry were performed for solid volume concentrations equal to: 10%; 20%, 30% and for water. As a result, temperature and velocity profiles were obtained.

Table 2. Thermal properties for limestone and liquid phase; D = 0.02 m; Tw = 293.15 K; q = −500 W/m.

Thermal Conduction for the Carrier Liquid W/(m K)	Thermal Conduction for Solid Particles W/(m K)	Specific Heat for the Carrier Liquid J/(kg K)	Specific Heat for Solid Particles J/(kg K)
0.598	1.221	4183.00	795.00

Figure 4a–c demonstrate the influence of solid concentration on the slurry temperature profiles at the same bulk velocity equal to 3.50 m/s. If the bulk velocity is constant, the temperature difference between the wall and the slurry increases significantly with increasing solid volume concentration. Therefore, considering Equation (15), it can be concluded that for the constant bulk velocity, the heat transfer coefficient decreases with increasing solid volume concentration because the density of heat flux and diameter of the pipe are constant. To confirm this, Figure 4d shows the influence of solid volume concentration on the Nusselt number for constant bulk velocity. It is assumed that if the calculations for C = 0% are presented, this means that there is a flow of pure water. At the same bulk velocity of the carrier liquid and slurry, it is seen that the highest rate of decrease in the Nusselt number is in the range of C = 0% to C = 10%, while for C > 10% the decrease in the Nusselt number is gradual.

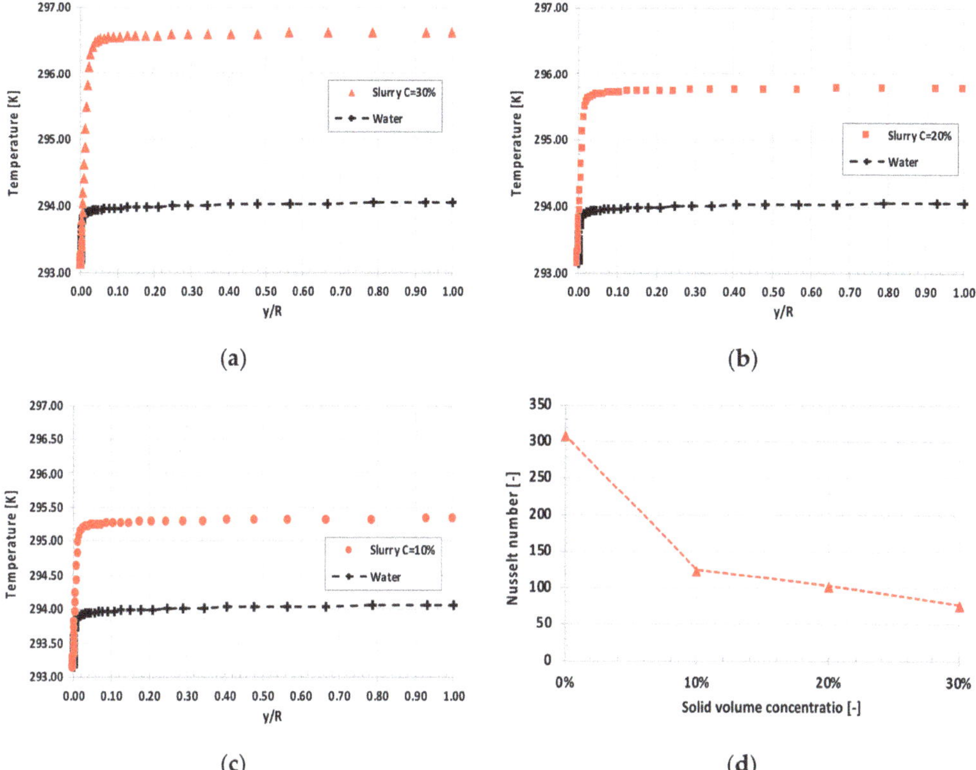

Figure 4. (**a**) Comparison of temperature profiles for slurry and water; C = 30%, U_b = 3.50 m/s; q = −500 W/m. (**b**) Comparison of temperature profiles for slurry and water; C = 20%; U_b = 3.50 m/s; q = −500 W/m. (**c**) Comparison of temperature profiles for slurry and water; C = 10%; U_b = 3.50 m/s; q = −500 W/m. (**d**) Influence of solid concentration on the Nusselt number; U_b = 3.50 m/s; q = −500 W/m.

If the slurry temperature distribution is known, the heat transfer coefficient can be found by calculating the density of heat flux at the slurry boundary as follows:

$$\dot{q} = -\lambda\,\frac{\partial \overline{T}}{\partial r} \tag{41}$$

The velocity distribution plays an important role in the heat exchange process and affects the Nusselt number. The viscous sublayer plays a crucial role for shear stress and heat transfer, while the buffer layer exhibits the highest generation of turbulence, which improves diffusion processes and, as a consequence, the effects on the heat transfer rate [83,84]. For this reason, the slurry velocity profile analysis is useful.

The velocity profiles for constant bulk velocity are presented in Figure 5a–c for different solid concentrations. Slurry velocity profiles are compared with carrier liquid profiles at the same bulk velocity and under the same thermal conditions (q = const; T_w = const) and for the same pipe diameter. However, it should be emphasized that it is useless to compare the velocity profiles of the slurry and water for the same Reynolds number. In such a case, the velocity profiles are substantially different because the viscosities of the slurry and water differ considerably. Figure 5a–c indicate that the velocity gradient at the pipe wall is lower for the slurry than for the carrier liquid. Differences are strongly dependent on the solid concentration. To illustrate what happens at the wall of the pipe, Figure 5d presents the logarithmic profiles of dimensionless turbulent stresses, calculated as a turbulent to laminar stress ratio (μ_t/μ_{app}), for slurry with C = 30% and for water. Again, predictions are made for a constant bulk velocity. It is seen that both profiles differ significantly. For R+ > 5, the dimensionless turbulent stresses are higher in the carrier liquid than in the slurry, which means that turbulent diffusion is greater in the water flow than in the slurry. This is a result of the high viscosity of the slurry. Taking into account the effect of velocity, we expect that for the same bulk velocity of the slurry and the carrier liquid, the heat transfer coefficient of the slurry should be lower than that of water.

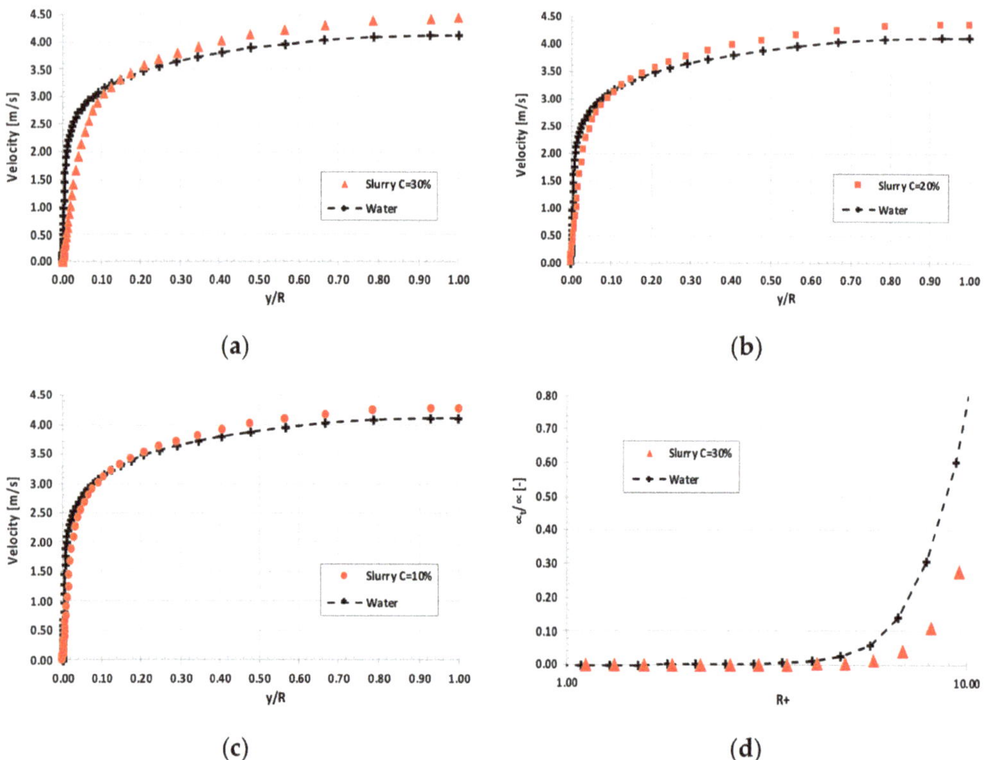

Figure 5. (a) Comparison of velocity profiles for slurry and water; C = 30%; U_b = 3.5 m/s; q = −500 W/m. (b) Comparison of velocity profiles for slurry and water; C = 20%; U_b = 3.5 m/s; q = −500 W/m. (c) Comparison of velocity profiles for slurry and water; C = 10%; U_b = 3.50 m/s; q = −500 W/m; (d) Comparison of logarithmic profiles of dimensionless turbulent stresses for slurry and water; C = 30%; U_b = 3.50 m/s; q = −500 W/m.

The most practical way to calculate the heat transfer coefficient is to use the Nusselt number. Figure 6a presents the influence of Reynolds number on the Nusselt number for slurry and water. The Nusselt number is seen to be higher for the slurry than for the water and increases with increasing solid concentration. However, the rate of increase in the Nusselt number from 20% to 30% seems to be lower than from 10% to 20%. Figure 6b presents the influence of the bulk velocity on the Nusselt number. It is seen that, for the same bulk velocity, the Nusselt number is much lower than the Nusselt number for the carrier liquid, and differences arise as the solid concentration increases. This is in line with earlier conclusions withdrawn by analyzing Figures 4a–c and 5a–c and with the conclusions of some researchers such as Rozenblit et al. [56], for instance.

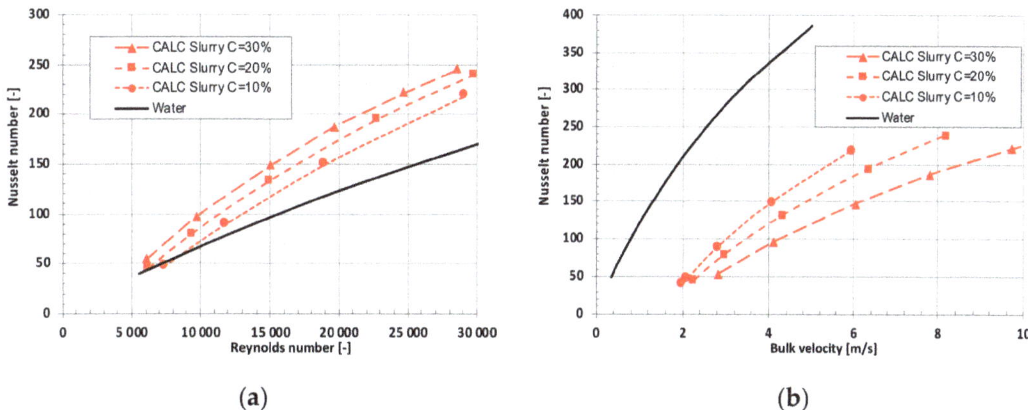

Figure 6. (**a**) Influence of Reynolds number on the Nusselt number for fine dispersive slurry and water; (**b**) Influence of bulk velocity on the Nusselt number for fine dispersive slurry and water.

From an engineering point of view, it is not practical to build a complex mathematical model to predict the heat transfer coefficient for the slurry flow. It is more practical for engineers to use a simple function, like the Nusselt number, which is defined as the ratio of convection to conduction of heat transfer. The Nusselt number allows for calculation of the heat transfer coefficient.

Based on simulations of thermally developed fine dispersive slurry flow, which exhibits yield stress and damping of turbulence in the range of solid volume concentration $C = (10$–$30)\%$ and for $T_w = $ const and $q = $ const, the following equation is developed for the Nusselt number:

$$Nu = 0.02296 \, Re^{0.8} \, Pr^{0.333} \, (1 - C)^{0.75} \left(1 - \frac{\tau_o}{\tau_w} \right)^{1.5} \tag{42}$$

Equation (42) was developed on the basis of the Nusselt number dedicated to water. The equation includes the properties of the slurry, such as solid volume concentration and dimensionless yield shear stress. Dimensionless yield shear stress was defined as the ratio of yield shear stress to wall shear stress. The wall shear stress can be easily calculated by Equation (32), for example. Equation (42) can be applied if $\tau_w > \tau_o$.

Figure 7 presents the calculation of a new Nusselt number, using the correlation expressed by Equation (42), versus the Nusselt number calculated from the mathematical model. The calculations using the new correlation (42) matched well with the numerical predictions. The discrepancy exists for low Nusselt numbers. As an example, for the lowest value of the Nusselt number and $C = 30\%$, the relative error is equal to -6%, while for the highest Nusselt number it is equal to -1%. Taking into account the influence of solid concentration on the accuracy of the prediction by Equation (42), the highest relative error was observed for $C = 10\%$ and $Nu = 44.3$ and is equal to -12%.

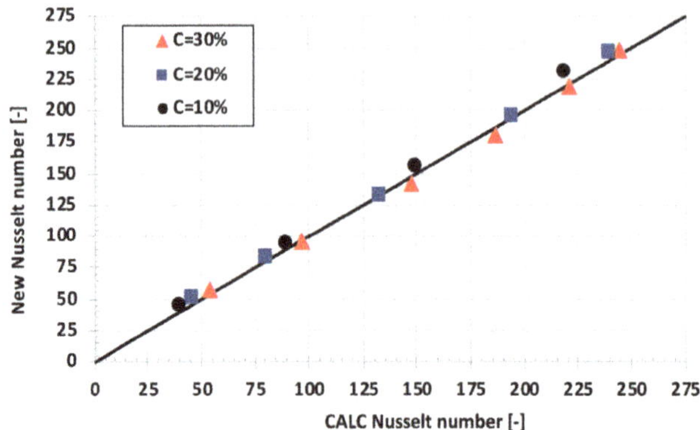

Figure 7. New Nusselt number expressed by (42) versus the Nusselt number predicted by the mathematical model for C = 10%; 20% and 30%.

7. Discussion and Conclusions

The heat transfer characteristics of the fine dispersive slurry, which exhibit yield stress and damping of turbulence, are not well understood. Experiments and modelling of such slurries are difficult. High-concentration slurry measurements are at risk of damage or contamination by intrusive probes. If optical methods are considered, attenuation of the light beam by solid particles occurs. Non-intrusive methods do not allow one to measure higher-order fluctuating parts of temperature and velocity. As a result of these difficulties, there are limited experiments and studies available in the literature.

Modelling of heat transfer in slurry requires special attention because not accurately predicted frictional head loss will affect the accuracy of heat transfer prediction. Therefore, the use of a properly designed mathematical model, which is successfully validated for isothermal flow, is crucial. Some researchers focused their simulations on the analogy between pressure drop and heat transfer in fluidized solid–liquid beds [54,85]. For example, Hashizume et al. [85] reasoned that the heat transfer coefficient predicted from the frictional pressure drop agreed fairly with the experimental data. The crucial point in this study is the hypothesis of Wilson and Thomas [25].

In this study, a simplified physical and mathematical model was developed to investigate the Nusselt number in fine dispersive slurry. Taking into account the assumptions made in the physical model, a new correlation is proposed for the Nusselt number, which depends on the Reynolds and Prandtl numbers, solid volume concentration, and dimensionless yield shear stress. The new Nusselt number is limited to fine dispersed slurries with a median particle diameter of approximately 20 μm, and for C = (10–30)%, Re = 6000–30,000; Pr = 7–75. The lowest limit of the Reynolds number was arbitrarily chosen because the accuracy of the k-ε turbulence model decreases for low Reynolds numbers. The highest limit of the Reynolds number was chosen on the basis of the bulk velocities reached. For example, for C = 30% and Re = 24,660, the bulk velocity of the slurry is 9.7 m/s. Such a high bulk velocity is impractical in engineering applications. The results of the computations confirmed that the new Nusselt number matches fairly well with the Nusselt number obtained from numerical predictions.

$$\lambda_{SL} = \lambda_L \left[1 + \frac{C\left(1 - \frac{\lambda_L}{\lambda_S}\right)}{\frac{\lambda_L}{\lambda_S} + 0.28(1-C)^{0.63\left(\frac{\lambda_S}{\lambda_L}\right)^{0.18}}} \right] \tag{43}$$

It is worth mentioning that some researchers calculate the heat conduction coefficient for the slurry differently from the one done in this study. Some authors use the correlation proposed by Etheram et al. [86], described by Equation (43). Correlation (43), compared with Equation (18), gives lower values of the heat conduction coefficient. However, the differences are not substantial. The relative difference between the calculations made using Equations (18) and (43) increases with increasing solid concentration, and for C = 30% it equals −8%, while for C = 10% it is only −2%. However, such differences do not affect the qualitative results of the predicted Nusselt number and also have a small influence on the qualitative results.

Based on the numerical simulation of heat transfer in a thermally developed turbulent flow of fine dispersive slurry, the following conclusions can be formulated:

1. The slurry velocity profile substantially influences the heat transfer process. Therefore, it is crucial that the mathematical model be validated for isothermal flow first.
2. The increase in the thickness of the viscous sublayer, which exists in fine dispersive slurry, depends on solid concentration and causes an increase in the resistance of the heat flux that acts from the slurry to the pipe wall.
3. For the same bulk velocity of the slurry, the Nusselt number decreases with solid volume concentration increase. The highest rate of decrease in the Nusselt number seems to be when $0 < C < 10\%$.
4. The new Nusselt number developed for the parameters studied depends on the solid concentration, the dimensionless yield shear stress, and the Reynolds and Prandtl numbers.

The new Nusselt number dedicated to fine dispersive slurries, which exhibit yield stress and turbulence damping, is in good agreement with numerical predictions and the highest relative error was observed for C = 10% and Nu = 44.3 and is equal to −12%.

The new Nusselt number is dedicated to specific slurries, which exhibit yield stress and turbulence damping, and requires validation. However, this needs reliable data, which are difficult to obtain. The new Nusselt number has a limitation, as the average particle diameter is not included. This is due to the fact that the mathematical model has been validated strictly for slurries with specific solid particle size distributions and averaged particle diameters, like it is in kaolin slurry, for instance.

More work needs to be done on examining the new Nusselt number for solid volume concentrations lower than 10%, greater than 30% and for different heat fluxes and for different pipe diameters. The influence of the Reynolds number, defined for slurries with yield stress, on the Nusselt number could be examined as well.

Funding: This research received no external funding.

Institutional Review Board Statement: Not applicable.

Informed Consent Statement: Not applicable.

Data Availability Statement: Not applicable.

Conflicts of Interest: The author declares that there is no conflict of interest.

Nomenclature

A	pipe cross section perpendicular to the symmetry axis (m^2)
A_*	outer surface of a pipe of unit length L (m^2)
a, b, c	empirical constants in the Taylor expansion (-)
C_i	constant in the Launder and Sharma turbulence model, I = 1, 2, (-)
C	solid volume concentration (cross sectional solid volume fraction) (%)
c_p	specific heat at constant pressure [J/(kg K)]
D	inner pipe diameter (m)
d_p	solid particle diameter, (μm)
f_μ	turbulence damping function at the pipe wall (-)
g	gravitational acceleration (m/s^2)

k	kinetic energy of turbulence (m^2/s^2)
L	unit length of the pipe (m)
m	mass flux, (kg/s)
Nu	Nusselt number (-)
Pr	Prandtl number (-)
p	static pressure (Pa)
Q	heat flux (W)
q	density of heat flux (W/m)
R	inner pipe radius (m)
r	distance from the symmetry axis (m)
Re	Reynolds number (-)
T	temperature (K)
t'	fluctuating part of the temperature (K)
U	velocity component in the ox direction (m/s)
u'	fluctuating component of velocity in the ox direction (m/s)
V	velocity component in or direction (m/s)
v'	fluctuating components of the velocity V in or direction (m/s)
W	velocity component in oθ direction (m/s)
X	coordinate, main flow direction (m)
y	coordinate for oy direction or distance from a pipe wall (m)
$\overline{}$	time-averaged quantity

Special characters

α	heat transfer coefficient (convective heat transfer) [W/(m^2 K)]
γ	shear rate (s^{-1})
δ_{ij}	Kronecker quantity, (-)
ε	rate of dissipation of kinetic energy of turbulence [m^2/s^3]
λ	coefficient of heat conduction [W/(m K)]
μ	coefficient of dynamic viscosity (Pa·s)
μ_{app}	slurry apparent viscosity (Pa·s)
μ_{PL}	plastic viscosity in the Bingham rheological model (Pa·s)
μ_t	turbulent viscosity (Pa·s)
ν	coefficient of kinematic viscosity (m^2/s)
ρ	density (kg/m^3)
σ_i	diffusion coefficients in the k-ε turbulence model, i = k, ε [-]
τ	shear stress [Pa]
τ_0	yield shear stress [Pa]
τ_w	wall shear stress [Pa]
Φ	general dependent variable, Φ = U, T, k, ε, λ, ρ, c$_p$

Subscript

b	bulk
i	ith direction, i = 1, 2, 3
j	jth direction, j = 1, 2, 3
L	liquid
S	solid
SL	slurry
t	turbulent
w	wall

References

1. Dai, Y.; Zhang, Y.; Li, X. Numerical and experimental investigations on pipeline internal solid-liquid mixed fluid for deep ocean mining. *Ocean. Eng.* **2021**, *220*, 108411. [CrossRef]
2. Wilson, K.C.; Clift, R.; Sellgren, A. Operating points for pipelines carrying concentrated heterogeneous slurries. *Powder Technol.* **2002**, *123*, 19–24. [CrossRef]
3. Tomareva, I.A.; Kozlovtseva, E.Y.; Perfilov, V.A. Impact of pipeline construction on air environment. *IOP Conf. Ser. Mater. Sci. Eng.* **2017**, *262*, 1–7. [CrossRef]
4. Michaels, A.S.; Bolger, J.C. The plastic flow behavior of flocculated Kaolin suspensions. *J. Ind. Eng. Chem. Fundam.* **1962**, *1*, 153–162. [CrossRef]

5. Roco, M.C.; Shook, C.A. Computational methods for coal slurry pipeline with heterogeneous size distribution. *Powder Technol.* **1984**, *39*, 159–176. [CrossRef]
6. Shook, C.A.; Roco, M.C. *Slurry Flow: Principles and Practice*; Butterworth–Heinemann: Boston, MA, USA, 1991. Available online: https://www.amazon.com/Slurry-Flow-Principles-Butterworth-Heinemann-Engineering/dp/0750691107 (accessed on 7 August 2021).
7. Gillies, R.G.; Schaan, J.; Sumner, R.J.; Mckibben, M.J.; Shook, C.A. Deposition velocities for Newtonian slurries in turbulent flow. *Can. J. Chem. Eng.* **2000**, *78*, 704–708. [CrossRef]
8. Doron, P.; Barnea, D. A three-layer model for solid liquid flow in horizontal pipes. *Int. J. Multiph. Flow* **1993**, *19*, 1029–1043. [CrossRef]
9. El-Nahhas, K.; El-Hak, N.G.; Rayan, M.A.; El-Sawaf, I. Flow behaviour of non-Newtonian clay slurries. In Proceedings of the Ninth International Water Technology Conference, IWTC9 2005, Sharm El-Sheikh, Egypt, 17–20 March 2005; pp. 627–640. Available online: https://citeseerx.ist.psu.edu/viewdoc/download?doi=10.1.1.302.772&rep=rep1&type=pdf (accessed on 7 August 2021).
10. Gopaliya, M.K.; Kaushal, D.R. Analysis of effect of grain size on various parameters of slurry flow through pipeline using CFD. *Part. Sci. Technol.* **2015**, *33*, 369–384. [CrossRef]
11. Messa, G.V.; Malavasi, S. Improvements in the numerical prediction of fully developed slurry flow in horizontal pipes. *Powder Technol.* **2015**, *270*, 358–367. [CrossRef]
12. Silva, R.; Garcia, F.A.P.; Faia, P.M.; Rasteiro, M.G. Settling suspensions flow modelling: A Review. *KONA Powder Part. J.* **2015**, *2015*, 41–56. [CrossRef]
13. Li, M.Z.; He, Y.P.; Liu, Y.D.; Huang, C. Pressure drop model of high-concentration graded particle transport in pipelines. *Ocean Eng.* **2018**, *163*, 630–640. [CrossRef]
14. Vlasák, P.; Matouek, V.; Chára, Z.; Krupicka, J.; Konfrst, J.; Keseley, M. Solid volume concentration distribution and deposition limit of medium-coarse sand-water slurry in inclined pipe. *J. Hydrol. Hydromech.* **2020**, *68*, 83–91. [CrossRef]
15. Shook, C.; Bartosik, A. Particle-wall stresses in vertical slurry flows. *Powder Technol. Elsevier Sci.* **1994**, *81*, 117–124. Available online: https://www.sciencedirect.com/science/article/abs/pii/003259109402877X (accessed on 7 August 2021). [CrossRef]
16. Wilson, K.C.; Thomas, A.D. Analytic model of laminar-turbulent transition for Bingham plastics. *Can. J. Chem. Eng.* **2006**, *84*, 520–526. [CrossRef]
17. Kelessidis, V.C.; Dalamarinis, P.; Maglione, R. Experimental study and predictions of pressure losses of fluids modeled as Herschel–Bulkley in concentric and eccentric annuli in laminar, transitional and turbulent flows. *J. Pet. Sci. Eng.* **2011**, *77*, 305–312. [CrossRef]
18. Talmon, A.M. Analytical model for pipe wall friction of pseudo-homogenous sand slurries. *Part. Sci. Technol. Int. J.* **2013**, *31*, 264–270. [CrossRef]
19. Cotas, C.; Asendrych, D. Numerical simulation of turbulent pulp flow: Influence of the non-Newtonian properties of the pulp and of the damping function. In Proceedings of the 8th International Conference for Conveying and Handling of Particulate Solids, Tel-Aviv, Israel, 3–7 May 2015; pp. 1–18. Available online: https://www.researchgate.net/publication/283345241 (accessed on 7 August 2021).
20. Cotas, C.; Asendrych, D.; Garcia, F.A.P.; Fala, P.; Rasteiro, M.G. Turbulent flow of concentrated pulp suspensions in a pipe—Numerical study based on a pseudo-homogeneous approach. In Proceedings of the COST Action FP1005 Final Conference, EU-ROMECH Colloquium 566, Trondheim, Norway, 9–11 June 2015. Available online: https://www.researchgate.net/publication/283342328 (accessed on 7 August 2021).
21. Cotas, C.; Silva, R.; Garcia, F.; Faia, P.; Asendrych, D.; Rasteiro, M.G. Application of different low-Reynolds k-ε turbulence models to model the flow of concentrated pulp suspensions in pipes. *Procedia Eng.* **2015**, *102*, 1326–1335. [CrossRef]
22. Rawat, A.; Singh, S.N.; Seshadri, V. Computational methodology for determination of head loss in both laminar and turbulent regimes for the flow of high concentration coal ash slurries through pipeline. *Part. Sci. Technol.* **2016**, *34*, 289–300. [CrossRef]
23. Kumar, N.; Gopaliya, M.K.; Kaushal, D.R. Experimental investigations and CFD modeling for flow of highly concentrated iron ore slurry through horizontal pipeline. *Part. Sci. Technol.* **2019**, *37*, 232–250. [CrossRef]
24. Mehta, D.; Radhakrishnan, A.K.T.; Van Lier, J.B.; Clemens, F.H.L.R. Assessment of numerical methods for estimating the wall shear stress in turbulent Herschel–Bulkley slurries in circular pipes. *J. Hydraul. Res.* **2020**, *58*, 196–213. [CrossRef]
25. Wilson, K.; Thomas, A. A new analysis of the turbulent flow of non-Newtonian fluids. *Can. J. Chem. Eng.* **1985**, *63*, 539–546. [CrossRef]
26. Zisselmar, R.; Molerus, O. Investigation of solid-liquid pipe flow with regard to turbulence modification. *Chem. Eng. J.* **1979**, *18*, 233–239. [CrossRef]
27. Hetsroni, G. Particles-turbulence interaction. *Int. J. Multiph. Flow* **1989**, *15*, 735–746. [CrossRef]
28. Gore, R.A.; Crowe, C.T. Modulation of turbulence by dispersed phase. *J. Fluids Eng.* **1991**, *113*, 304–307. [CrossRef]
29. Jianren, F.; Junmei, S.; Youqu, Z.; Kefa, C. The effect of particles on fluid turbulence in a turbulent boundary layer over a cylinder. *Acta Mech. Sin.* **1997**, *13*, 36–43. Available online: https://link.springer.com/article/10.1007/BF02487829 (accessed on 7 August 2021). [CrossRef]
30. Eaton, J.K.; Paris, A.D.; Burton, T.M. Local distortion of turbulence by dispersed particles. In Proceedings of the 30th Fluid Dynamics Conference, Norfolk, VA, USA, 28 June–1 July 1999. [CrossRef]

31. Fessler, J.R.; Eaton, J.K. Turbulence modification by particles in a backward-facing step flow. *J. Fluid Mech.* **1999**, *394*, 97–117. Available online: http://citeseerx.ist.psu.edu/viewdoc/download?doi=10.1.1.575.8674&rep=rep1&type=pdf (accessed on 7 August 2021). [CrossRef]

32. Li, D.; Luo, K.; Fan, J. Modulation of turbulence by dispersed solid particles in a spatially developing flat-plate boundary layer. *J. Fluid Mech.* **2016**, *802*, 359–394. [CrossRef]

33. Li, D.; Luo, K.; Wang, Z.; Xiao, W.; Fan, J. Drag enhancement and turbulence attenuation by small solid particles in an unstably stratified turbulent boundary layer. *Phys. Fluids* **2019**, *31*, 1–19. [CrossRef]

34. Sumner, R.J. Concentration Variations and Their Effects on in Flowing Slurries and Emulsions. Ph.D. Thesis, University of Saskatchewan, Saskatoon, SK, Canada, 1992.

35. Xu, J.; Gillies, R.; Small, M.; Shook, C.A. Laminar and turbulent flow of Kaolin slurries. In Proceedings of the Hydrotransport 12, Cranfield, UK, 28–30 September 1993; pp. 595–613. Available online: https://www.src.sk.ca/sites/default/files/resources/pipe%2520flow%2520technology%2520papers%25201991-2007.pdf (accessed on 7 August 2021).

36. Fathinia, F.; Parsazadeh, M.; Heshmati, A. Turbulent forced convection flow in a channel over periodic grooves using nanofluids. *Int. J. Mech. Mechatron. Eng.* **2012**, *6*, 12–2782. [CrossRef]

37. Faruk, O.C.; Celik, N. Numerical investigation of the effect of fow and heat transfer of a semi-cylindrical obstacle located in a channel. *Int. J. Mech. Aerosp. Ind. Mechatron. Manuf. Eng.* **2013**, *7*, 891–896. Available online: https//www.researchgate.net/publication/285116236 (accessed on 7 August 2021).

38. Sen, D.; Ghosh, R. A Computational study of very high turbulent flow and heat transfer characteristics in circular duct with hemispherical inline baffles. *Int. J. Mech. Aerosp. Ind. Mechatron. Manuf. Eng.* **2015**, *9*, 1046–1051. Available online: https://www.researchgate.net/publication/312134104 (accessed on 7 August 2021).

39. Zong, Y.; Bai, D.; Zhou, M.; Zhao, L. Numerical studies on heat transfer enhancement by hollow-cross disk for cracking coils. *Chem. Eng. Process. Process. Intensif.* **2019**, *135*, 82–92. [CrossRef]

40. Nakhchi, M.E.; Esfahani, J.A. Numerical investigation of heat transfer enhancement inside heat exchanger tubes fitted with perforated hollow cylinders. *Int. J. Therm. Sci.* **2020**, *147*, 106153. [CrossRef]

41. Pandey, L.; Singh, S. Numerical Analysis for Heat Transfer Augmentation in a Circular Tube Heat Exchanger Using a Triangular Perforated Y-Shaped Insert. *Fluids* **2021**, *6*, 247. [CrossRef]

42. Bartosik, A. Numerical Modelling of Fully Developed Pulsating Flow with Heat Transfer. Ph.D. Thesis, Kielce University of Technology, Kielce, Poland, 1989.

43. Hagiwara, Y. Effects of bubbles, droplets or particles on heat transfer in turbulent channel flows. *Flow Turbul. Combust.* **2011**, *86*, 343–367. [CrossRef]

44. Bayat, H.; Majidi, M.; Bolhasan, M.; Karbalaie Alilou, A.; Mirabdolah, A.; Lavasani, A. Unsteady flow and heat transfer of nanofluid from circular tube in cross-flow. *Int. Sch. Sci. Res. Innov.* **2015**, *9*, 2078–2083. [CrossRef]

45. Hamed, H.; Mohhamed, A.; Khalefa, R.; Habeeb, O. The effect of using compound techniques (passive and active) on the double pipe heat exchanger performance. *Egypt. J. Chem.* **2021**, *64*, 2797–2802. [CrossRef]

46. Harada, E.; Toda, M.; Kuriyama, M.; Konno, H. Heat transfer between wall and solid-water suspension flow in horizontal pipes. *J. Chem. Eng. Jpn.* **1985**, *18*, 33–38. [CrossRef]

47. Harada, E.; Kuriyama, M.; Konno, H. Heat transfer with a solid liquid suspension flowing through a horizontal rectangular duct. *Heat Transf. Jpn. Res.* **1989**, *18*, 79–94. [CrossRef]

48. Wang, X.; Xu, X.; Choi, S.U.S. Thermal conductivity of nanoparticle-fluid mixture. *J. Thermophys. Heat Transf.* **1999**, *13*, 474–480. [CrossRef]

49. Ku, J.H.; Cho, H.H.; Koo, J.H.; Yoon, S.G.; Lee, J.K. Heat transfer characteristics of liquid-solid suspension flow in a horizontal pipe. *KSME Int. J.* **2000**, *14*, 1159–1167. Available online: https://link.springer.com/article/10.1007%2FBF03185070 (accessed on 7 August 2021). [CrossRef]

50. Amoura, M.; Alloti, M.; Mouassi, A.; Zeraibi, N. Study of heat transfer of nanofluids in a circular tube. *Int. J. Phys. Math. Sci.* **2013**, *7*, 1464–1469. [CrossRef]

51. Bubbico, R.; Celata, G.P.; D'Annibale, F.; Mazzarotta, B.; Menale, C. Comparison of the Heat Transfer Efficiency of Nanofluids. *Chem. Eng. Trans.* **2015**, *43*, 703–708. [CrossRef]

52. Zakaria, I.A.; Mohamed, W.A.; Mamat, A.M.; Rahman, S. Thermal analysis of heat transfer enhancement and fluid flow for low concentration of Al2O3 water-ethylene glycol mixture nanofluid in a single PEMFC cooling plate. *Energy Procedia* **2015**, *79*, 259–264. [CrossRef]

53. Hamad, F.A.; He, S. Heat transfer from a cylinder in cross-flow of single and multiphase flows. *Int. J. Mech. Aerosp. Ind. Mechatron. Manuf. Eng.* **2017**, *11*, 370–374. Available online: https://www.researchgate.net/publication/313824654 (accessed on 7 August 2021).

54. Jacimovski, D.; Garic-Grulovic, R.; Grbavcic, Z.; Boškovic-Vragolovic, N. Analogy between momentum and heat transfer in liquid-solid fluidized beds. *Powder Technol.* **2015**, *274*, 213–216. [CrossRef]

55. Chavda, N.K.; Patel, G.V.; Bhadauria, M.R.; Makwana, M.N. Effect of nanofluid on friction factor of pipe and pipe fittings: Part ii effect of copper oxide nanofluid. *Int. J. Res. Eng. Technol.* **2015**, *4*, 697–700. Available online: https://www.slideshare.net/esatjournals/effect-of-nanofluid-on-friction-factor-of-pipe-and-pipe-fittings-part-ii-effect-of-copper-oxide-nanofluid (accessed on 7 August 2021).

56. Rozenblit, R.; Simkhis, M.; Hetsroni, G.; Barnea, D.; Taitel, Y. Heat transfer in horizontal solid-liquid pipe flow. *Int. J. Multiph. Flow* **2000**, *26*, 1235–1246. [CrossRef]

57. Bartosik, A. Simulation of heat transfer to Kaolin slurry which exhibits enhanced damping of turbulence. In Proceedings of the 10th International Conference on Heat Transfer, Fluid Mechanics and Thermodynamics, Orlando, FA, USA, 14–16 July 2014; pp. 2318–2325. Available online: http://hdl.handle.net/2263/44670 (accessed on 7 August 2021).

58. Bartosik, A. Simulation of a yield stress influence on Nusselt number in turbulent flow of Kaolin slurry. In Proceedings of the ASME Summer Heat Transfer Conference, Washington, DC, USA, 10–14 July 2016; p. 2. [CrossRef]

59. Ramisetty, K. *Prediction of Concentration Profiles of a Particle-Laden Slurry Flow in Horizontal and Vertical Pipes*; Jawaharlal Nehru Technological University Hyderabad: Hyderabad, India, 2010; p. 113. Available online: https://shareok.org/bitstream/handle/11244/10042/Ramisetty_okstate_0664M_11139.pdf?sequence=1 (accessed on 7 August 2021).

60. Reynolds, O. On the dynamical theory on incompressible viscous fluids and the determination of the criterion. *Philos. Trans. R. Soc. Lond.* **1895**, *186*, 123–164. [CrossRef]

61. Boussinesque, J. Theorie de l'ecoulement tourbillant. *Mem. Acad. Sci.* **1877**, *23*, 46.

62. Blom, J. An experimental determination of the turbulent Prandtl number in a developing temperature boundary layer. *Eindh. Univ. Technol.* **1970**. [CrossRef]

63. Launder, B.E.; Sharma, B.I. Application of the energy-dissipation model of turbulence to the calculation of flow near a spinning disc. *Lett. Heat Mass Transfer.* **1974**, 131–138. Available online: https://www.sciencedirect.com/science/article/abs/pii/0094454874901507 (accessed on 7 August 2021).

64. Spalding, D.B. *Turbulence Models for Heat Transfer*; Report HTS/78/2; Department of Mechanical Engineering, Imperial College London: London, UK, 1978.

65. Spalding, D.B. *Turbulence Models—A Lecture Course*; Report HTS/82/4; Department of Mechanical Engineering, Imperial College London: London, UK, 1983.

66. Mathur, S.; He, S. Performance and implementation of the Launder–Sharma low-Reynolds number turbulence model. *Comput. Fluids* **2013**, *79*, 134–139. [CrossRef]

67. Abir, I.A.; Emin, A.M. A comparative study of four low-Reynolds-number k-e turbulence models for periodic fully developed duct flow and heat transfer. *Numer. Heat Transf. Part B Fundam.* **2016**, *69*, 234–248. [CrossRef]

68. Hedlund, A. *Evaluation of RANS Turbulence Models for the Simulation of Channel Flow*; Teknisk-naturvetenskaplig Fakultet UTH-enheten: Upsala, Sweden, 2014; p. 26. Available online: https://www.diva-portal.org/smash/get/diva2:771689/FULLTEXT01.pdf (accessed on 7 August 2021).

69. Davidson, L. *An Introduction to Turbulence Models*; Chalmers University of Technology: Goteborg, Sweden, 2018.

70. Lawn, C.J. The determination of the rate of dissipation in turbulent pipe flow. *J. Fluid Mech.* **1971**, *48*, 477–505. Available online: http://www.tfd.chalmers.se/~{}lada/postscript_files/kompendium_turb.pdf (accessed on 7 August 2021). [CrossRef]

71. Metzner, A.B.; Reed, J. Flow of non-Newtonian fluids-correlation of the laminar, transition and turbulent flow regions. *AIChEJ* **1955**, *1*, 434–440. [CrossRef]

72. Biswas, P.K.; Gidiwalla, K.M.; Sanyal, S.C.D. A simple technique for measurement of apparent viscosity of slurries: Sand-water system. *Mater. Des.* **2002**, *23*, 511–519. [CrossRef]

73. Bartosik, A. *Simulation and Experiments of Axially-Symmetrical Flow of Fine- and Coarse-Dispersive Slurry in Delivery Pipelines*; Monograph M-11; Kielce University of Technology: Kielce, Poland, 2009.

74. Roache, P.J. Computational Fluid Dynamics. *Hermosa Publ. Albuq.* **1982**, *206*, 332.

75. Hermandes-Peres, V.; Abdulkadir, M.; Azzopardi, B.J. Grid generation issues in the CFD modelling of two-phase flow in a pipe. *J. Comput. Multiph. Flows* **2011**, *3*, 13–26. [CrossRef]

76. Bartosik, A. Application of rheological models in prediction of turbulent slurry flow. *Flow Turbul. Combust.* **2010**, *84*, 277–293. [CrossRef]

77. Salamone, J.S.; Newman, M. Water suspensions of solids. *Ind. Eng. Chem.* **1955**, *47*, 283–288. [CrossRef]

78. Ozbelge, T.A.; Somer, T.G. A heat transfer correlation for liquid-solid flows in horizontal pipes. *Chem. Eng. J.* **1994**, *55*, 39–44. [CrossRef]

79. Dittus, F.W.; Boelter, L.M.K. Heat transfer in automobile radiators of the tubular type. *Int. Commun. Heat Mass Transf.* **1985**, *12*, 3–22. [CrossRef]

80. Bergman, T.L.; Lavine, A.S.; Incropera, F.P.; DeWitt, D.P. *Fundamentals of Heat and Mass Transfer*, 8th ed.; John Willey & Sons: Hoboken, NJ, USA, 2018. Available online: https://www.wiley.com/en-us/ES81119320425 (accessed on 7 August 2021).

81. Sparrow, E.M.; Abraham, J.; Gorman, J. *Advances in Heat Transfer*; Academic Press: Cambridge, MA, USA; Elsevier Inc.: Amsterdam, The Netherlands, 2017; Volume 49, pp. 1–323. Available online: https://www.elsevier.com/books/advances-in-heat-transfer/sparrow/978-0-12-812411-6 (accessed on 7 August 2021).

82. Slatter, P.T. Transitional and Turbulent Flow on Non-Newtonian Slurries in Pipes. Ph.D. Thesis, University of Cape Town, Cape Town, South Africa, 1994.

83. Prandtl, L. Bemerkung uber den warmeubergang in rohr. *Phys. Z.* **1928**, *29*, 487.

84. Cebeci, T.; Smith, A.M.O. *Analysis of Turbulent Boundary Layers*; Academic Press: Cambridge, MA, USA, 1974. Available online: https://www.amazon.com/Analysis-Turbulent-Boundary-Layers-Tuncer/dp/B009DKC2YK (accessed on 7 August 2021).

85. Hashizume, K.; Kimura, Y.; Morita, S. Analogy between pressure drop and heat transfer in liquid-solid circulating fluidized beds *Trans. Jpn. Soc. Mech. Eng.* **2008**, *74*, 2014–2019. [CrossRef]
86. Etheram, H.; Arani, A.A.; Sheikhzadeh, G.A.; Aghaei, A.; Malihi, A.R. The effect of various conductivity and viscosity models considering Brownian motion on nanofluids mixed convection flow and heat transfer. *Trans. Phenom. Nano Micro Scales* **2016**, *4*, 78–81. [CrossRef]

Article

Experimental and Numerical Study of Heat Pipe Heat Exchanger with Individually Finned Heat Pipes

Grzegorz Górecki *, Marcin Łęcki, Artur Norbert Gutkowski, Dariusz Andrzejewski, Bartosz Warwas, Michał Kowalczyk and Artur Romaniak

Faculty of Mechanical Engineering, Institute of Turbomachinery, Lodz University of Technology, 219/223 Wolczanska Street, 90-924 Lodz, Poland; marcin.lecki@p.lodz.pl (M.Ł.); artur.gutkowski@p.lodz.pl (A.N.G.); dariusz.a.zsp@gmail.com (D.A.); bartosz.warwas@dokt.p.lodz.pl (B.W.); michal.kowalczyk.1@dokt.p.lodz.pl (M.K.); artur.romaniak@dokt.p.lodz.pl (A.R.)
* Correspondence: grzegorz.gorecki@p.lodz.pl; Tel.: +48-42-631-23-20

Abstract: The present study is devoted to the modeling, design, and experimental study of a heat pipe heat exchanger utilized as a recuperator in small air conditioning systems (airflow $\approx$ 300–500 m^3/h), comprised of individually finned heat pipes. A thermal heat pipe heat exchanger model was developed, based on available correlations. Based on the previous experimental works of authors, refrigerant R404A was recognized as the best working fluid with a 20% heat pipe filling ratio. An engineering analysis of parametric calculations performed with the aid of the computational model concluded 20 rows of finned heat pipes in the staggered arrangement as a guarantee of stable heat exchanger effectiveness $\approx$ 60%. The optimization of the overall cost function by the "brute-force" method has backed up the choice of the best heat exchanger parameters. The 0.05 m traversal (finned pipes in contact with each other) and 0.062 m longitudinal distance were optimized to maximize effectiveness (up to 66%) and minimize pressure drop (less than 150 Pa). The designed heat exchanger was constructed and tested on the experimental rig. The experimental data yielded a good level of agreement with the model—relative difference within 10%.

Keywords: heat pipe heat exchanger; wickless heat pipe; heat transfer; individually finned tubes

Citation: Górecki, G.; Łęcki, M.; Gutkowski, A.N.; Andrzejewski, D.; Warwas, B.; Kowalczyk, M.; Romaniak, A. Experimental and Numerical Study of Heat Pipe Heat Exchanger with Individually Finned Heat Pipes. *Energies* **2021**, *14*, 5317. https://doi.org/10.3390/en14175317

Academic Editor: Artur Bartosik

Received: 9 July 2021
Accepted: 21 August 2021
Published: 27 August 2021

Publisher's Note: MDPI stays neutral with regard to jurisdictional claims in published maps and institutional affiliations.

1. Introduction

The heat pipe (HP) has no moving parts and uncomplicated construction. It is a reliable, and passive heat transfer device—working fluid transport within HP occurs naturally, without additional energy input. Its heat conductance could be higher than any known solid material [1], so it can efficiently transfer heat along significant distances. The separate heat pipe (SHP) is a type of HP, that has a transport or adiabatic section, which separates the evaporator and the condenser section [2]. Because of the existence of the transport section, SHPs are commonly used as heat transferring elements in heat exchangers (HEXs). The type of HEX in which heat pipes are utilized is called a heat pipe heat exchanger (HPHE). Recently, HPHEs have gained popularity in heat recovery applications, e.g., in air conditioning, technological processes, etc. [3,4]. Much attention is focused on testing the capability of HPHEs as recuperators in residential ventilation and air conditioning systems [5]. HPHE intended for heat recuperation in the area of air conditioning, made from a finned bundle of horizontal heat pipes, was tested in [6]. Heat pipes were filled with R-11 refrigerant, and it was found that the peak effectiveness of HPHE is approximately 0.5 (for a range of volumetric flow of air: 1200–3000 m^3/h). Baghban and Majideian [7] investigated HPHE as a recuperator for surgery rooms in hospitals. They chose methanol as the SHP's working fluid and tested it for volumetric flow approx. 400 m^3/h. The very small effectiveness which they obtained (ε = 0.16) was a result of the utilization of bare HP without any fins. Wu et al. [8] proved experimentally that HPHE can be used effectively as a dehumidifier in air conditioning systems. Their

HPHE consisted of continuously finned copper SHPs with R-22 used as working fluid. Longo et al. [9] made an experimental and theoretical analysis of HPHE consisting of finned HPs with internal micro-fins. For the volumetric airflow range, 400–1000 m^3/h HPHE effectiveness peaked at approximately 0.55. Working fluids, in this case, were modern refrigerants: R-134a, R-1234yf, and R-1234ze. Small, continuously finned HPHE was tested in [10] for air volumetric flows close to 40 m^3/h (water as working fluid). The highest, recorded effectiveness was $\varepsilon = 0.6$. Rajski et al. [11] performed a theoretical analysis of wickless HP-based HPHE, which worked as an indirect evaporative cooler. For the considered air flows up to 450 m^3/h, high coefficients of performance of evaporative cooler were achieved. Yau and Ahmadzadehtalatapeh [12] conducted an interesting experiment on the effects of working fluid charge ratio on HPHE effectiveness. They have shown that optimal effectiveness is attained for a filling ratio exceeding slightly the amount to saturate the HP wick. The above-mentioned HPHEs are based on HPs with wicking structure (operate horizontally as well as vertically) or wickless HP (operate at an inclination that ensures a gravity-assisted return of the condensed working fluid). Although, there are many promising, novel HPHE types, which are made from other, more unusual types of HPs, such as pulsating or micro HPs. Their usage in heat exchangers was presented and analyzed by Vasiliev [13]. A flat micro-fins HP array was investigated as air to the air heat exchanger by Yang et al. [14]. Fresh air flow was kept at 1000 m^3/h and outflow air at 1500 m^3/h. Maximum recuperation efficiency was found to be 0.83. Recently, Yang et al. [15] have tested the performance of the pulsating heat pipe heat exchanger using deionized water and HFE-7000 as working fluids. For the airflow range, up to 300 m^3/h, the maximum effectiveness obtained was approx. 50%. Even for HPs meant to be working as pulsating HPs, the chosen diameter was too large to induce the oscillating motion of fluid, and each of HP pass worked as an individual wickless HP. A few papers were focused strictly on modeling of HPHEs. Brough et al. [16] successfully used TRNSYS software to simulate HPHE response to transient input conditions. The same program (TRNSYS) was applied for prediction of yearly energy recovery and dehumidification intensification resulting from HPHE installation in an air conditioning system in tropics conditions [17]. Yu et al. [18] developed an optimization procedure for segmented separate type HPHE. Righetti et al. [19] compared their own computational model with experimental data for six-rows HPHE. Excellent agreement was obtained for heat transfer rate and pressure drop. It convinced the authors of the present work that the most accurate results could be produced by our own, customized numerical model of HPHE. From the above literature review one can draw the conclusion that examined HPHEs were mainly based on the wick-assisted and wickless HPs. Air heat transfer resistance was reduced mainly by the application of aluminum continuous fins (plate-fin tube assembly). In most of the available research, volumetric flows above 1000 m^3/h were considered. The present theoretical study is the only one where HPs are individually finned, based on the literature review carried out by the authors. Theoretical analyses were conducted for a smaller air flow range (370–530 m^3/h) than usually stated in the similar experimental works. It makes the present study important in the view of broadening the knowledge about the design of air-to-air HPHEs.

2. Thermal Calculations of Heat Pipe Heat Exchanger

Usage of heat pipes as basic heat exchanging elements for a heat exchanger forces two alternative designs:

- Continuous fin tube HEX;
- Individually finned tube bundle HEX.
- Individually finned tube bundle HEX was chosen because of the following reasons:
- Freedom of tube spacing variation. Continuous fins are fabricated with standard tube spacing and its optimization is limited to few spacing options;
- Individually finned tubes are more resistant to mechanical damage and withstand higher pressure differences without deformation;

- Construction enables easy cleaning and disassembly for replacement of damaged HP in the bundle.

2.1. Computational Model Description

The computational model was developed to design the HPHEs parameters. As finned HPs are basic heat transfer components, heat transfer through HPs is calculated according to the formula:

$$\dot{Q} = \frac{\Delta T_m}{\sum R} \tag{1}$$

The overall thermal resistance of finned heat pipe:

$$\sum R = R_{ca} + 2R_{tube} + R_{hp} + R_{ha}, \tag{2}$$

where thermal resistance of conduction through tube wall:

$$R_{tube} = \frac{\ln \frac{d_o}{d_i}}{2 \cdot \pi \cdot k \cdot L}, \tag{3}$$

thermal resistance of cold air flow over finned surface:

$$R_{ca} = \frac{1}{h_{ca} \cdot \epsilon_o \cdot A_s}, \tag{4}$$

thermal resistance of hot air flow over finned surface

$$R_{ha} = \frac{1}{h_{ha} \cdot \epsilon_o \cdot A_s}. \tag{5}$$

The heat transfer coefficient on the airside was calculated according to [20]. The maximal average velocity of air through finned tube bundle:

$$w_{max} = \frac{\dot{V}}{A_o}. \tag{6}$$

Reynolds number:

$$Re_d = \frac{W_{max} \cdot d_r}{v} \tag{7}$$

Nusselt number:

$$Nu = \frac{h \cdot d_r}{k} \tag{8}$$

Figure 1 shows the finned tube arrangement with characteristic geometrical dimensions.

Minimum airflow area:

$$A_o = \left[\left(\frac{L_3}{X_t} - 1 \right) z' + (X_t - d_r) - \left(d_f - d_r \right) t_f N_f \right] L_1, \tag{9}$$

where N_f is the number of fins per meter:

$$N_f = 1/S, \tag{10}$$

$$z' = 2x' \; if \; 2x' < 2y', \tag{11}$$

$$z' = 2y' \; if \; 2y' < 2x', \tag{12}$$

in which $2x'$ and y' are given by:

$$2x' = (X_t - d_r) - \left(d_f - d_r \right) t_f N_f, \tag{13}$$

$$y' = \left[\left(\frac{X_t}{2} \right)^2 + (X_l)^2 \right]^{0,5} - d_r - \left(d_f - d_r \right) t_f N_f. \tag{14}$$

Dimensions $2x'$ and y' are depicted in Figures 1 and 2.

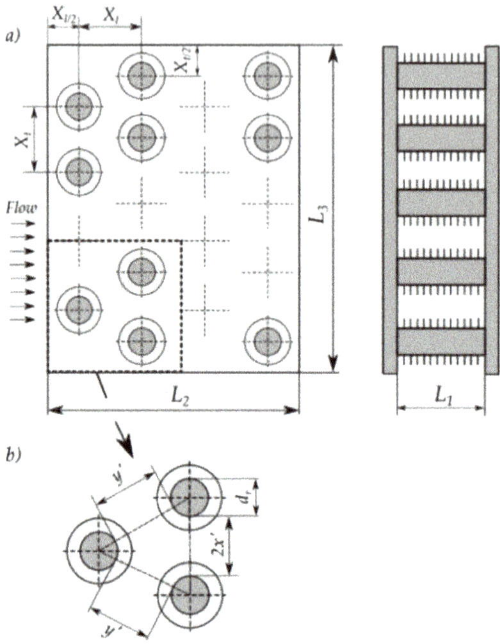

Figure 1. Finned heat exchanger tube arrangement: (**a**) airflow through the heat pipes, (**b**) minimum airflow area (according to [20]).

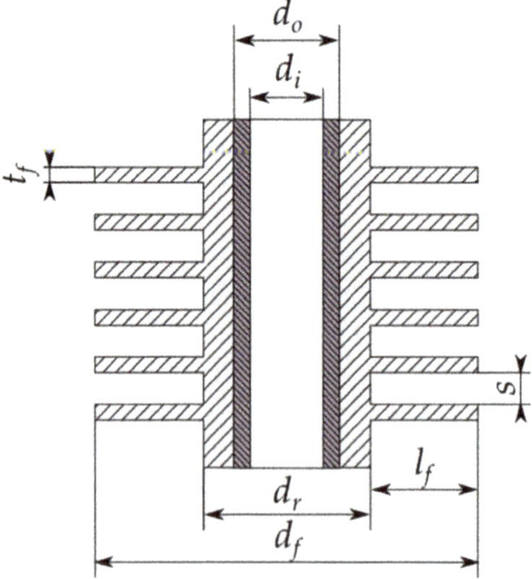

Figure 2. Geometrical dimensions of finned HP.

For high-fin tube banks, the correlation based on experimental heat transfer data is [21]:

$$Nu = 0.1387 Re_d^{0.718} Pr^{1/3} \left(\frac{s}{l_f} \right)^{0.296}, \tag{15}$$

with a standard deviation of 5.1%.

For an equilateral triangular pitch with high-finned tubes, the friction coefficient can be calculated by [21]:

$$f = 9.465 \, Re_d^{-0.316} \left(\frac{X_t}{d_r} \right)^{-0.927} \tag{16}$$

with a standard deviation of 7.8%. This is applicable for the following parameter definitions:

$$Re_d = 2000 - 50{,}000$$
$$P_t / d_r = 1.687 - 4.50$$

For isosceles triangular layout [22]:

$$f = 9.465 Re_d^{-0.316} \left(\frac{X_t}{d_r} \right)^{-0.927} \left(\frac{X_t}{X_l} \right)^{0.515} \tag{17}$$

This is applicable for:

$$d_r = 18.6 - 40.9 \; mm,$$
$$l_f / d_r = 0.35 - 0.56,$$
$$l_f / s = 4.5 - 5.3,$$
$$X_t / d_r = 1.8 - 4.6,$$
$$X_l / d_r = 1.8 - 4.6,$$
$$N_t \geq 6,$$

where N_t is the number of heat exchanger tube rows.

The pressure drop of airflow through finned tube bundles;

$$\Delta p = 2 f N_t \rho w_{max}^2. \tag{18}$$

The efficiency of the annular fin (rectangular profile with adiabatic tip) is obtained from correlation [23]:

$$\eta_f = \frac{\frac{2 r_i}{m}}{r_o^2 - r_i^2} \frac{K_1(mr_i) I_1(mr_o) - I_1(mr_i) K_1(mr_o)}{K_0(mr_i) I_1(mr_o) - I_0(mr_i) K_1(mr_o)}, \tag{19}$$

where $r_o = d_f/2$, $r_i = d_r/2$, I_0—modified Bessel function of order 0, I_1—modified Bessel function of order 1, K_0—modified Bessel function K of order 0, K_1—modified Bessel function K of order 1, m coefficient for annular fin:

$$m = \sqrt{\frac{2 \cdot h}{k \cdot t_f}} \tag{20}$$

Finned surface area is obtained by the following equation:

$$A_s = A_r + A_f, \tag{21}$$

where unfinned area of tube:

$$A_r = \left(S - t_f \right) \pi \, d_r \, N_f, \tag{22}$$

finned area of tube:

$$A_f = \left(2\,\pi\,\left(r_o^2 - r_i^2\right) + \pi\,d_f\,t_f\right)N_f. \tag{23}$$

Overall finned surface efficiency:

$$\varepsilon_o = \frac{A_r + A_f \cdot \eta_f}{A_s}. \tag{24}$$

The thermal resistance of the wickless heat pipe is taken from the regression of experimental data from previous authors' work [22], where the thermal performance of various diameters of wickless HPs were investigated for different working fluids filling ratios. The experiment parameters are summarized in Table 1. Various diameters of HPs were tested to obtain thermal resistance characteristics. On the condenser and evaporator sections of HPs, jacket HEXs were installed. The condenser section HEX was fed with cold water, while through the evaporator section HEX hot water flowed. Because of the low volumetric flow of cooling and heating water, the range of heat fluxes obtained for HPs was nearly identical as predicted in the case of the air-to-air HPHE. The lower convective heat transfer coefficients of the air were compensated for by the area enlargement due to the finned surface (the enlargement factor was around 20). The other aim of the study was to choose the best working fluid of the examined substances. Refrigerant R404A was recognized as the best working fluid (giving the highest thermal throughput) from other tested fluids (R134a, R410A, and R407C). A twenty percent volumetric filling ratio was chosen consecutively as the best of four considered in the study (10%, 20%, 30%, and 40%). Thermal resistance versus heat transfer rate for a 32 mm outer diameter heat pipe is shown in Figure 3. The relative uncertainty of the HPs' thermal resistance is nearly identical to the relative uncertainty of the heat transfer rate stated in [22] (typically 10–25%). Thermal resistances are nearly identical for filling ratios from 20–40%. They are considerably lower for 10%, although this filling ratio was rejected because of dry-out limit occurrence. Thermal resistance increases sharply for a 10% filling ratio near to the highest heat throughput (near 150 W) because the falling film of the refrigerant was broken, and the dry patches at the evaporator section impeded the heat flow.

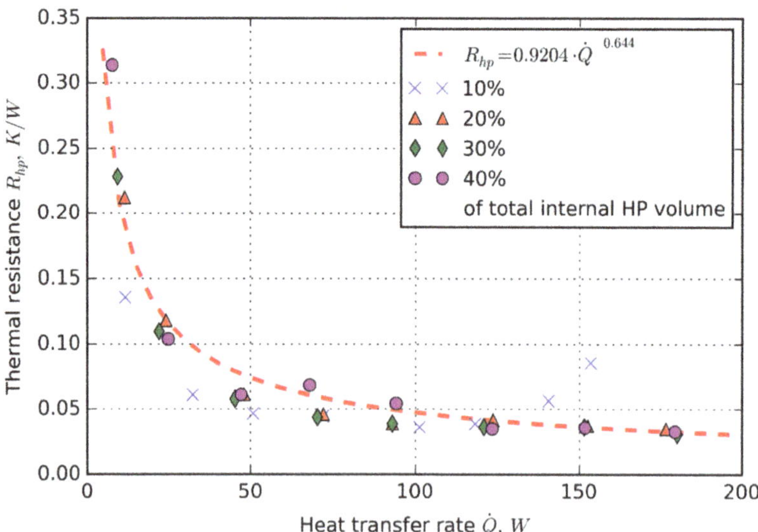

Figure 3. Thermal resistance of 32 mm outer diameter heat pipe vs. heat transfer rate for different filling ratios [22].

Table 1. Parameters of the experiment [24].

Parameter Name	Value
Evaporator length	250 mm
Condenser length	250 mm
Adiabatic section length	50 mm
Cooling water inlet temperature	10 °C
Heating water inlet temperature	20, 25, 30, 35, 40, 45, 50 °C
Volumetric flow of cooling and heating water	15 l/h
Material of HP container	Copper
Tested diameters	20, 32 mm

The next lowest filling ratio was considered the best. Even its thermal resistance was comparable with 30 and 40%; lower filling cuts expenses on working fluid and reduces the total weight of HPHE. Experimental data for 20 and 32 mm heat pipes were fitted by a correlation (25) in Figure 4. The correlation takes into account various diameters of HPs. As it can be seen in Figure 4, Equation (25) successfully fits the experimental data [24] for two examined HPs' outer diameters: 20 and 32 mm.

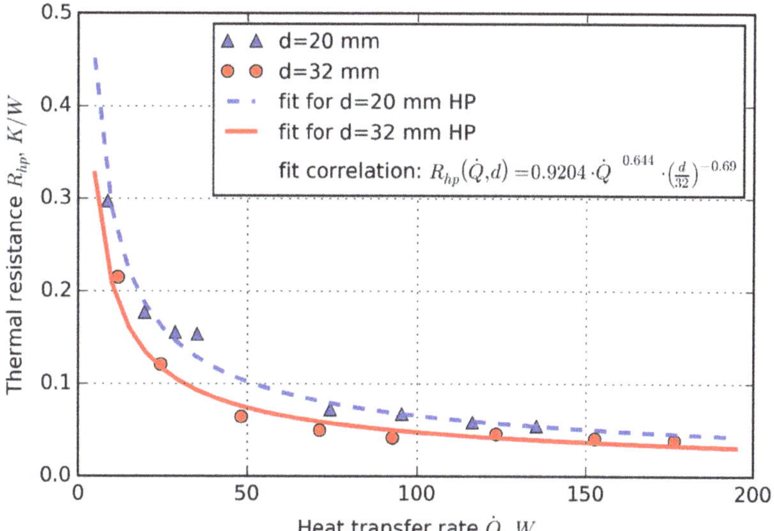

Figure 4. Thermal resistance vs. heat transfer rate for 20% filling ratio, for *d* = 20 mm, and *d* = 32 mm HP. Experimental data were fitted with (25) [23].

Correlation (25) was used to predict the overall thermal resistance of HPs in the computational model:

$$R_{hp}\left(d, \dot{Q}\right) = 0.9204 \cdot \dot{Q}^{-0.644} \left(\frac{d}{32}\right)^{-0.69} \tag{25}$$

The computational model uses Equations (1)–(25) for the iterative calculation of heat transfer rate, and air streams outlet temperatures. The initial value of the heat transfer rate is guessed and after each iteration, it is updated until the residual becomes lower than 1 W. The iterative technique to solve the non-linear set of equations is the Gauss–Seidel method with an under-relaxation factor of 0.4. This algorithm is used to find a solution (heat transfer rate) for every HPHE row. The HEX is divided into control volumes according to Figure 5. Nodes were at the inlet and outlet planes to each row. Each node represents an unknown temperature, besides the known values at the inlets (blue nodes for cold air temperatures: T_{c0}–T_{cN}, red nodes for hot air temperatures: T_{h0}–T_{hN}). After all of the temperatures and heat transfer rates $\dot{Q}_0 - \dot{Q}_N$ were updated, and the maximum residual of all of the temperatures is calculated. If it is not lower than 0.001 °C the temperatures and heat transfer rates are updated again. The iterative technique is also Gauss–Seidel but without the under-relaxation.

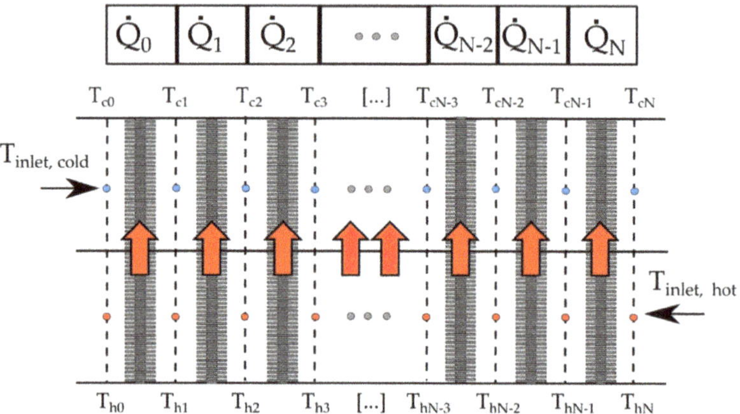

Figure 5. Scheme of the dividing of HPHE into control volumes.

2.2. Thermal Design Parameters

The design process of HPHE starts with the specification of assumed design parameters which result from HPHE type and its utilization. HPHE will be utilized as an air-to-air recuperator for air conditioning systems. This heat exchanger type implies the following design parameters:

- Working fluid is air at (approximately) atmospheric pressure;
- Effective range of working fluid temperatures is from −20 to 50 °C.

HPHE will be used in very small air conditioning systems (ventilation of up to 160–180 m^2 spaces). Both airstreams' volumetric flow rates are taken as typical for small systems: $V = 300$ m^3/h.

Computational model. The output parameter, which is the main indicator of the efficiency of heat transfer, is HEX temperature effectiveness:

$$\varepsilon = \frac{t_{hi} - t_{ho}}{t_{hi} - t_{ci}} \tag{26}$$

where t_{hi}—hot air stream inlet temperature, t_{ho}—hot air stream outlet temperature, and t_{ci}—cold air stream inlet temperature. A hundred percent recuperation efficiency would occur if the hot stream temperature would drop at the outlet to the level of cold stream inlet temperature.

The goal of optimization is to design HPHE with thermo-hydraulic characteristics (thermal effectiveness, pressure drop) similar to or exceeding most of the existing HPHE reported earlier in the literature review. Comparative parameters are:

- Pressure drop: 40 Pa (not greater than 200 Pa);
- Thermal efficiency: 60%.

The geometry of the finned tube was chosen to ensure the ease of HPHE prototype manufacture. It was selected from the bimetallic high-finned tubes manufacturer catalog—CEMAL company [24]. Heat exchanger channel height was L_1 = 0.245 m (a consequence of finned heat pipe geometry), and width L_3 = 0.245 m was a result of the assumption of a square channel. Geometrical parameters of the finned aluminum tube applied on the standard copper tube: ø 22 × 1 mm are given in Table 2 and presented in Figure 6. Fins were cold-rolled on the outer aluminum tube.

Table 2. Geometrical parameters for the high-finned tube.

Parameter Name	Value
S	2.5 mm
d_i	20 mm
d_o	22 mm
d_f	50 mm
l_f	13 mm
t_f	0.8 mm
δ_1	1.2 mm
δ_o	0.4 mm
d_r	24 mm

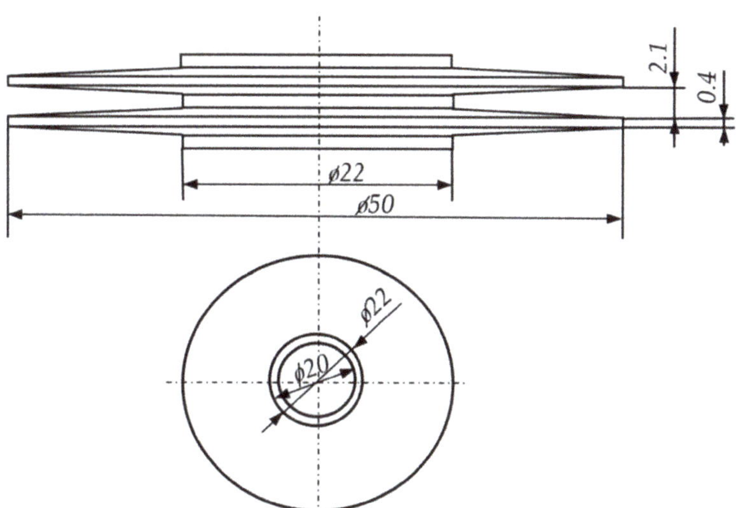

Figure 6. Finned surface heat pipe heat exchanger dimensions.

2.3. Parametric Optimization of HPHE

The initial arrangement of finned heat pipes is shown in Figure 7 (three first rows). It is a start positioning for the further thermal and flow analysis carried out with the use of the computational model. This arrangement is considered for the first set of plots: Figures 8–10. Volumetric flow rates for hot and cold airstreams are held constant (300 m^3/h). Inlet cold air temperature is kept at 10 °C and hot air inlet temperature at 30 °C. The counterflow arrangement was chosen as it produces a higher mean temperature difference between streams than parallel flow. A staggered HPs arrangement was also selected over the inline as it ensures better mixing and temperature uniformity of air streams. Initially, the isosceles layout was assumed due to a minimal pressure drop [20]. Figure 8 shows the heat transfer rate for inline and staggered finned HPs arrangement as a function of the number of HPHE rows. It is a validation of the correctness of the computational model calculations. As expected in the whole, considered range of the number of rows, the staggered arrangement produces a higher heat transfer rate than inline. As the number of rows increases, thermal effectiveness grows non-linearly (Figure 9)—the shape of the function is similar as in the case of heat transfer rate. The previously stated goal is 60% effectiveness (red horizontal line), and it corresponds to a 17-rows HPHE. Because of thermal calculations uncertainty, 20 rows were chosen (three extra rows) to obtain the assumed effectiveness. Pressure drop increases linearly with the number of rows and exceeds 40 Pa at approximately nine rows (Figure 10). For 20 rows HPHE, ΔP of 100 Pa is not exceeded, which makes that acceptable value for small air conditioning systems (it can be handled by a typical air duct fan). Further increase in effectiveness by adding more rows (more than 20) is not viable, because the addition of two rows produces approximately 10 Pa extra pressure drop (which is 10% relative increase) while only 2–3% raise of effectiveness. It supports 20 rows as the final choice, taking into account that a pressure drop increases linearly, while the effectiveness slope becomes less steep for increasing row number (adding more rows becomes even less viable in terms of effectiveness vs. ΔP change as row number increases).

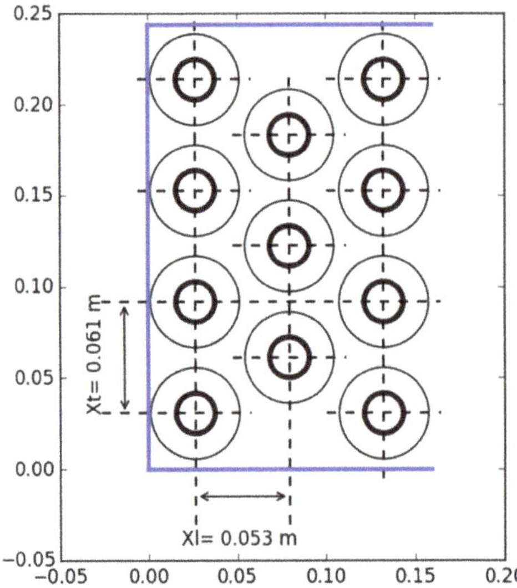

Figure 7. The initial arrangement of the HPs in HPHE (staggered).

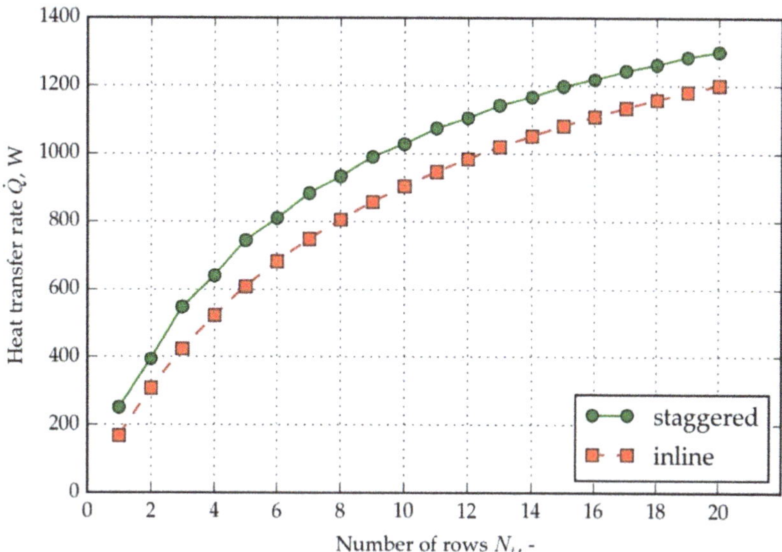

Figure 8. Heat transfer rate for inline vs. staggered heat pipe arrangement, for inlet cold air temperature 10 °C and hot air inlet temperature 30 °C.

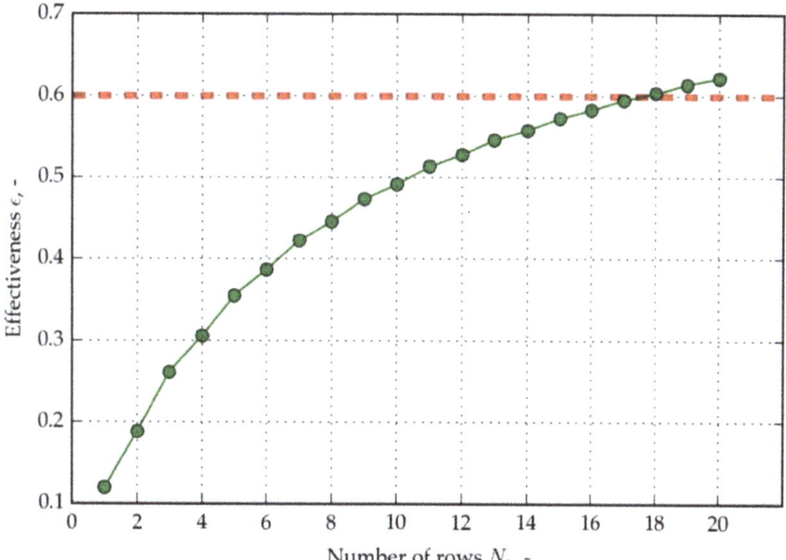

Figure 9. Thermal effectiveness of HPHE as a function of number of rows, for inlet cold air temperature 10 °C and hot air inlet temperature 30 °C.

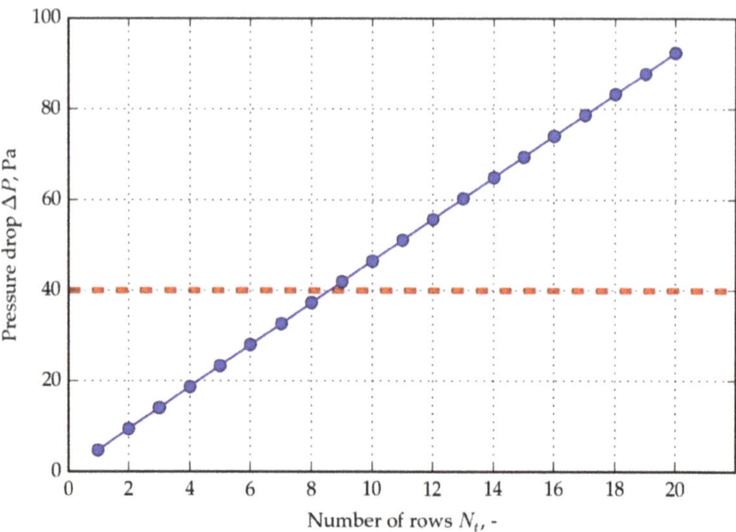

Figure 10. Pressure drop for HPHE vs. number of rows, for inlet cold air temperature 10 °C and hot air inlet temperature 30 °C.

The dependence of parameters on hot air inlet temperature is shown in Figures 11–13. All computations for the plots were made with the assumption of 20 HPs rows and 0 °C cold air stream inlet temperature. Figure 11 shows the dependence of the Reynolds number, which is decreasing for increasing hot air inlet temperature. It is an effect of an increase in air kinematic viscosity with temperature. There is approximately a 15% Reynolds number decrease relative to the value for the lowest considered temperature: 8 °C. Figure 12 presents the pressure drop as a function of the hot air inlet temperature. The pressure drop decreases linearly by 10% for the assumed temperature range—it is caused mainly by a decrease in air density with temperature. The combined effect of air thermophysical properties changes with temperature (variation with pressure is negligible) and mean temperature rise impacts HPHE effectiveness, yet very weakly for hot air temperatures greater than 30 °C (Figure 13). The relative change from 30 to 42 °C is just a fraction of a percent. The above analysis shows that the assumption of approximately 60% effectiveness is valid for a broad range of temperatures for 20 rows of HPHE. There is a potential benefit in reducing a pressure drop for higher temperatures, although it is calculated for a constant volumetric flow of air, which could decrease in a real situation, where a fan is inducing the flow. The working point of a fan can change for lower air density and flow conditions could vary.

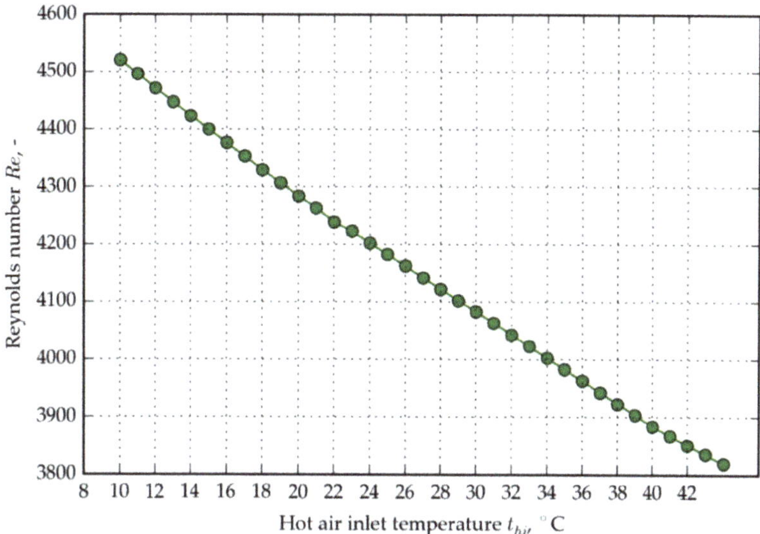

Figure 11. Reynolds number vs. hot air inlet temperature, for cold stream inlet temperature 0 °C and 20 HPs rows.

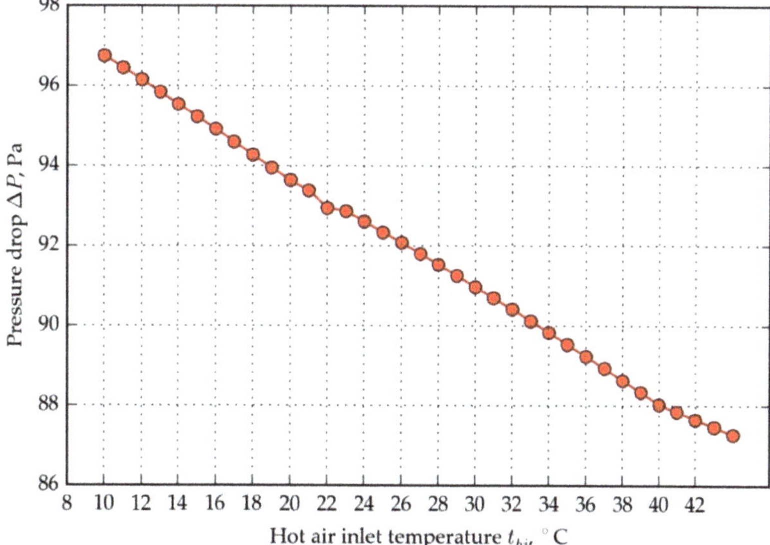

Figure 12. Pressure drop vs. hot air inlet temperature, for cold stream inlet temperature 0 °C and 20 HPs rows.

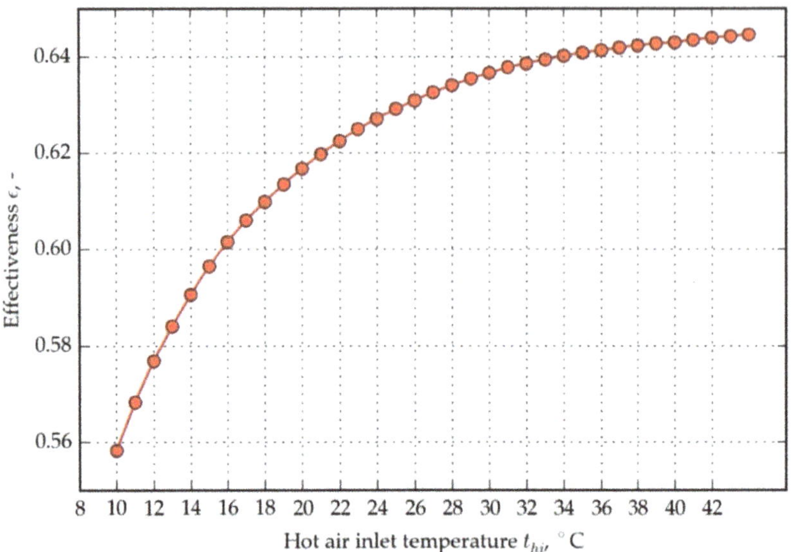

Figure 13. Effectiveness vs. hot air inlet temperature, for cold stream inlet temperature 0 °C and 20 HPs rows.

Figure 14 shows hot air heat transfer coefficient dependence on the number of rows (inlet hot air temperature: 30 °C, inlet cold air temperature: 10 °C). The computational model allows very small dependency of the heat transfer coefficient as equation (15) is valid for the number of rows greater than six; therefore, the calculated heat transfer coefficients for a smaller number of rows are uncertain. This inaccuracy, however, is not very important, as the practical number of rows which produces reasonable HPHE efficiency ($\varepsilon > 50\%$) is greater than 10. The dependence of the hot air heat transfer coefficient on the hot air inlet temperature, for cold air inlet temperature 0 °C, is shown in Figure 15. For the broad range of temperatures, heat transfer coefficient changes only slightly—it drops by 5%. This proves the stability of the working point of HPHE under various temperature conditions. Three-dimensional graphs are introduced for defining the optimal spacing between the center of the tubes—longitudinal and traversal in the respect to the airflow direction. The subsequent analysis was carried out with an assumption of constant hot and cold air inlet temperatures (10 and 30 °C). Figure 16 presents the exchanged heat transfer rate as a function of traversal and longitudinal spacings. Heat transfer rate changes stepwise at $X_t = 0.062$ m because below that spacing it is possible to add one extra HP in the row. Further decrease in traversal distance results in weak growth. At minimal traversal spacing $X_t = 0.05$ m (contact of finned HPs), heat transfer rate attains the maximal value. Longitudinal spacing does not affect the heat transfer rate. HPHE effectiveness, shown in Figure 17, exhibits similar behavior to heat transfer rate, with the peak value of 64% for minimal traversal spacing. The pressure drop rises more sharply with the decrease in X_t than with the reduction in X_l. As longitudinal spacing does not affect heat transfer significantly, moving away HPs in the longitudinal direction is advised, yet excessive spacing could cause an unacceptable increase in the total length of HPHE. The peak of a pressure drop is noted for minimal spacings (approx. 160 Pa).

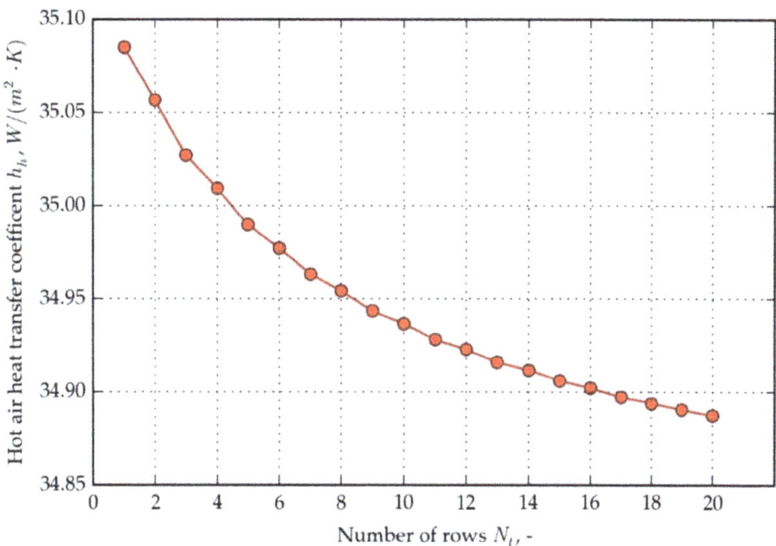

Figure 14. Hot air heat transfer coefficient vs. the number of rows, for hot air inlet temperature 30 °C and cold air inlet temperature 10 °C.

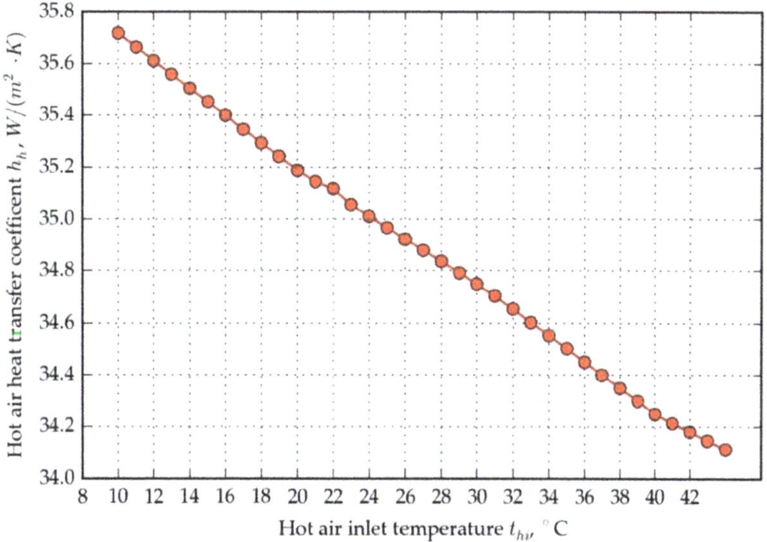

Figure 15. Hot air heat transfer coefficient vs. hot air inlet temperature, for cold air inlet temperature 0 °C.

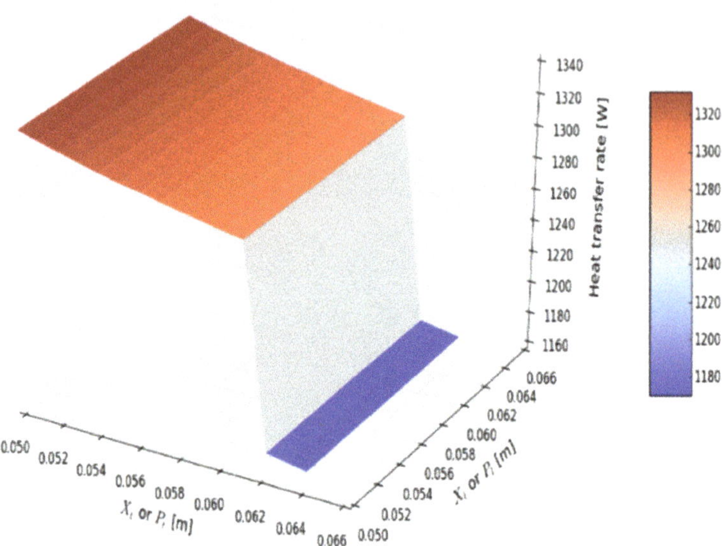

Figure 16. Heat transfer rate vs. traversal and longitudinal spacing of the tubes.

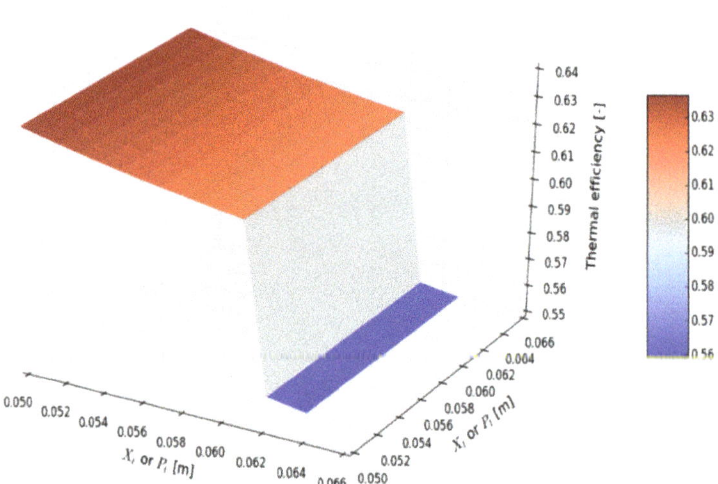

Figure 17. Thermal efficiency vs. traversal and longitudinal spacing of the tubes.

The hot air heat transfer coefficient grows by 20% with traversal spacing reduction (Figure 18). Enhancement by altering the X_l is negligible. Figure 19 presents the dependence of HPHE effectiveness on X_t and hot air inlet temperature. A maximal value of 66% is attained for minimal X_t and the maximal temperature difference between hot and cold airstreams. As expected for higher temperature differences, HPHE performs better.

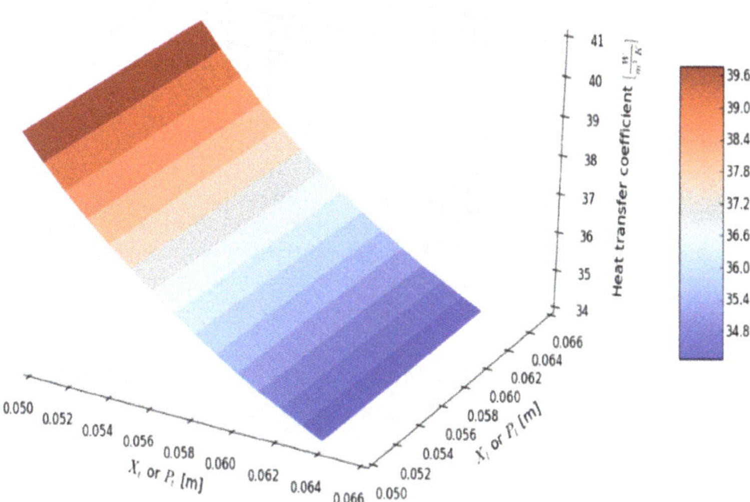

Figure 18. Heat transfer coefficient vs. traversal and longitudinal spacing of the pipes.

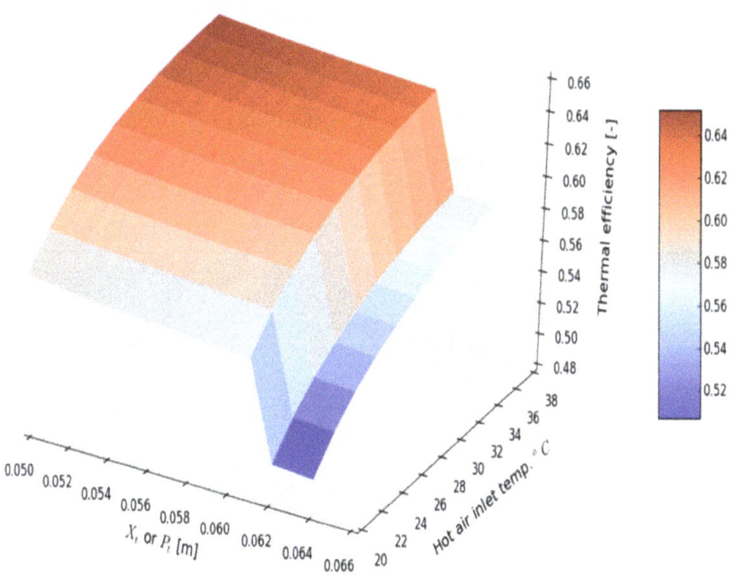

Figure 19. Thermal efficiency vs. traversal spacing of the pipes and hot air inlet temperature.

3. Validation of Computational Model by the Experiment

The prototype of HPHE was manufactured for the validation of the computational model. The HPHE tests were carried out in three successive measurement series. Each series was characterized by different values of air volumetric flows in the ducts caused by the change in rotational speed of the radial fans. The rotational speed of the fans was varied by fan motor speed controllers (based on TRIAC's electronic circuits). The measurement series consisted of three tests, during which the volumetric flow rates of the fluid flowing in the channels were kept steady.

The first measurement series was carried out at average volumetric flows of approx. 530 m^3/h (hot air duct) and approx. 400 m^3/h (cold air), respectively.

The second measurement series was carried out with average volumetric flows of approx. 520 m^3/h (hot air duct) and approx. 460 m^3/h (cold air), respectively. This series is distinguished by a high-volume flow rate in the hot air duct.

The third measurement series was carried out at average volume flows of approximately 380 m^3/h (hot air duct) and 370 m^3/h (cold air), respectively. This series was characterized by the lowest average volumetric flows in the air channels. The range of airflows corresponds to average velocities: 1.7–2.5 m/s, which is typical for small air-conditioning ducted systems (for systems with capacities < 1000 m^3/h, 3 m/s velocity is considered maximal, to avoid high pressure drops). HEX studies were conducted at the experimental rig shown in Figure 20.

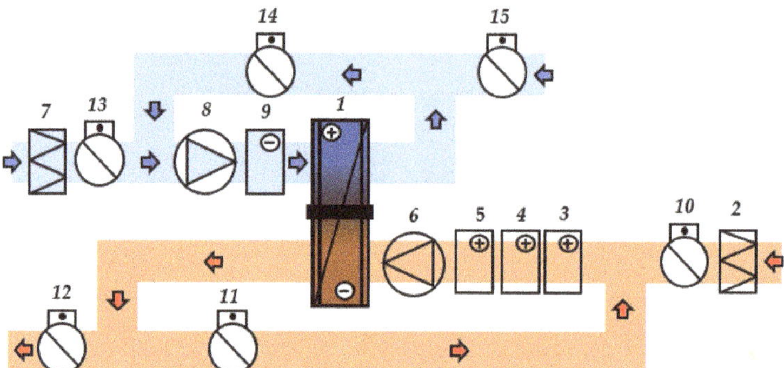

Figure 20. Diagram of the HPHE test rig: 1—HPHE; 2, 7—air filters; 3–5—electric preliminary heaters; 6, 8—air fans; 9—air cooler; 10–15—manual dampers.

The subject of the test is a HEX (1) made of finned SHP (Figure 21) for the recovery of heat from exhaust air.

Figure 21. Photo of the finned heat pipe.

Two air streams (hot—exhaust and cold—fresh air) pass through the HEX. The airflow arrangement is countercurrent. The HEX is made up of 32 HPs in a staggered arrangement. Each of the HP's is filled with R404A refrigerant. Based on previous research [22] the optimal amount of refrigerant in HP has been determined as 20% of its volume. Air flows through steel air ducts with a diameter of d = 0.2 m, while in the part adjacent to HPHE they transform into ducts with a rectangular flow area: 0.24 m × 0.25 m. Streams of the exhaust (hot) and fresh (cold) air flow in closed circuits. Airflow in the ducts is forced, utilizing fans (6, 8). There is one fan for exhaust and one for fresh air. The speeds of both fans are individually regulated. The air parameters are regulated by a cooler (9) and heaters (3–5). Cooler is an element of a split refrigeration system (refrigerant R404A). The air-cooled condenser, which is expelling the heat absorbed from an air stream, is placed outside the building. The cooling capacity is regulated by manually adjusting the automatic expansion valve opening, which is metering a refrigerant flow to the cooler. Heaters are simple electric air duct heaters. There is a temperature probe upstream of every heater for the feedback to the automation system controlling the power output for obtaining the pre-set temperature. Air is drawn in through filters (2, 7) and the flow direction is controlled by mechanical dampers. The experimental rig arrangement allows for cooling only the top channel of HPHE and heating the bottom channel (Figure 19).

The air system can operate in both closed and open circuit (drawing or releasing air to outside). The experimental tests were carried out in conditions where supply and fresh air flowed in closed circuits without contact with the external air. Both HPHE flow channels were divided into 27 equal areas. In the centers of these areas, velocity and temperature measurements were taken. In each of the measurement planes, the average velocity and temperature were calculated based on 27 measurements. The temperature and velocity measurements were performed utilizing the "climate meter" probe LB-580 [25], which has the functions of a thermal anemometer, thermometer (Pt1000), and capacitive hygrometer. The mass flow rate was calculated from the average value of the velocity at the measurement planes. In each of the airflow ducts (fresh and exhaust air) there were two measurement planes—directly upstream and downstream of the HPHE. The differences between the average velocities of a single airstream did not exceed 5%. The average relative humidity of air was also checked at the inlet and outlet of HPHE, but in any of the cases no change was recorded, so no water vapor condensation occurred on the heat exchange surfaces.

The graph shown in Figure 22 specifies the values of the average heat transfer rates depending on the set temperature of the fresh (cold) air at the HPHE inlet for three measuring ranges, differing in the average air volumetric flow rates in the ducts. The experimental values were shown with error bars and computed values from the numerical model. The temperature of the fresh air at the inlet to the HPHE in the condenser section changed according to Figure 3 and in the evaporator section, it was kept at 24 °C. The charts were made for the air flows shown in Table 3.

Table 3. Exhaust and fresh air flows.

Measurement Series Number	I	II	III
fresh air (m^3/h)	400	460	370
exhaust air (m^3/h)	530	520	380

The characteristics presented in Figure 22 show that with the increase in the temperature at the inlet of the fresh air, the heat transfer rates became lower. The decrease in the average air volume flows in individual measurement series is accompanied by a decrease in the average heat transfer rates. The highest average experimental heat transfer rate, 1773 W, was obtained for the second case, for which the average temperature measurement of the fresh air at the heat exchanger inlet was equal to 1.5 °C. The lowest average heat flux of 592 W was obtained for the third case and the fresh air temperature of 14.70 °C. There is a good agreement between experimental heat transfer rates and those obtained from the model described in the following section—the average difference between measured and computed values is approximately 10%. In the high heat transfer rates region, the model output is within the estimated measurement uncertainty. Relative uncertainty is in the range of 5–8%. The model is more differing from the experimental data for the smaller heat fluxes. As can be seen in Figure 22, as fresh air inlet temperature increases, the measurement, and numerical curves diverge. The highest relative difference is 20%, the lowest is 1%. The reason for the higher discrepancy for lower heat fluxes can be the higher uncertainty of the heat pipe thermal resistance for low throughputs. The same argument could be made for the airside convective heat transfer correlation.

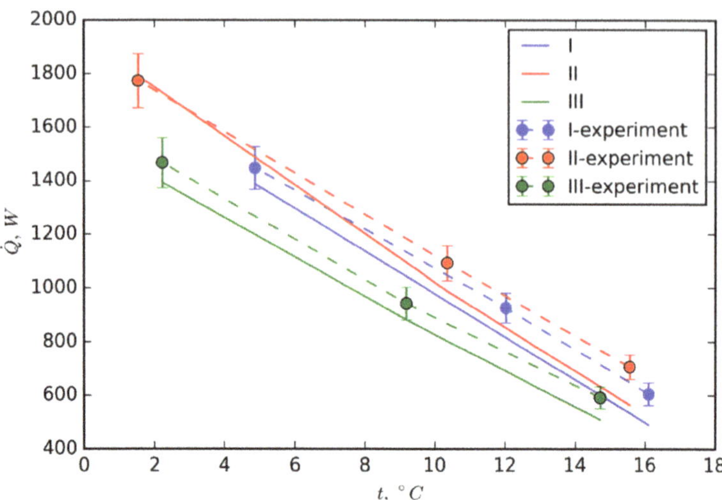

Figure 22. Average HPHE heat transfer rates $\dot{Q}$ from model and experimental for varying volumetric airflow rates.

Figure 23 shows the values of the experimental HEX efficiency η defined as the ratio of the heat transfer rate absorbed by the fresh air to the value of the heat transfer rate released by the exhaust air. These values depend on the average inlet temperature of the fresh air. The graph shows the characteristics for three cases, characterized by different average fluid volumetric flow rates.

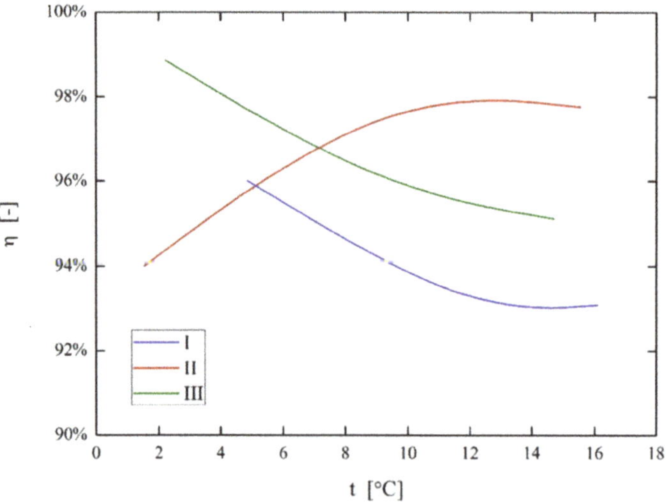

Figure 23. HEX efficiency η versus the fresh air temperature at the inlet to HEX for different volumetric flow rates.

Based on the above diagram, it can be concluded that the efficiency of the HEX η decreases in most cases with increasing temperature of the fresh air at the inlet. The exception is the second case, for which the average volume flows were 520 m^3/h (the exhaust air duct) and 460 m^3/h (the fresh air duct). At a lower average temperature

(1.54 °C), a lower efficiency (94%) was obtained than for similar measurements at higher temperatures (97.8%). The lowest efficiency (93%) was obtained for the temperature of 16.09 °C, which leads to the conclusion that it is worth ensuring that the HEX works with air at possibly the lowest temperatures. The highest efficiency (98.9%) was obtained for the third measurement of the third series (2.22 °C). In the entire tested range, the efficiency of the HEX was in the range of approx. 93–98%.

Figure 24 shows the values of the temperature effectiveness of the heat exchanger ε. Effectiveness depends on the fresh air inlet temperature. The graph shows the characteristics for three cases, characterized by different average volumetric flow rates in the air-ducts.

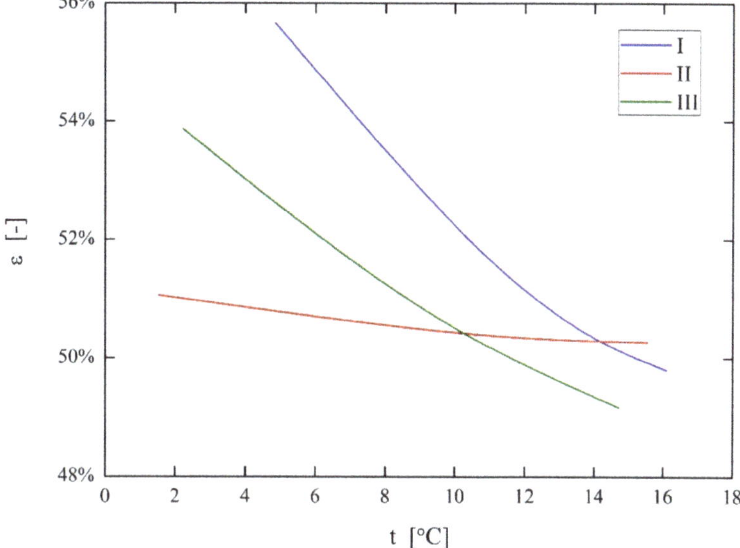

Figure 24. HEX temperature effectiveness vs. the fresh air inlet temperature for different volumetric flow rates.

Based on the characteristics in Figure 24, it is possible to determine the ranges of the temperature effectiveness in the case of the tested HEX. The highest temperature effectiveness was recorded for the first measurement series (56%). The lowest temperature effectiveness of 49.2% was observed for measurement series characterized by the lowest volumetric airflow rates. The recorded temperature effectiveness was in the range of 49–56%. The temperature effectiveness decreased with increasing the fresh air temperature for all measurement series.

4. Discussion

The present work was aimed at the parametric design of HPHE aided by the computational model. The model was validated against the experiment on a nine-row prototype HPHE. There was a good agreement between the model and measured heat transfer rates of the prototype HPHE. Process and design specifications were made for HPHE utilized as an air-to-air recuperator for air conditioning systems. This HEX type implies the following design parameters:

The HP container is a 1 mm thick 20 mm outer diameter copper tube, and the finned surface is made of aluminum. Finned tubes are manufactured by cold rolling technology. Aluminum was chosen because of its lower weight than copper, and considerably lower price.

The computational model was used to obtain the heat transfer rates, effectiveness, and pressure drop for varying parameters. The first set of calculations with assumed constant inlet and outlet air temperatures was analyzed to choose a number of rows that fulfills the 60% thermal efficiency goal: 20 rows were chosen (including three extra HPS as a safety factor in calculations). In this arrangement (for 300 m^3/h volumetric flow of hot and cold airstreams), the pressure drop does not exceed 100 Pa, increasing the maximal temperature difference of HPHE decreases in ΔP, with a relatively slight increase in effectiveness. The penalty for a reduction in temperature difference is not great as effectiveness drops by 4% over a 20 °C decrease. To sum up, a 20-rows HPHE exhibits stable effectiveness over a broad temperature range with an acceptable pressure drop for small air conditioning applications.

Further parametric computations were made to choose the optimal spacing between the HPs—3D plots show the dependence of efficiency and pressure drop on X_l and X_t. Traversal heat pipes spacing has the greatest influence on heat transfer rate; upon reduction in traversal spacing there is a steep increase in heat transfer rate and effectiveness for $X_t \approx 61$ mm caused by the increase in the number of HPs in rows. Four and three pipe arrangement can be used (70 HPs total) instead of three and two (50 HPs total). $X_t = 50$ was chosen as the most effective transverse spacing between HPs, which corresponds to the physical contact of HPs. Smaller X_l means a shorter heat exchanger but also a higher pressure drop (Figure 23). To minimize the pumping power of the fan longitudinal spacing $X_l = 61$ mm was chosen. It corresponds to the total length of the heat exchanger, $L_2 = 1.22$ m.

The final dimensions of the designed heat pipe heat exchanger are:

Length $L_2 = 1.22$ m; Height $L_1 = 0.245$ m, Width $L_3 = 0.245$ m.

The final finned heat pipe arrangement is shown in Figure 25. Tables 4 and 5 summarize HPHE working parameters for typical winter and summer conditions. The present best HPHE geometrical parameters choice is supported by the optimization of the overall cost function, which could be simplified to:

$$overall\ cost = HP\ cost \cdot N + cost\ of\ fan\ operation - savings\ due\ to\ heat\ recuperation \qquad (27)$$

where N—a number of HPs in an HPHE. The cost of the manufacture of one heat pipe was estimated as USD 21, and the value of the working fluid inside one HP is approx. USD 1.5. The cost of the fan operation is estimated, assuming a fixed price of 1 kWh of electricity (0.2 USD/kWh). The first two terms in Equation (27) generate cost—initial cost of manufacture of HPHE plus the consumed electric energy for the fans' operation. The savings come from the last term which takes into account the heat recuperation—in the cold months, the recuperated amount of thermal energy is priced as electrical energy (use of electrical heater), whereas in hot months, recuperated energy is divided by the Coefficient of Performance of the refrigeration device hypothetically used for airstream cooling. Based on the heat transfer rates, pressure drops, and air volumetric flow rate from the numerical model, the overall cost of HPHE in operation for 4 years is plotted in Figure 26. The chosen independent parameters were number of rows and traversal spacing X_t. The longitudinal spacing does not affect heat transfer or pressure drop significantly, so it was excluded from the analysis. The optimization method used to minimize the overall cost function is the so-called "brute force" method. The function's value is computed at each point of a multidimensional grid of points, to find the global minimum. An obvious disadvantage of this approach is the high computing power requirements, but the advantage is a certainty of obtaining the minimum in the range of the specified parameters. The minimum is marked in Figure 26, being USD −4218 (a negative value means that the savings are greater than the cost of the investment) for 18 rows HPHE, and optimal spacing $Xt = 0.051$ m. It is very similar to the final dimensions of the HPHE chosen according to previous engineering analysis.

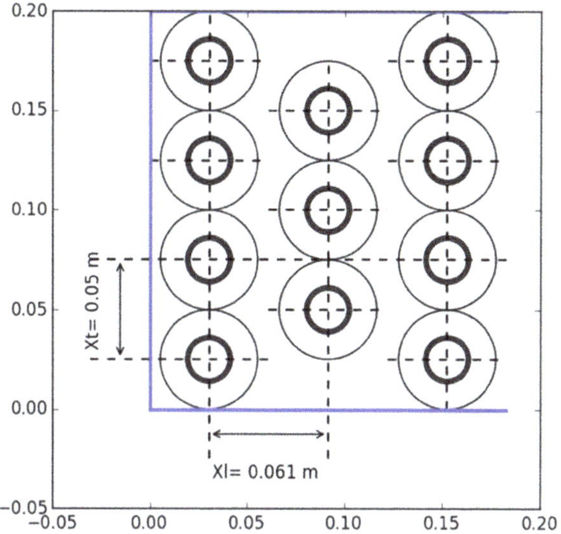

Figure 25. Final heat pipe heat exchanger arrangement.

Table 4. Final HPHE working parameters for typical summer conditions.

$\dot{Q}$ (W)	ε (%)	t_{hi} (°C)	t_{ho} (°C)	t_{ci} (°C)	t_{co} (°C)	Δp (Pa)	U (W/(m^2 K))
446.9	55.6	30.0	25.4	22.0	26.5	160	5.74

Table 5. Final HPHE working parameters for typical winter conditions.

$\dot{Q}$ (W)	ε (%)	t_{hi} (°C)	t_{ho} (°C)	t_{ci} (°C)	t_{co} (°C)	Δp (Pa)	U (W/(m^2 K))
2334	64.9	22.0	−1.27	−10.0	10.8	147	10.53

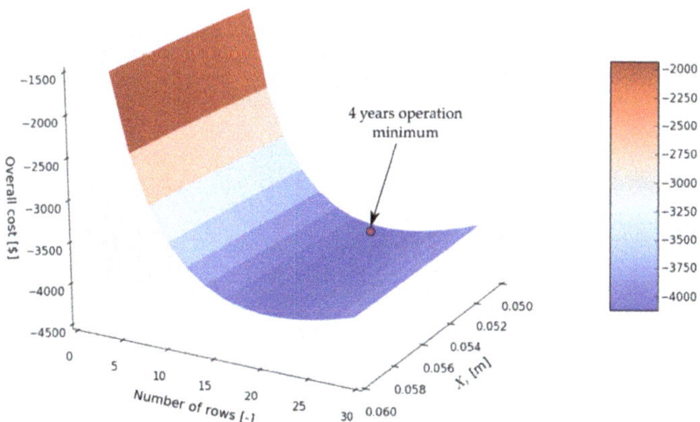

Figure 26. The overall cost of 4 years of operation of HPHE as a function of number of rows and traversal HPs spacing.

5. Conclusions

The final parameters given in Tables 4 and 5 prove that the designed 20-row individually finned HPHE is characterized by high recuperation effectiveness (55–65%). The effectiveness and heat transfer rate is higher for winter conditions, where higher temperature differences between airstreams exist.

The designed HPHE made of an individually finned HPs bundle is a competitive solution to the continuous plate-fine HPHE. The continuous fin HEX is a more thermally efficient construction (smaller number of rows for the same heat transfer rate), although the individual finning has its advantages in HPHE:

- It is more durable—can withstand higher pressure differences on the airside;
- Circular fins are less susceptible to mechanical damage than continuous fins, which can be easily deformed;
- Individual finning allows for simple identification of damaged HPs in the bundle and their replacement;
- HPHE with individually finned HPs can be more easily cleaned than continuously finned, therefore it is more robust in flue gas recuperation.

Even though more common continuous finned HPHEs outperform individually finned constructions, taking into account the above advantages, the construction proposed by authors can be more desired, especially from the exploitation and maintenance perspective.

The heat transfer intensification by the utilization of turbulators in an individually finned tube bundle (between HPs) is an attractive research direction, as it is not a widely recognized topic in the literature. It could improve the HPHE effectiveness significantly.

6. Patents

Polish patent number: P.415828, "Two-phase thermosiphon heat exchanger", date 16-08-2018.

Author Contributions: Conceptualization, M.Ł. and G.G.; Methodology, M.Ł.; Software, M.Ł., A.N.G., and A.R.; Supervision, G.G.; Visualization, G.G., D.A., and B.W.; Writing—original draft, M.Ł., G.G., and M.K.; Writing—review and editing, A.N.G. All authors have read and agreed to the published version of the manuscript.

Funding: Research funded by the National Centre for Research and Development, research project LIDER (grant no. *LIDER/08/42/L-3/11/NCBR/2012*) "Intensification of the heat transfer process in innovative heat exchanger—investigation with the utilization of PIV method".

Conflicts of Interest: The authors declare no conflict of interest.

Nomenclature

A	heat transfer surface area, m^2
A_f	finned area of the tube, m^2
A_o	minimum flow area, m^2
A_r	root (unfinned) area of the tube, m^2
A_s	area of finned surface, m^2
d_f	circular fin diameter, m
d_o, d_i	outside and inside HP diameter, m
d_r	fin root diameter, m
f	friction coefficient, -
h	heat transfer coefficient, W/(m^2·K)
h_{ca}, h_{ha}	cold and hot air heat transfer coefficients, W/(m^2·K)
I_0, I_1	modified Bessel function of order 0 and 1
k	thermal conductivity, W/(m·K)
K_0, K_1	modified Bessel function K of order 0 and 1
l_f	fin length, m
L	length of HP, m
L_1, L_2, L_3	height, length, and width of HEX, m

m	the coefficient for fin efficiency—Equation (20), -
N	number of HPs in HPHE bundle, -
N_f	number of fins per unit length, $1/m$
N_t	number of heat exchanger tube rows, -
Nu	Nusselt number, -
p	pressure, Pa
$\dot{Q}$	heat transfer rate through heat pipe, W
r_o, r_i	outer and inner diameter of a circular fin, m
R	thermal resistance, K/W
R_{ca}	convective thermal resistance of cold air flow, K/W
R_{ha}	convective thermal resistance of hot air flow, K/W
R_{hp}	inside resistance of HP, K/W
R_{finned}	overall thermal resistance of the finned surface, K/W
R_{tube}	thermal resistance of tube, K/W
Re_d	Reynolds number based on diameter, -
s	fin spacing, m
t_f	fin thickness, m
t_{co}, t_{ci}	outlet and inlet cold air stream temperature, $^\circ$C
t_{ho}, t_{hi}	outlet and inlet hot air stream temperature, $^\circ$C
T	temperature, K
T_m	enthalpy mixed-mean-temperature, K
U	overall heat transfer coefficient, $W/(m^2{\cdot}K)$
$\dot{V}$	volumetric flow rate, m^3/s
w_{max}	maximal average velocity through a finned tube bundle, m/s
x, y, z	cartesian coordinates, m
x', y', z'	the distance between HPs according to Figure 6 and Equations (11) and (12)
X_t	traversal spacing between tube centers, m
X_l	longitudinal spacing between tube centers, m

Greek Symbols

Δp	a pressure drop, Pa
ε_o	overall finned surface efficiency, -
η_f	fin efficiency, -
ν	kinematic viscosity, m^2/s
ρ	density, kg/m^3

Abbreviations

HEX	heat exchanger
HP	heat pipe
HPHE	heat pipe heat exchanger
SHP	separate heat pipe

References

1. Reay, D.A.; David, A.; Kew, P.A.; Peter, A.; Dunn, P.D.; Peter, D. *Heat Pipes*; Butterworth-Heinemann: Oxford, UK, 2006; ISBN 9780080464770.
2. Liu, D.; Tang, G.F.; Zhao, F.Y.; Wang, H.Q. Modeling and experimental investigation of looped separate heat pipe as waste heat recovery facility. *Appl. Therm. Eng.* **2006**, *26*, 2433–2441. [CrossRef]
3. Srimuang, W.; Amatachaya, P. A review of the applications of heat pipe heat exchangers for heat recovery. *Renew. Sustain. Energy Rev.* **2012**, *16*, 4303–4315. [CrossRef]
4. Tian, E.; He, Y.L.; Tao, W.Q. Research on a new type waste heat recovery gravity heat pipe exchanger. *Appl. Energy* **2017**, *188*, 586–594. [CrossRef]
5. Hughes, B.R.; Chaudhry, H.N.; Calautit, J.K. Passive energy recovery from natural ventilation air streams. *Appl. Energy* **2014**, *113*, 127–140. [CrossRef]
6. Abd El-Baky, M.A.; Mohamed, M.M. Heat pipe heat exchanger for heat recovery in air conditioning. *Appl. Therm. Eng.* **2007**, *27*, 795–801. [CrossRef]

7. Noie, S.H. Investigation of thermal performance of an air-to-air thermosyphon heat exchanger using ε-NTU method. *Appl. Therm. Eng.* **2006**, *26*, 559–567. [CrossRef]

8. Wu, X.P.; Johnson, P.; Akbarzadeh, A. Application of heat pipe heat exchangers to humidity control in air-conditioning systems. *Appl. Therm. Eng.* **1997**, *17*, 561–568. [CrossRef]

9. Righetti, G.; Zilio, C.; Longo, G.A. Comparative performance analysis of the low GWP refrigerants HFO1234yf, HFO1234ze(E) and HC600a inside a roll-bond evaporator. *Int. J. Refrig.* **2015**, *54*, 1–9. [CrossRef]

10. Jadhav, T.S.; Lele, M.M. Experimental performance and parametric analysis of heat pipe heat exchanger for air conditioning application integrated with evaporative cooling. *Heat Mass Transf. und Stoffuebertragung* **2017**, *53*, 3287–3293. [CrossRef]

11. Rajski, K.; Danielewicz, J.; Brychcy, E. Performance Evaluation of a Gravity-Assisted Heat Pipe-Based Indirect Evaporative Cooler. *Energies* **2020**, *13*, 200. [CrossRef]

12. Yau, Y.H.; Ahmadzadehtalatapeh, M. Performance Analysis of a Heat Pipe Heat Exchanger Under Different Fluid Charges. *Heat Transf. Eng.* **2014**, *35*, 1539–1548. [CrossRef]

13. Vasiliev, L.L. Heat pipes in modern heat exchangers. *Appl. Therm. Eng.* **2005**, *25*, 1–19. [CrossRef]

14. Yang, J.; Zhao, Y.; Chen, A.; Quan, Z. Thermal Performance of a Low-Temperature Heat Exchanger Using a Micro Heat Pipe Array. *Energies* **2019**, *12*, 675. [CrossRef]

15. Yang, K.-S.; Jiang, M.-Y.; Tseng, C.-Y.; Wu, S.-K.; Shyu, J.-C. Experimental Investigation on the Thermal Performance of Pulsating Heat Pipe Heat Exchangers. *Energies* **2020**, *13*, 269. [CrossRef]

16. Brough, D.; Ramos, J.; Delpech, B.; Jouhara, H. Development and validation of a TRNSYS type to simulate heat pipe heat exchangers in transient applications of waste heat recovery. *Int. J. Thermofluids* **2021**, *9*, 100056. [CrossRef]

17. Yau, Y.H.; Ahmadzadehtalatapeh, M. Predicting yearly energy recovery and dehumidification enhancement with a heat pipe heat exchanger using typical meteorological year data in the tropics. *J. Mech. Sci. Technol.* **2011**, *25*, 847–853. [CrossRef]

18. Yu, Z.T.; Hu, Y.C.; Cen, K.F. Optimal design of the separate type heat pipe heat exchanger. *J. Zhejiang Univ. Sci.* **2005**, *6*, 23–28. [CrossRef]

19. Righetti, G.; Zilio, C.; Mancin, S.; Longo, G.A. Heat Pipe Finned Heat Exchanger for Heat Recovery: Experimental Results and Modeling. *Heat Transf. Eng.* **2018**, *39*, 1011–1023. [CrossRef]

20. Thulukkanam, K. *Heat Exchanger Design Handbook*, 2nd ed.; Mechanical Engineering, CRC Press: Boca Raton, Florida, USA, 2013; ISBN 1439842124.

21. Briggs, D.E.; Young, E.H. Convection heat transfer and pressure drop of air flowing across triangular pitch banks of finned tubes. *Proc. Chem. Eng. Prog. Symp. Ser.* **1963**, *59*, 1–10.

22. Gorecki, G. Investigation of two-phase thermosyphon performance filled with modern HFC refrigerants. *Heat Mass Transf.* **2018**, *54*, 2131–2143. [CrossRef]

23. Łęcki, M.; Górecki, G. Different approaches to FVM method fluid flow and heat transfer simulation inside thermosyphon. In Proceedings of the 15th International Heat Transfer Conference, Kyoto, Japan, 10–15 August 2014.

24. Rura Bimetalowa Wysokożebrowana RBW—Cemal.com.pl—Najwyższej Jakości rury Żebrowane. Available online: http://cemal.com.pl/oferta/rura-bimetalowa-wysokozebrowana-rbw/ (accessed on 15 December 2020).

25. Climate Meter LB-580—Multiparameter | ClimateLoggers.com. Available online: https://climateloggers.com/en/lb580-climate-meter.measuring-systems.html (accessed on 16 August 2021).

Article

Heat Transfer Analysis for Non-Contacting Mechanical Face Seals Using the Variable-Order Derivative Approach

Slawomir Blasiak

Department of Manufacturing Engineering and Metrology, Faculty of Mechatronics and Mechanical Engineering, Kielce University of Technology, Aleja Tysiaclecia Panstwa Polskiego 7, 25-314 Kielce, Poland; sblasiak@tu.kielce.pl; Tel.: +48-41-34-24-756

Abstract: This article presents a variable-order derivative (VOD) time fractional model for describing heat transfer in the rotor or stator in non-contacting mechanical face seals. Most theoretical studies so far have been based on the classical equation of heat transfer. Recently, constant-order derivative (COD) time fractional models have also been used. The VOD time fractional model considered here is able to provide adequate information on the heat transfer phenomena occurring in non-contacting face seals, especially during the startup. The model was solved analytically, but the characteristic features of the model were determined through numerical simulations. The equation of heat transfer in this model was analyzed as a function of time. The phenomena observed in the seal include the conduction of heat from the fluid film in the gap to the rotor and the stator, followed by convection to the fluid surrounding them. In the calculations, it is assumed that the working medium is water. The major objective of the study was to compare the results of the classical equation of heat transfer with the results of the equations involving the use of the fractional-order derivative. The order of the derivative was assumed to be a function of time. The mathematical analysis based on the fractional differential equation is suitable to develop more detailed mathematical models describing physical phenomena.

Keywords: heat transfer analysis; non-contacting mechanical face seal; variable order derivative integral transform

Citation: Blasiak, S. Heat Transfer Analysis for Non-Contacting Mechanical Face Seals Using the Variable-Order Derivative Approach. *Energies* **2021**, *14*, 5512. https://doi.org/10.3390/en14175512

Academic Editor: Chi-Ming Lai

Received: 20 July 2021
Accepted: 31 August 2021
Published: 3 September 2021

Publisher's Note: MDPI stays neutral with regard to jurisdictional claims in published maps and institutional affiliations.

1. Introduction

Over the last few years, there have been many studies using fractional-order differential and integral operators to generalize classical differential and integral calculus with the aim of further understanding the nature of complex systems. Currently, attempts are being made to apply fractional calculus to solve various physical, mechanical, biological, or chemical problems. While classical integer-order operators are dependent only on the local behavior of the function, fractional-order operators accumulate all the information about the function. Another fundamental feature of fractional derivatives is that they are defined along a segment, not at a point, as is the case with classical derivatives. This feature ensures a more accurate and effective analysis of different phenomena. One of the shortcomings that most models have to overcome is that, mathematically, velocity is an instantaneous velocity defined at a point. Thus, it can be seen from the literature that differential calculus is used in the heat transfer theory.

It has been found, for example, that variable thermal conductivity is a key physical property of materials, especially when it is dependent on temperature. Variable thermal conductivity is of significance in a wide range of applications, including modern physics and mechanical engineering. It is taken into consideration primarily to determine the effect of temperature on the performance of machine elements. Special attention is paid to this property when rapid changes in temperature occur, as they may affect the operation of mechanical assemblies and subassemblies, and consequently the whole machine. The relationship between variable thermal conductivity and fractional differential calculus

99

has been investigated extensively for problems related to fluid mechanics [1,2], viscous elasticity [3], relaxation [4], thermal elasticity and conductivity [5,6], control theory [7,8], and many others [9–11]. Over the years, researchers have extended the classical theory of elasticity and introduced the theory of thermoelasticity.

The mechanical properties, especially the relaxation, of additively manufactured materials were analyzed, for example, in [4]. A mathematical model was developed using fractional differential calculus to further fit the relaxation curves with the experimental data.

Warbhe et al. [5] used the quasi-static approach to the fractional-order theory of thermoelasticity to solve a two-dimensional problem for a thin circular plate with zero temperature across the lower surface and constant and evenly distributed temperature across the insulated upper surface. The integral transform technique was employed to solve the physical problem, while the displacement potential function was applied to calculate thermal stresses. Povstenko and Kyrylych [6] studied the interface between two solids. They considered generalized boundary conditions of a non-ideal thermal contact to solve the heat conduction equation of the fractional order as a function of time using the Caputo derivative.

The extensive monograph by Kaczorek [7] presented the theoretical and practical aspects of the application of fractional calculus to fractional discrete-time linear systems.

The research by Nowacki, described for instance in [12–14], was a breakthrough in this field. His findings are still valid and thus widely cited, as they can be applied to solve a large number of theoretical and practical engineering problems.

Knowledge of all theories available in this area is crucial to understand the deformations of elastic materials. Their thermal and mechanical behavior needs to be taken into account whenever thermoelastic deformations result in leaks, as is the case with non-contacting face seals, where the gap height may range from one to several micrometers.

Fractional differential calculus has proved well-suited to predict many physical phenomena associated with elastic media, e.g., thermal conductivity, heat transfer, and viscoelasticity. The classical differential equations describing physical phenomena are modified using the variable-order derivative (VOD) time fractional approach. Recently, there has been much research focusing on extending the applications of fractional differential calculus, e.g., [15–18]. These studies provide an insight into many important aspects associated with heat conduction. In [15], for example, Povstenko presented fundamental solutions to the time-fractional advection diffusion equation for two cases: the plane and the half-plane. He also considered Cauchy problems, inverse source problems, and Dirichlet problems. The solutions were expressed in terms of Bessel integral functions combined with Mittag–Leffler functions. In his other articles [16,17], Povstenko discussed solutions for heat transfer in composite media, in which he employed the time-fractional heat conduction equation with the Caputo derivative of fractional order to describe heat transfer in both constituent materials ($0 < \alpha \leq 2$ and $0 < \beta \leq 2$, respectively). The problem was solved under ideal contact conditions. This suggests that the temperature of the two materials and the heat fluxes were the same.

Liu et al. [19] proposed an approach to solving the time fractional nonlinear heat conduction equation. Similar solutions were proposed by Koca and Lotfy [20,21].

Another issue is an inverse problem in heat transfer. Liu and Feng [22], for instance, solved it using the fractional differential equation for a time-dependent derivative. The solution to the 2D time-fractional inverse problem of diffusion was based on an improved kernel technique. Maximum a posteriori estimation was required. Convergence was achieved using the regularization term and deriving a priori probability.

The primary aim of this article was to show how the results obtained by solving the classical equation of heat transfer differ from those obtained for the same physical phenomenon by employing the fractional differential equation. The study involved using the derivative order as a time-dependent function. The results from the two approaches were compared graphically. The analytical solutions of the main physical quantities were developed using the Marchi–Zgrablich transform, the finite Fourier cosine transform, and, finally, the Laplace transform.

2. Mathematical Model

The classical Fourier heat equation can be written as:

$$\mathbf{q} = -\lambda \nabla \vartheta,$$ (1)

where $\mathbf{q}$—heat flux vector, $\vartheta = T - T_o$ and λ—thermal conductivity.

When combined with the energy conservation principle, the heat equation can be used to calculate the local deformation:

$$-\nabla \mathbf{q}(t) = \rho\, C_p\, \frac{\partial \vartheta(t)}{\partial t},$$ (2)

with ρ being the density and C_p the specific heat capacity.

Many researchers have examined the application of differential calculus or integral calculus involving the use of derivatives of fractional orders. Fractional calculus is a natural extension of the notions of differentials and integrals used in classical differential calculus and integral calculus, respectively. Recently, there has been much interest in the use of fractional differential calculus to explain and model many physical phenomena. For example, fractional differential calculus has been used to design and/or model $PI^\lambda D^\mu$ controllers, heat conduction [23–25], thermoelasticity [26], complex nonlinear systems [7], supercapacitors, electrical and mechanical systems [27–29], electrical filters, dielectric relaxation, diffusion, and viscoelasticity [30].

The time fractional heat conduction equation for the rotor can be written as [31]:

$$\frac{\partial^\alpha \vartheta}{\partial t^\alpha} = \kappa\, \Delta\vartheta \qquad 0 < \alpha \le 2 \text{ and } \kappa = \frac{K}{\rho\, C_p}.$$ (3)

For the case considered here, it was assumed that $\alpha = \alpha(t)$. The function $\alpha(t)$ was defined as a function of time (see Figure 1).

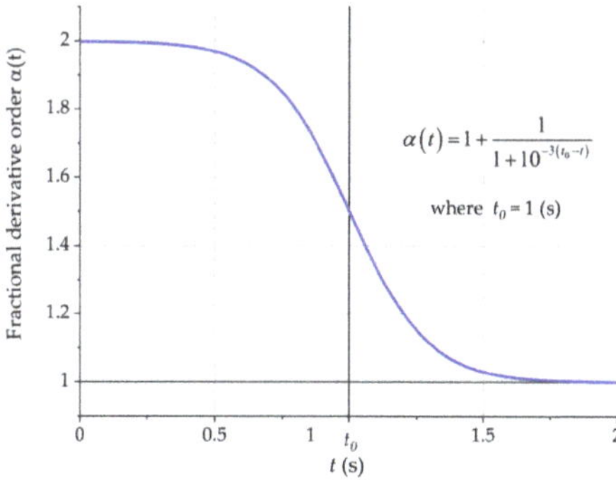

Figure 1. Fractional derivative order α vs. time.

The function describing the change in the coefficient $\alpha(t)$ can be written using the following formula:

$$\alpha(t) = \begin{cases} 1 + \dfrac{1}{1+10^{-3(t-t_0)}} & t \in (0, 2) \\ 1 & t > 2 \end{cases},$$ (4)

The transition to the time fractional heat conduction equation with the boundary conditions, was described, for example, in [32] as:

$$\frac{\partial^\alpha \vartheta}{\partial t^\alpha} = \kappa \left(\frac{1}{r} \frac{\partial \vartheta}{\partial r} + \frac{\partial^2 \vartheta}{\partial r^2} + \frac{\partial^2 \vartheta}{\partial z^2} \right), \text{ for } r_i \leq r \leq r_o; \ 0 \leq z \leq L; \ t > 0, \tag{5}$$

where L is the thickness of the sealing rings.

The Caputo derivative of the fractional order can be defined as described in [15]:

$$\frac{\partial^\alpha f(t)}{\partial t^\alpha} = \begin{cases} \frac{1}{\Gamma(n-\alpha)} \int\limits_0^t (t-\tau)^{n-\alpha-1} \frac{d^n f(\tau)}{d\tau^n} d\tau, \ n-1 < \alpha < n, \\ \frac{d^n f(t)}{dt^n}, \ \alpha = n, \end{cases} \tag{6}$$

where $\Gamma(\alpha)$ is the gamma function, with the initial conditions being: $t = 0$ $\vartheta = 0$, $0 < \alpha \leq 2$, $t = 0$ $\frac{\partial \vartheta}{\partial t} = 0$, $1 < \alpha \leq 2$.

The heat conduction equation for the stator is thus rewritten as:

$$\frac{1}{r} \frac{\partial \theta}{\partial r} + \frac{\partial^2 \theta}{\partial r^2} + \frac{\partial^2 \theta}{\partial z^2} = 0, \text{ for } r_i \leq r \leq r_o; \ 0 \leq z \leq L, \tag{7}$$

The heat transfer model analyzed here also needs to take into consideration the heat flux generated in the gap, which can be described with a relationship based on the simplified energy equation. A similar method was used in [32]:

$$\mu \left(\frac{\partial v_\phi}{\partial z} \right)^2 + \lambda^f \frac{\partial^2 T^f}{\partial z^2} = 0 \tag{8}$$

The velocity of the fluid particles v_φ is linearly variable. Ranging from zero on the stator surface to the value of (ωr) on the rotor surface, it can be described as:

$$\frac{\partial v_\phi}{\partial z} = \frac{\omega r}{h} \tag{9}$$

The energy Equation (8) was solved taking into account relationship Equation (9) to calculate the distribution of temperature in the gap.

Considering sealing rings with unmodified face surfaces and neglecting other factors affecting their geometry, e.g., mechanical deformations, we can assume that the height of the gap is constant: $h = const$. The gap height considered by other researchers is that discussed, for example, in reference [33].

As dynamic viscosity is a temperature-dependent parameter, here it can be defined using the formula proposed by Li et al. [1]:

$$\mu = \mu_o \exp(-b \left(T_m - T_o \right)). \tag{10}$$

The average temperature of the medium in the gap can be determined from:

$$T_m = \frac{1}{h(r)} \int\limits_0^{h(r)} T^f dz. \tag{11}$$

Equation (10) is a function describing the distribution of dynamic viscosity $\mu(r)$ in the radial direction.

The seal performance is largely affected by the forces acting on the stator and the rotor. The closing force produced by the spring ensures the leak tightness of the device in the OFF mode and the contact of the stator and the rotor. However, when the device operates (is ON), the opening force is generated by the pressure of the medium in the gap. The distribution of pressure can be described using the one-dimensional Reynolds equation:

$$\frac{d}{dr}\left(r\frac{\rho h^3}{\mu}\frac{dp}{dr}\right) = 0, \tag{12}$$

where the boundary conditions are:

$$p(r)|_{r=r_i} = p_i, \ \ p(r)|_{r=r_o} = p_o \tag{13}$$

Equation (12) is the simplified Reynolds equation describing the changes in pressure of the incompressible medium in the gap only in the radial direction.

3. Boundary Conditions

The schematic diagram in Figure 2 illustrates the non-contacting face seal with the flexibly mounted stator (1) and the rotor (2) connected to the shaft (6) of the turbo machine. The gap between the stator and the rotor, with a height ranging from one micrometer to several or even more than ten micrometers, is filled with a sealing medium, e.g., water. The distribution of temperature on the stator and rotor surfaces is dependent mainly on the heat flux in the gap.

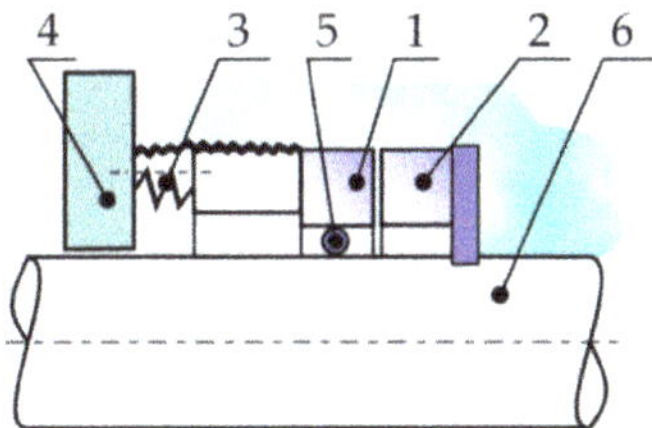

Figure 2. Non-contacting face seal made up of the flexibly mounted stator (1), the rotor (2), the spring (3), the housing (4), the O-ring (5), and the shaft (6).

The boundary conditions are represented graphically in Figure 3.

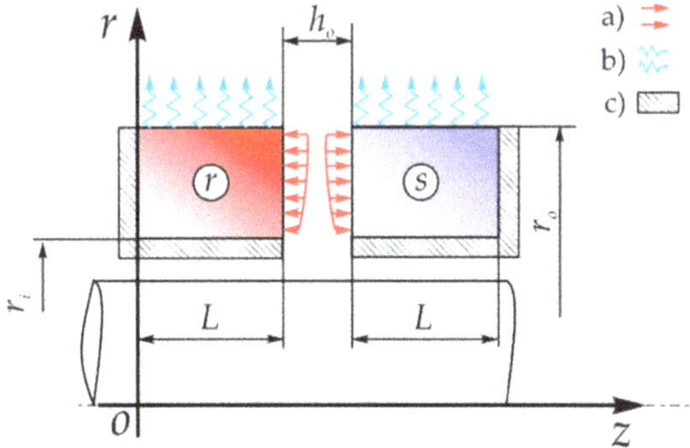

Figure 3. Boundary conditions for the heat transfer in the non-contacting face seal. (a) Heat flux, (b) convection, and (c) insulated surfaces.

In this analysis, it is assumed that the inner and bottom surfaces of the rotor (r_i) and the inner and top surfaces of the stator are not in contact with the surroundings; thus, heat transfer for these surfaces takes the general form: $\frac{\partial \theta}{\partial n} = \frac{\partial \theta}{\partial n} = 0$. In a real system, these surfaces are in contact with other elements of the seal characterized by different physical properties. Under certain conditions, heat transfer taking place between these elements can be assumed to be negligible. Another reason to introduce the boundary conditions ($\frac{\partial \vartheta}{\partial n} = 0$) for the surfaces is the considerable simplification of the calculations for the analytically solved model.

As the gap is limited by the ring faces, the boundary conditions for the rotor and the stator are as follows:

$$\lambda \frac{\partial \vartheta}{\partial z}\bigg|_{z=L} = \lambda^f \frac{\partial \theta^f}{\partial z}\bigg|_{z=0} \quad \text{and } \vartheta = \theta^f, \tag{14}$$

$$\lambda^f \frac{\partial \theta^f}{\partial z}\bigg|_{z=h} = \lambda \frac{\partial \vartheta}{\partial z}\bigg|_{z=0} \quad \text{and } \theta^f = \theta, \text{ respectively} \tag{15}$$

For the outer surface of the rotor (r_o), heat transfer is assumed to be by convection, and it can be expressed by:

$$-\lambda \frac{\partial \vartheta}{\partial r}\bigg|_{r=r_o} = \alpha^f \, \vartheta|_{r=r_o}, \text{ for the rotor} \tag{16}$$

$$-\lambda \frac{\partial \theta}{\partial r}\bigg|_{r=r_o} = \alpha^f \, \theta|_{r=r_o}, \text{ for the stator.} \tag{17}$$

where α^f is the convection coefficient.

All the above boundary conditions are necessary to solve the system of three differential equations.

4. Problem Solution

For the case considered here, a cylindrical coordinate system corresponding to the geometry of the physical model was used, and thus Equation (3) can be written as:

$$\frac{\partial^\alpha \vartheta}{\partial \tau^\alpha} = \frac{1}{r}\frac{\partial \vartheta}{\partial r} + \frac{\partial^2 \vartheta}{\partial r^2} + \frac{\partial^2 \vartheta}{\partial z^2}, \text{ where } \tau = \kappa t \tag{18}$$

Equation (18) was solved by adopting the Marchi–Zgrablich transform, the finite Fourier cosine transform, and the Laplace transform. The general form of the finite integral transform by Marchi-Zgrablich [34] is:

$$\mathcal{H}(f(x)) = \overline{f}(n) = \int_{r_i}^{r_o} r\,f(x)\,S_p(\lambda, \alpha, k_n\,r)\,dr \tag{19}$$

The inverse integral transform proposed by Marchi-Zgrablich can be expressed as:

$$\mathcal{H}^{-1}\left(\overline{f}(x)\right) = f(x) = \sum_{n=1}^{\infty} a_n\,S_p(\lambda, \alpha, k_n \cdot r), \tag{20}$$

where

$$a_n = \frac{\overline{f}_p(n)}{C_n}, \tag{21}$$

$$C_n = \int_{r_i}^{r_o} r\left[S_p(\lambda, \alpha, k_n\,r)\right]^2 dr. \tag{22}$$

Thus, Equation (18) is written as:

$$\frac{\partial^\alpha \overline{\vartheta}}{\partial \tau^\alpha} = \left(\frac{\partial^2 \overline{\vartheta}}{\partial z^2} - k^2{}_n \overline{\vartheta}\right). \tag{23}$$

The finite Fourier cosine transform is described in the general form as:

$$\mathcal{F}_c[g(x)] = g^*(m) = \int_0^L g(x) \cos\left(\frac{m\pi x}{L}\right) dx, \tag{24}$$

where $m = 1, 2, 3, \ldots$

$$\mathcal{F}_c\left[\frac{d^2 g(x)}{dx^2}\right] = (-1)^m \left.\frac{dg(x)}{dx}\right|_{x=L} - \left.\frac{dg(x)}{dx}\right|_{x=0} - \frac{m^2\pi^2}{L^2} \underbrace{g^*(m)}_{g^*(n)=\mathcal{F}_c[g(x)]} . \tag{25}$$

The inverse transform is given in the general form as:

$$\mathcal{F}_c^{-1}[g^*(m)] = g(x) = \frac{g^*(m=0)}{L} + \frac{2}{L}\sum_{m=1}^{\infty} g^*(m) \cos\left(\frac{m\pi x}{L}\right). \tag{26}$$

Applying the finite Fourier cosine transform to write Equation (23) and assuming the boundary conditions given in Equation (14), we obtain:

$$\frac{\partial^\alpha \overline{\vartheta}^*}{\partial \tau^\alpha} = \left(\frac{\overline{\varphi}_q}{\lambda} - \frac{m^2\pi^2}{L^2}\overline{\vartheta}^*(m) - k^2{}_n\overline{\vartheta}^*(m)\right). \tag{27}$$

With the Laplace transform [25] being:

$$\mathcal{L}\{D_C^\alpha f(t)\} = s^\alpha \hat{f}(s) - \sum_{k=0}^{n-1} f^{(k)}(0^+) s^{\alpha-1-k}, n-1 < \alpha < n, \tag{28}$$

we can rewrite Equation (27) as:

$$s^\alpha \hat{\overline{\vartheta}}^*(s) = \frac{\overline{\varphi}_q}{\lambda} - \frac{m^2\pi^2}{(L^r)^2}\hat{\overline{\vartheta}}^*(s) - k^2{}_n\hat{\overline{\vartheta}}^*(s). \tag{29}$$

Once the transformations are completed, Equation (18) has the following form:

$$\hat{\overline{\vartheta}}^*(s) = \frac{\overline{\varphi}_q}{\lambda}\frac{1}{\omega^2}\left(\frac{1}{s} - \frac{s^{\alpha-1}}{(s^\alpha + \omega^2)}\right) \text{ where } \left(\frac{m^2\pi^2 + L^2 k^2{}_n}{L^2}\right) = \omega^2 \text{ and } \omega_0{}^2 = k^2{}_n \tag{30}$$

After the three inverse integral transforms are employed, the equation describing the temperature distribution for the rotor takes the final form:

$$T^r = T_o + \left(\begin{array}{l} \dfrac{1}{L}\sum_{n=1}^{\infty} \dfrac{\int_{r_i}^{r_o} r\,\overline{\varphi}_q\, S_p\left(\lambda^s,\alpha^f, k_n r\right) dr}{\lambda\int_{r_i}^{r_o} r\left[S_p(\lambda,\alpha, k_n r)\right]^2 dr}\, S_p(\lambda,\alpha, k_n r)\left(\dfrac{1}{\omega_0{}^2} - \dfrac{1}{\omega_0{}^2}E_\alpha\left(-\kappa\omega_0{}^2 t^\alpha\right)\right) + \\[2em] +\dfrac{2}{L}\sum_{n=1}^{\infty} \dfrac{\int_{r_i}^{r_o} r\,\overline{\varphi}_q\, S_p\left(\lambda^s,\alpha^f, k_n r\right) dr}{\lambda\int_{r_i}^{r_o} r\left[S_p(\lambda,\alpha, k_n r)\right]^2 dr}\, S_p(\lambda,\alpha, k_n r)\sum_{m=1}^{\infty} \cos\left(\dfrac{m\pi x}{L}\right)\left(\dfrac{1}{\omega^2} - \dfrac{1}{\omega^2}E_\alpha\left(-\kappa\omega^2 t^\alpha\right)\right) \end{array}\right), \tag{31}$$

where $E_{\alpha,\beta}(z)$ is the function of the Mittag–Leffler type [35,36]:

$$E_{\alpha,\beta}(z) = \sum_{v=1}^{\infty} \frac{z^v}{\Gamma(\alpha v + \beta)} \quad \alpha > 0,\ \beta > 0. \tag{32}$$

For $\alpha, \beta = 1$, $E_{\alpha,\beta}(z) = e^z$, and thus Equation (31) is solved using the classical heat equation with predetermined initial and boundary conditions.

$$S_p(\lambda, \alpha, k_n \cdot r) = (-k_n Y_1(k_n r_i) + \alpha Y_0(k_n r_o) - \lambda k_n Y_1(k_n r_o)) J_0(k_n r) \\ + (k_n J_1(k_n r_i) - \alpha J_0(k_n r_o) + \lambda k_n J_1(k_n r_o)) Y_0(k_n r) \tag{33}$$

Equation (33) is dependent on the boundary conditions assumed for the rotor or stator model. The distribution of temperature in the stator is defined as:

$$T^s = T_0 + \sum_{n=1}^{\infty} \frac{\cosh(k_n z) \int_{r_i}^{r_o} r\, \theta^f\, S_p\left(\lambda^s, \alpha^f, k_n\, r\right) dr}{\cosh(k_n L) \int_{r_i}^{r_o} r\left[S_p(\lambda^s, \alpha^f, k_n \cdot r)\right]^2 dr} S_p\left(\lambda^s, \alpha^f, k_n\, r\right). \tag{34}$$

where θ^f is the excess temperature of the medium in the gap, calculated as $\theta^f = T^f - T_0$.

5. Operating Conditions

The parameters characterizing the material used for the rotor and the stator are provided in Table 1.

Table 1. Properties of the rotor and stator materials.

Properties / Material	Density ρ (kg/m^3)	Poisson's Ratio ν	Thermal Conductivity λ (W/mK)	Thermal Expansion τ (1/°C)	Specific Heat C_p (J/kgK)
Silicon carbide	3100	0.18	130	$5 \cdot 10^{-6}$	750
Resin-impregnated carbon	1860	0.20	15	$4 \cdot 10^{-6}$	–

The parameters concerning the seal geometry and performance are shown in Table 2. It is assumed that the surrounding fluid is water with a temperature of 20 °C.

Table 2. Parameters related to seal geometry and performance.

Seal Geometry		Seal Performance	
Inner radius r_i (m)	0.050	Convection coefficient (water) α^f (W/m^2 K)	18,000
Outer radius r_o (m)	0.055	Temperature of the surrounding fluid T_0 (°C)	20
Rotor/Stator thickness L (m)	0.010	Angular speed ω (rad/s)	800
Gap height h (m)	$2 \cdot 10^{-6}$		

The analytical solution to the mathematical model considered here shows the distribution of temperature in the cross-section of the seal when α in Equation (18) has a fractional order value of $1 \leq \alpha \leq 2$. Moreover, the parameter α is time-dependent, which is illustrated graphically in Figure 1.

6. Results and Discussion

The simulations were performed to determine the influence of the factor α on the distribution of temperature in the seal (Figure 4).

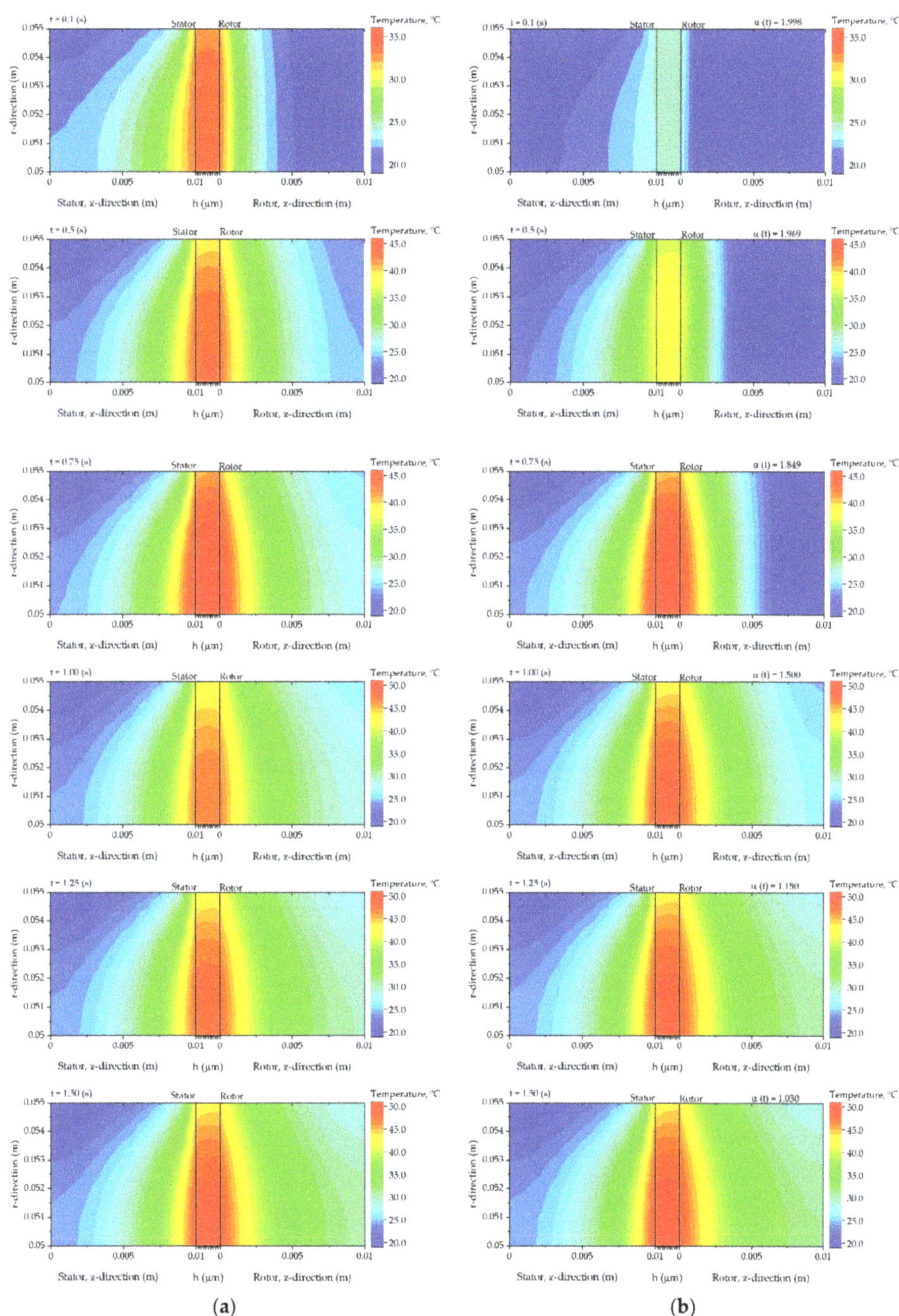

Figure 4. Distribution of temperature in the cross-section of the non-contacting face seal at different values of α. (**a**) for α = 1, (**b**) for α(t) = var.

Figure 4 illustrates the distributions of temperature in the rotor over time, ranging from 0.1 s to 1.5 s. When the classical heat equation is used (Figure 4a), the temperature of the rotor increases gradually because of the heat flux generated in the gap. The greatest difference between the diagrams is observed for the time range 0.1–1.25 s. With the assumption that $\alpha(t)$, Figure 1 suggests that if values higher than $t_0 = 1$ s are used, the order of the derivative tends toward unity.

The heat flux provided by the sealing ring face at a given moment of time for $t = 0.05$ (s) causes a local increase in temperature of 2.1 °C, which may result in thermal deformations of the sealing ring in contact with the gap. In the case of the fractional differential equation (for $t = 0.05$ (s)), the equation is hyperbolic in nature, as illustrated in Figure 5b.

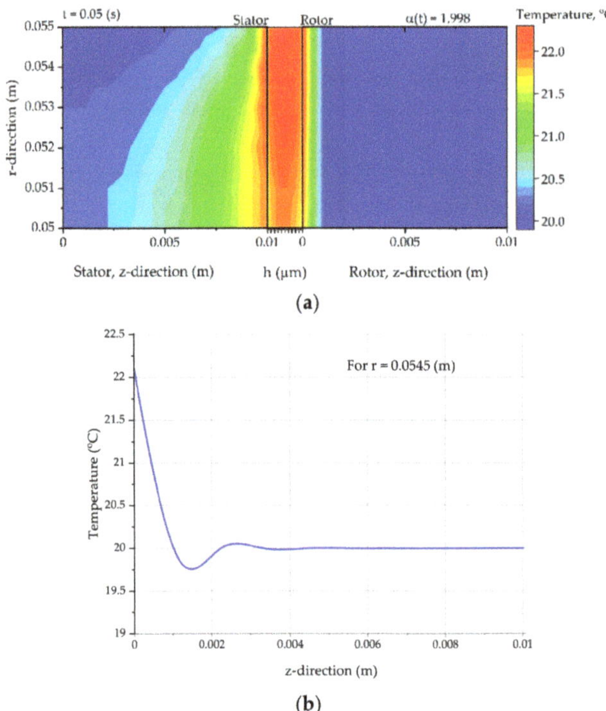

Figure 5. (**a**) Distribution of temperature in the seal for $t = 0.05$ (s) and $\alpha = 1.998$, (**b**) temperature vs. the rotor thickness for $r = 0.0545$ (m).

The solution based on the variable-order derivative (VOD) time fractional model provides a better correlation between the predicted data and the observed results than the classical approach.

7. Validation of Results

The results were validated using Ansys Workbench software (Figure 6).

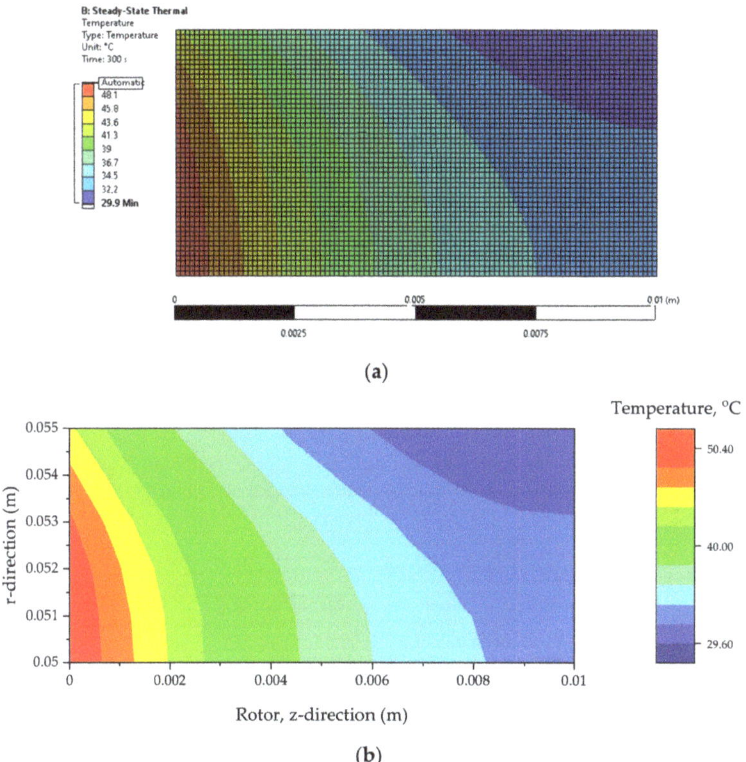

(a)

(b)

Figure 6. Temperature distribution in the rotor calculated by (**a**) the Ansys Workbench model and (**b**) the proposed model for t = 300 s.

Figure 6 depicts the temperature distributions calculated using the Ansys Workbench model (Figure 6a) and those obtained analytically (Figure 6b).

Figure 7 compares the distributions of temperature along the radius (r = 0.0525 m).

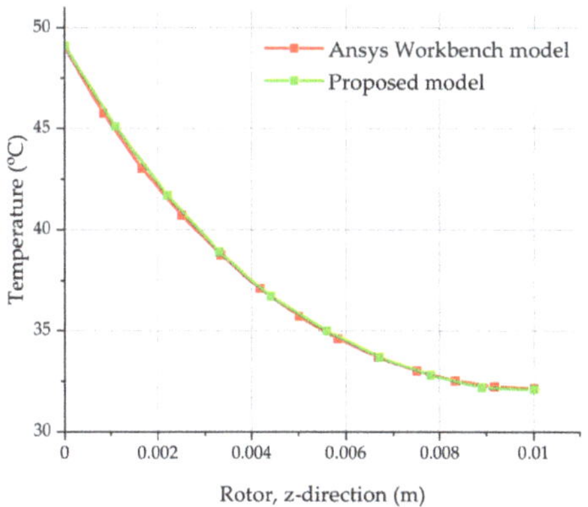

Figure 7. Distributions of temperature along the rotor height (thickness) for r = 0.0525 m.

From Figure 7, it is apparent that the results obtained using both calculation methods are in good agreement.

8. Conclusions

This article has presented a mathematical model to determine the heat transfer phenomena occurring in the non-contacting face seal used in a turbo machine. The problem was solved analytically by applying three integral transforms. The resulting relationships described the changes in temperature in the face seal cross-section. The boundary conditions were taken into consideration. The classical Fourier law was extended by applying the time fractional derivative where $\alpha = \alpha(t)$. This model can be used to describe unstable heat transfer conditions or thermal shock.

Even small local changes in temperature of an order of 2–10 °C can cause thermoelastic deformations of the stator and the rotor. As a result, there is a change in the gap geometry. During the first moments of the startup, unstable conditions can be observed because the seal operates under dry friction conditions; thus, the heat flux generated between the rotor and the stator is much greater than during the non-contact operation of the seal assumed in this model. The deformations may contribute to a greater leak.

It should be mentioned that the mathematical analysis based on the fractional differential equation is suitable to develop more accurate models to analyze similar physical phenomena.

Funding: This research received no external funding.

Institutional Review Board Statement: Not applicable.

Informed Consent Statement: Not applicable.

Data Availability Statement: Not applicable.

Conflicts of Interest: The author declares no conflict of interest.

References

1. Li, C.; Zheng, L.; Zhang, X.; Chen, G. Flow and heat transfer of a generalized Maxwell fluid with modified fractional Fourier's law and Darcy's law. *Comput. Fluids* **2016**, *125*, 25–38. [CrossRef]
2. Bartosik, A. Application of Rheological Models in Prediction of Turbulent Slurry Flow. *Flow Turbul. Combust.* **2010**, *84*, 277–293. [CrossRef]
3. Kozior, T. The Influence of Selected Selective Laser Sintering Technology Process Parameters on Stress Relaxation, Mass of Models, and Their Surface Texture Quality. *3D Print. Addit. Manuf.* **2020**, *7*, 126–138. [CrossRef]
4. Bochnia, J.; Blasiak, S. Fractional relaxation model of materials obtained with selective laser sintering technology. *Rapid Prototyp. J.* **2019**, *25*, 76–86. [CrossRef]
5. Warbhe, S.D.; Tripathi, J.J.; Deshmukh, K.C.; Verma, J. Fractional Heat Conduction in a Thin Circular Plate with Constant Temperature Distribution and Associated Thermal Stresses. *J. Heat Transf.* **2017**, *139*, 44502. [CrossRef]
6. Povstenko, Y.; Kyrylych, T. Fractional heat conduction in solids connected by thin intermediate layer: Nonperfect thermal contact. *Contin. Mech.* **2019**, *31*, 1719–1731. [CrossRef]
7. Kaczorek, T. *Selected Problems of Fractional Systems Theory*; Springer Berlin Heidelberg: Berlin/Heidelberg, Germany, 2011; ISBN 9783642205026.
8. Laski, P.A. Fractional-order feedback control of a pneumatic servo-drive. *Bull. Pol. Acad. Sci.-Tech. Sci.* **2019**, *67*, 53–59. [CrossRef]
9. Zmarzly, P.; Gogolewski, D.; Kozior, T. Design guidelines for plastic casting using 3D printing. *J. Eng. Fibers Fabr.* **2020**, *15*. [CrossRef]
10. Adamczak, S.; Zmarzly, P.; Kozior, T.; Gogolewski, D. Assessment of Roundness and Waviness Deviations of Elements Produced by Selective Laser Sintering Technology. In *Engineering Mechanics 2017*; ACAD SCI Czech Republic, INST Thermomechanics: Prague, Czech Republic; ISBN 978-80-214-5497-2.
11. Takosoglu, J.E.; Laski, P.A.; Blasiak, S. Innovative Modular Pneumatic Valve Terminal with Self-Diagnosis, Control and Network Communications. In *Engineering Mechanics 2014*; Fuis, V., Ed.; ACAD SCI Czech Republic, INST Thermomechanics: Prague, Czech Republic; pp. 644–647; ISBN 978-80-214-4871-1.
12. Nowacki, W. Problems of thermoelasticity. *Prog. Aerosp. Sci.* **1970**, *10*, 1–63. [CrossRef]
13. Nowacki, W. *Dynamic Problems of Thermoelasticity*; Noordhoof International: Leyden, CO, USA, 1976; ISBN 978-90-286-0045-4.
14. Nowacki, W. *Thermoelasticity*, 2nd ed.; Elsevier Science: Kent, UK, 2014; ISBN 978-1-4831-6248-5.

15. Povstenko, Y.Z. Fundamental Solutions to Time-Fractional Advection Diffusion Equation in a Case of Two Space Variables. *Math. Probl. Eng.* **2014**, *2014*, 705364. [CrossRef]
16. Povstenko, Y. Fractional Heat Conduction in an Infinite Medium with a Spherical Inclusion. *Entropy* **2013**, *15*, 4122–4133. [CrossRef]
17. Povstenko, Y. Fractional heat conduction in a semi-infinite composite body. *Commun. Appl. Ind. Math.* **2014**, *6*. [CrossRef]
18. Raslan, W.E. Application of fractional order theory of thermoelasticity to a 1D problem for a spherical shell. *J. Theor. Appl. Mech.* **2016**, *54*, 295–304. [CrossRef]
19. Liu, J.-G.; Yang, X.-J.; Feng, Y.-Y. On integrability of the time fractional nonlinear heat conduction equation. *J. Geom. Phys.* **2019**, *144*, 190–198. [CrossRef]
20. Koca, I. Modeling the heat flow equation with fractional-fractal differentiation. *Chaos Solitons Fractals* **2019**, *128*, 83–91. [CrossRef]
21. Lotfy, K. Analytical solution of fractional order heat equation under the effects of variable thermal conductivity during photothermal excitation of spherical cavity of semiconductor medium. *Waves Random Complex Medium* **2021**, *31*, 239–254. [CrossRef]
22. Liu, S.; Feng, L. An Inverse Problem for a Two-Dimensional Time-Fractional Sideways Heat Equation. *Math. Probl. Eng.* **2020**, *2020*, 5865971. [CrossRef]
23. Povstenko, Y. Time-fractional radial heat conduction in a cylinder and associated thermal stresses. *Arch. Appl. Mech.* **2012**, *82*, 345–362. [CrossRef]
24. Povstenko, Y.Z. Axisymmetric Solutions to Time-Fractional Heat Conduction Equation in a Half-Space under Robin Boundary Conditions. *Int. J. Differ. Equ.* **2012**, *2012*, 154085. [CrossRef]
25. Jiang, X.; Xu, M. The time fractional heat conduction equation in the general orthogonal curvilinear coordinate and the cylindrical coordinate systems. *Phys. A* **2010**, *389*, 3368–3374. [CrossRef]
26. Povstenko, Y. *Fractional Thermoelasticity*; Springer International Publishing: Cham, Switzerland, 2015; ISBN 978-3-319-15334-6.
27. Mainardi, F. *Fractional Calculus and Waves in Linear Viscoelasticity: An Introduction to Mathematical Models*; Imperial College Press: London, UK; Hackensack, NJ, USA, 2010; ISBN 978-1-84816-329-4.
28. Blasiak, M.; Blasiak, S. Application of Fractional Calculus in Harmonic Oscilator. In *Engineering Mechanics 2017*; ACAD SCI Czech Republic, INST Thermomechanics: Prague, Czech Republic, 2017; pp. 146–149; ISBN 978-80-214-5497-2.
29. Blasiak, M.; Blasiak, S. The Application of Integral Transforms to Solving Partial Differential Equations of the Fractional Order. In *Engineering Mechanics 2017*; ACAD SCI Czech Republic, INST Thermomechanics: Prague, Czech Republic, 2017; pp. 150–153; ISBN 978-80-214-5497-2.
30. Li, Z.L.; Sun, D.G.; Han, B.H.; Sun, B.; Zhang, X.; Meng, J.; Liu, F.X. Response of viscoelastic damping system modeled by fractional viscoelastic oscillator. *Proc. Inst. Mech. Eng. Part C J. Mech. Eng. Sci.* **2017**, *231*, 3169–3180. [CrossRef]
31. Qi, H.-T.; Xu, H.-Y.; Guo, X.-W. The Cattaneo-type time fractional heat conduction equation for laser heating. *Comput. Math. Appl.* **2013**, *66*, 824–831. [CrossRef]
32. Blasiak, S. Time-fractional heat transfer equations in modeling of the non-contacting face seals. *Int. J. Heat Mass Transf.* **2016**, *100*, 79–88. [CrossRef]
33. Tournerie, B.; Danos, J.C.; Frene, J. Three-Dimensional Modeling of THD Lubrication in Face Seals. *J. Tribol.* **2001**, *123*, 196–204. [CrossRef]
34. Ghonge, B.E.; Ghadle, K.P. Deflection of transient thermoelastic circular plate by Marchi-Zgrablich and Laplace integral transform technique. *Appl. Mech. Lett.* **2012**, *2*, 21004. [CrossRef]
35. Haubold, H.J.; Mathai, A.M.; Saxena, R.K. Mittag-Leffler Functions and Their Applications. *J. Appl. Math.* **2011**, *2011*, 298628. [CrossRef]
36. Rogosin, S. The Role of the Mittag-Leffler Function in Fractional Modeling. *Mathematics* **2015**, *3*, 368–381. [CrossRef]

Article

Dependence of Conjugate Heat Transfer in Ribbed Channel on Thermal Conductivity of Channel Wall: An LES Study

Joon Ahn [1,*], **Jeong Chul Song** [2] **and Joon Sik Lee** [2]

[1] School of Mechanical Engineering, Kookmin University, Seoul 02707, Korea
[2] School of Mechanical and Aerospace Engineering, Seoul National University, Seoul 08826, Korea; sjc7595@snu.ac.kr (J.C.S.); jslee123@snu.ac.kr (J.S.L.)
* Correspondence: jahn@kookmin.ac.kr

Abstract: A series of large eddy simulations was conducted to analyze conjugate heat transfer characteristics in a ribbed channel. The cross section of the rib is square and the blockage ratio is 0.1. The pitch between the ribs is 10 times the rib height. The Reynolds number of the channel is 30,000. In the simulations, the effect of the thermal resistance of the solid wall of the channel on convective heat transfer was observed in the turbulent flow regime. The numerical method used was based on the immersed boundary method and the concept of effective conductivity is introduced. When the conductivity ratio between the solid wall and the fluid (K^*) exceeded 100, the heat transfer characteristics resembled those for an isothermal wall, and the cold core fluid impinging and flow recirculation mainly influenced the convective heat transfer. For $K^* \leq 10$, the effect of the cold core fluid impinging became weak and the vortices at the rib corners strongly influenced the convective heat transfer; the heat transfer characteristics were therefore considerably different from those for an isothermal wall. At $K^* = 100$, temperature fluctuations at the upstream edge of the rib reached 2%, and at $K^* = 1$, temperature fluctuations in the solid region were similar to those in the fluid region. The rib promoted heat transfer up to $K^* = 100$, but not for $K^* \leq 10$. The Biot number based on the channel wall thickness appears to adequately explain the variation of the heat transfer characteristics with K^*.

Keywords: ribbed channel; large eddy simulation; immersed boundary method; conjugate heat transfer; thermal conductivity ratio

Citation: Ahn, J.; Song, J.C.; Lee, J.S. Dependence of Conjugate Heat Transfer in Ribbed Channel on Thermal Conductivity of Channel Wall: An LES Study. *Energies* **2021**, *14*, 5698. https://doi.org/10.3390/en14185698

Academic Editor: Artur Bartosik

Received: 31 July 2021
Accepted: 6 September 2021
Published: 10 September 2021

Publisher's Note: MDPI stays neutral with regard to jurisdictional claims in published maps and institutional affiliations.

1. Introduction

Gas turbines are mainly used as prime movers for aircraft propulsion and natural gas power generation. Their thermal efficiency and output increase with their inlet temperature [1]. Current state-of-the-art turbine engines operate at inlet temperatures (1700 °C) well above the melting point of the material (1000 °C), and hence, the turbine blades are cooled by compressor bleed air (700 °C). Cooling air is supplied to the internal cooling passage of the turbine blade (Figure 1a), where ribs are installed to promote heat transfer. If the predicted temperature of the blade is out of 30 °C, the blade life can be halved, and therefore, local heat transfer by the rib and the blade temperature should be accurately predicted [2].

The effect of promoting heat transfer by using ribs has been studied under various conditions and by considering various variables over the past several decades. It has been reported that the channel blockage ratio, rib pitch (=p) [3], rib angle of attack [4], and rib shape [5,6] have a significant effect on performance. In the recirculating flow region behind the rib, heat transfer is not active. When the blockage ratio is increased, the pressure drop increases significantly, so the optimal values of blockage ratio and pitch exist [7]. It is recommended that the rib height (=e) be 10% of the channel height and the pitch be 10 times the rib height [8].

113

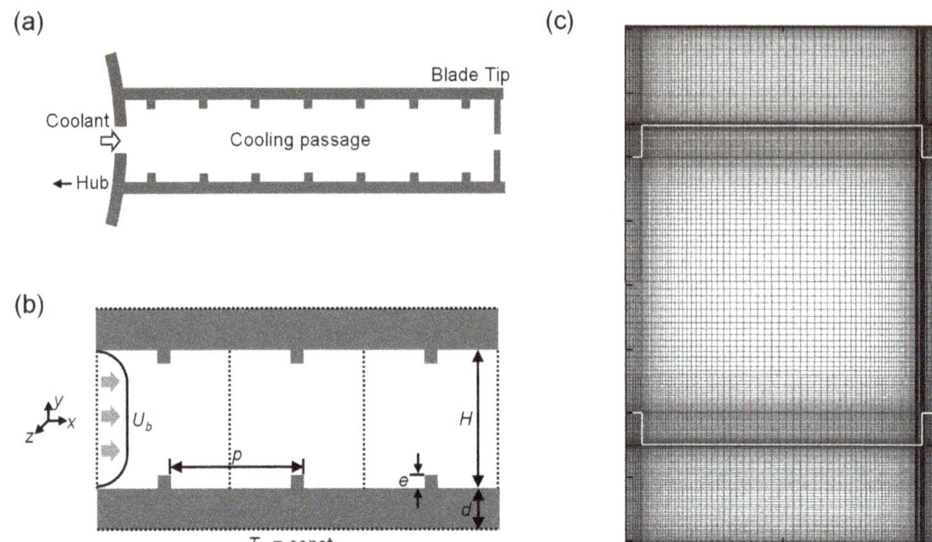

Figure 1. Computational domain and grid system: (**a**) schematic diagram of the internal cooling passage; (**b**) the computational domain; and (**c**) the grid system.

At a typical pitch of p/e = 10, since 90% of the channel wall is a bare surface, many studies have dealt with the heat transfer of the channel wall. Although rib heat transfer has received relatively little attention, on-rib heat transfer significantly contributes to the total heat transfer [6,9]. Most of the past studies have been performed under isothermal or iso-flux thermal boundary conditions [10]. In actual cooled turbine blades, solid conduction exists and is important for predicting the blade temperature [10,11]. In particular, heat transfer by the rib, where the conduction effect is concentrated, is not purely convective [12,13].

Iaccarino et al. [13] performed a Reynolds-averaged Navier–Stokes (RANS) simulation by adopting the k-ε $v2f$ model and considering the conduction of the rib. They also imposed an iso-flux condition on the channel wall, despite conduction between the rib and the channel wall existing in actual cooled blades. Subsequently, studies conducted experiments on conjugate heat transfer including channel walls [14,15]. When conduction was considered, the average heat transfer coefficient decreased by 26% in the rib and 10% in the channel wall compared with the case of pure convection [15].

Using k-ε $v2f$ improves [16], but in the ribbed channel problem, RANS underpredicts the heat transfer and does not accurately predict local peaks. Large eddy simulations (LESs) can solve these problems [17,18]. Furthermore, mechanisms for generating local peaks can be elucidated by observing instantaneous flow and thermal fields [18,19]. For studying the conjugate heat transfer of ribbed channels [14,15], Scholl et al. [20,21] performed a LES. The LES predicted that the heat transfer of the channel wall was in good agreement with the experiment, but the heat transfer coefficient at the front edge and back of the rib was higher than those observed in the experiment [20].

The difference observed between experimental and LES results at the edge of the rib indicates the possibility of an optical problem involving infrared camera images [22]. Additionally, the possibility of solid/fluid time disparity [23] caused by Scholl et al. [20,21] using the conduction–convection linkage as the heat transfer coefficient forward temperature back method was raised. Recently, Oh et al. [24] performed a fully coupled LES using the IBM (Immersed Boundary Method) for the same problem [13,14]. While the thermal response of a solid and a fluid has been shown to affect temporal changes in the temperature, it does not significantly affect the time-averaged heat transfer [24].

In the above studies [14,15,20,21,24], the blockage ratio was 0.3 and the thickness of the channel wall was the same as the rib height. In actual turbine blades, the typical blockage ratio is about 0.1, and the wall thickness is usually 2 to 7 times the rib height [25,26]. Ahn et al. [27] performed a fully coupled LES of a ribbed channel with a blockage ratio of 0.1 by applying the IBM. They also set the channel wall thickness to be thrice the ribs to consider an appropriate length scale when defining the Biot number (Bi). They found that Bi was below 0.1, and the amount of reduction in the conjugate heat transfer compared with pure convection was predicted to be 3%.

The reasons for the effect of conduction being small (3%) are the blockage ratio and the thermal conductivity of the blade material, which is 500 to 600 times that of air [27]. On the basis of dimensional analysis, Cukurel and Arts [15] showed that the heat transfer characteristics of a ribbed channel with a conducting wall can be expressed as

$$\mathrm{Nu} = f(\mathrm{Re}, \mathrm{K}^*, \mathrm{Bi}), \tag{1}$$

where the thermal conductivity ratio K^* $(=k_s/k_f)$ is important for conjugate heat transfer. The blade material and coolant (air) are predetermined and therefore not actively considered. Recently, as 3D printing has been reviewed as a production technique for turbine blades [28,29], it has become possible to have a turbine blade with different thermal conductivities. Recently, as a new gas turbine cycle [30,31] has been studied as a countermeasure against global warming, cooling fluids with thermal conductivities different from that of air, such as carbon dioxide and water vapor, are being studied [32,33].

The effect of the thermal conductivity ratio has been partially addressed in previous studies related to ribbed channels. Iaccarino et al. [13] compared cases with $\mathrm{K}^* = 0.1, 1$, and 100, but their results were RANS results and they did not consider wall conduction. Oh et al. [24] only verified their code for laminar flow over a flat plate in the range of $\mathrm{K}^* = 0.01$ to 1, and they did not report the effect of K^* in the ribbed channel. In the current study, we observed the effect of the thermal conductivity ratio on the turbulent heat transfer of the ribbed channel for $\mathrm{K}^* = 1, 10, 100$, and 566.26.

In this study, a channel wall with a thickness thrice the rib height was considered in the calculation area and a fully coupled LES was performed using the IBM. The thermal conductivity ratio was analyzed for 1, 10, and 100 cases with 566, which is a typical value for a gas turbine blade, and the results were compared. On the basis of the time average temperature field and heat transfer distribution, it was examined whether the mechanism responsible for promoting heat transfer under pure convection conditions is valid even when the conductivity ratio is different. Furthermore, conjugate heat transfer characteristics for different K^* values were observed through the instantaneous flow and thermal fields. Finally, the thermal performance of a ribbed channel according to the thermal conductivity ratio and the Biot number is discussed.

2. Numerical Methods

In the simulation, an in-house code was used. This code was based on the finite volume incompressible Navier–Stokes solver [31] at Stanford University, USA. All spatial derivatives were discretized by the central difference with second-order accuracy. The semi-implicit fractional step method based on the Crank–Nicolson and Runge–Kutta methods with third-order accuracy was used as the time integration method. The code was revised to deal with solids in the flow field with the IBM [32] and supplemented to solve the convective heat transfer between solids and fluids [33]. Later, it was revised twice, once to perform an LES of turbulent heat transfer [18,19], and again to handle conjugate heat transfer (CHT) [34].

The computational domain analyzed in this study is presented in Figure 1b. Ribs with a square cross-section and a height (e) that was 0.1 of the channel height (H) are arranged at intervals of 10 e on the upper and lower sides of the channel. By comparing the computational domain, including three periods in the x direction and one including single period, the same time average result was obtained, and finally, the computational area was set to include single period as shown in Figure 1c. The spanwise domain was set to be 2.5 π e, for which zero fall-off was observed with two-point correlation in a smooth channel simulation [18,19]. The thickness of the solid wall was set to be three times e, which corresponds to 30% of H.

Periodic boundary conditions were imposed in the main flow (x) and spanwise (z) directions. In the wall normal direction (y), no slip conditions or isothermal conditions were imposed on the upper and lower surfaces in the computational domain. The grid system comprised 128, 256, and 48 meshes in the x, y, and z directions (Figure 1c). A nonuniform grid was used in the x and y directions, while a uniform grid was used in the z direction. This was similar to the resolution obtained by [35] for a grid independent solution by performing a resolution test on the LES of a ribbed duct.

The grid-filtered incompressible Navier–Stokes equation and energy equation were adopted as the dimensionless governing equations, and they can be expressed as follows [34]:

$$\frac{\partial \overline{u}_i}{\partial x_i} - ms = 0, \tag{2}$$

$$\frac{\partial \overline{u}_i}{\partial t} + \frac{\partial \overline{u}_i \overline{u}_j}{\partial x_j} = -\frac{d\overline{p}}{dx_i} + \frac{1}{Re} \frac{\partial^2 \overline{u}_i}{\partial x_j \partial x_j} + \frac{\partial \tau_{ij}}{\partial x_j} + f_i, \tag{3}$$

$$\frac{\partial \overline{\theta}}{\partial t} + \omega \frac{\partial}{\partial x_j} \left(\overline{u}_j \overline{\theta} \right) = \frac{C^* K^*}{Re\, Pr} \frac{\partial^2 \overline{\theta}}{\partial x_j \partial x_j} + \frac{\partial q_j}{\partial x_j} + \xi. \tag{4}$$

Under the assumption of fully developed flow, the mean streamwise pressure and temperature gradient were decoupled as follows to impose periodic boundary conditions in the streamwise direction [36]:

$$P(\mathbf{x}, t) = -\beta x + p(\mathbf{x}, t), \tag{5}$$

$$T(\mathbf{x}, t) = \gamma x + \theta(\mathbf{x}, t), \tag{6}$$

where β and γ are the mean streamwise pressure and temperature gradients, respectively. These two parameters were determined to satisfy the conservation of global momentum and energy, respectively [36].

In the energy equation, the thermal properties of the solid were distinguished from those of the fluid by defining the heat capacity ratio C^* and the thermal conductivity ratio K^*, and by introducing the concept of effective thermal conductivity (k_e). The effective thermal conductivity was determined to satisfy the continuity of temperature and heat flux at the interface [34]. The parameter ω was a convection correction factor, and it was 0 in the cell containing the solid–fluid boundary and 1 in the remaining cells when conduction between the solid and the fluid was considered. In Equation (4), ξ was introduced to maintain second-order accuracy in the cell containing the interface [34]. The code was verified through the CHT problem involving a ribbed duct and a circular cylinder [34].

Turbulent flow was analyzed using an LES. In Equations (3) and (4), τ_{ij} and q_j are the sub-grid scale turbulent stress and turbulent heat flux, respectively, and τ_{ij} was determined as a dynamic sub-grid model by using scale similarity and setting a test filter around the grid [37,38]. The dynamic sub-grid model provided better results than the constant model for the ribbed channel problem [35]. Similar to τ_{ij}, q_j was determined dynamically; this approach yields better results in problems where the flow and heat transfer are dissimilar [39]. The simulation were performed for 10,000 time steps to reach a steady state. After that, additional 10,000 time steps ($t\, U_b/D_h = 5$) were carried out to obtain the statistics.

Numerical analysis was performed with thermal conductivity ratios of 1, 10, 100, and 566.26 to examine the thermal resistance effect of the solid wall. The value 566.26 is the thermal conductivity ratio between the gas turbine blade material and air [22]. The remaining flow and geometry conditions are summarized in Table 1. In this study, an LES was performed to obtain the instantaneous flow field and turbulence statistics along with the time average flow and temperature field. Furthermore, the turbulent heat transfer on the fluid side and changes in temperature fluctuations on the solid side according to the thermal conductivity ratio were observed.

Table 1. Parameters related to conjugate heat transfer.

	Present Study	**Liou et al. [6]**	**Cukurel et al. [14,15]**	**Scholl et al. [20,21]**
Method	LES	Hologram	IR Camera	LES
Re	30,000	10,200	40,000	40,000
Pr	0.71	0.71	0.71	0.71
K^*	566.36	1368.8	618.32	618.32
C^*	0.00031	0.00038	0.00031	0.00031
d/e	3	0.75	1	1

3. Results and Discussion

3.1. Code Validation Study

Figure 2a shows the time-averaged streamline in the ribbed channel. Similar to the streamlines [40] obtained with experimental data, recirculating cells exist behind the rib, and there is a secondary vortex between the recirculation flow and the back of the rib. There is a vortex on the front side of the rib and there is a separation bubble on the top side of the rib. While the LES-predicted flow field was almost identical to that observed in the experiment, the corner vortices occurring before and after the rib were predicted to be slightly larger compared with those in the experiment. This appears to be because of the difference between the blockage ratio and the channel aspect ratio. In the time-averaged temperature distribution under isothermal conditions (Figure 2b), a high-temperature region was formed behind the rib since heat transfer was not active.

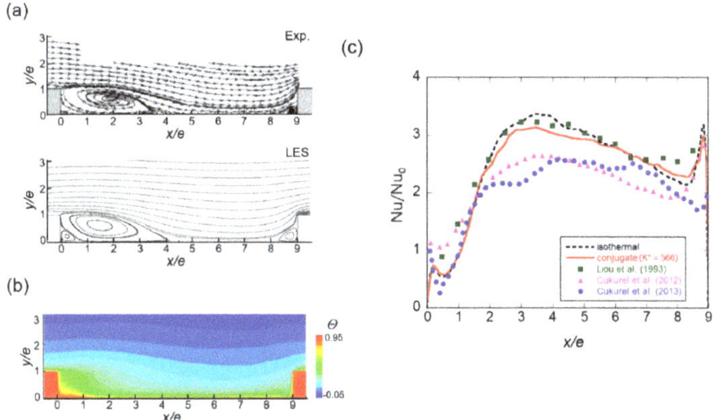

Figure 2. Time-averaged flow and thermal fields for the isothermal wall: (**a**) comparison of time-averaged streamlines with particle image velocimetry measurement data of Casarsa et al. [40], (**b**) the time-averaged temperature field, and (**c**) the Nusselt number ratio on the channel wall between ribs.

Figure 2c shows a comparison of the heat transfer coefficient distribution at the channel wall with those in the literature (see Table 1). The heat transfer coefficient was compared with the Nusselt number ratio, where Nu_0 is the Nusselt number of a smooth channel wall obtained using the following Dittus–Boelter correlation:

$$Nu_0 = 0.023 \, Re^{0.8} \, Pr^{0.4}. \tag{7}$$

When wall conduction was considered (red solid line in Figure 2c), the local heat transfer change slightly decreased compared with that obtained through locally isothermal analysis (black dotted line in Figure 2c). Among the experimental data, it was in good agreement with that of Liou et al. [6] (green square), who adopted the same blockage ratio (=0.1). The results of Cukurel et al. [14,15] compared together had a blockage ratio of 0.3, and the pink triangle and blue circle show iso-flux and conjugate results, respectively. Although not as quantitatively consistent as [6], the tendency of the convective heat transfer to be higher than the conjugate heat transfer in the range of 2 < *x/e* < 4 and the reversal of this tendency downstream are identical to those in the present study.

3.2. Time-Averaged Thermal Fields and Heat Transfer

The time-averaged temperature fields for different thermal conductivity ratios (K*) are compared in Figure 3. For a high conductivity ratio (K* $\geq$ 100), the temperature distributions are similar to those for the isothermal wall condition (Figure 3a,b). The fluid region accounts for most of the thermal resistance, and therefore, the temperature distribution inside the solid wall is uniform and the temperature in the fluid region near the solid–fluid interface varies.

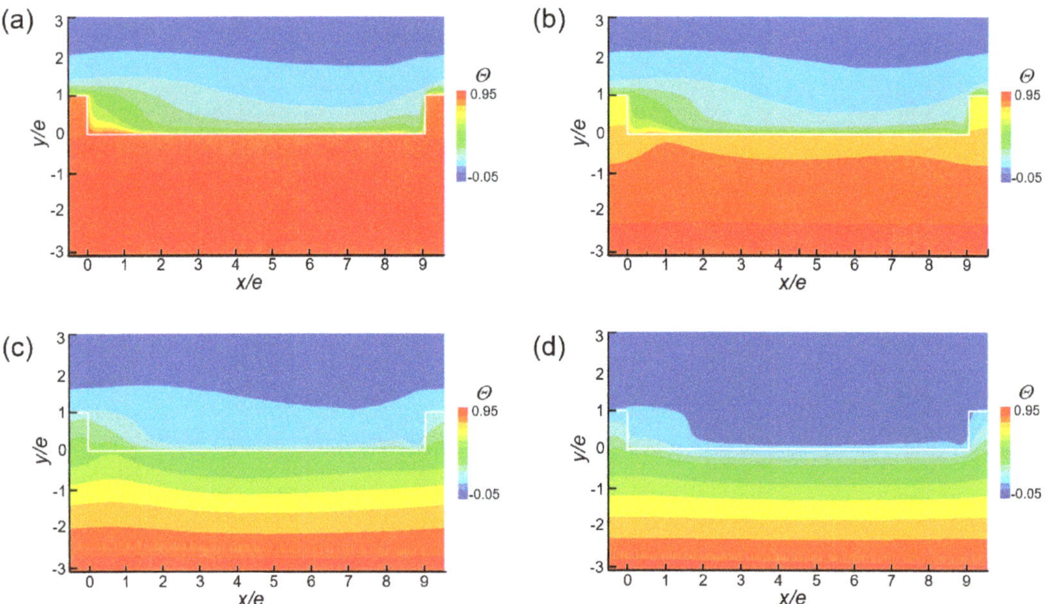

Figure 3. Time-averaged thermal fields: (**a**) K* = 566.26, (**b**) K* = 100.00, (**c**) K* = 10.00, and (**d**) K* = 1.00.

On the other hand, in the case of low conductivity ratios (K* $\leq$ 10), the contour lines are narrowly packed in the solid region and the temperature distribution in the fluid region is relatively uniform since the thermal resistance inside the solid wall is greater than that of the fluid (Figure 3c,d).

Figure 4 shows the heat flux inside the solid wall for each thermal conductivity ratio; the contours are isotherms. In general, the thermal resistance of a fluid varies greatly spatially because of the recirculation of the flow and the collision of the cold core fluid [18,19]. On the other hand, the thermal resistance inside the solid wall is constant. When the thermal conductivity ratio is large (K* $\geq$ 100), the thermal resistance in the fluid dominates. Therefore, the heat flux vector inside the solid wall is concentrated on the rib and directed toward both edges of the rib (Figure 4a,b).

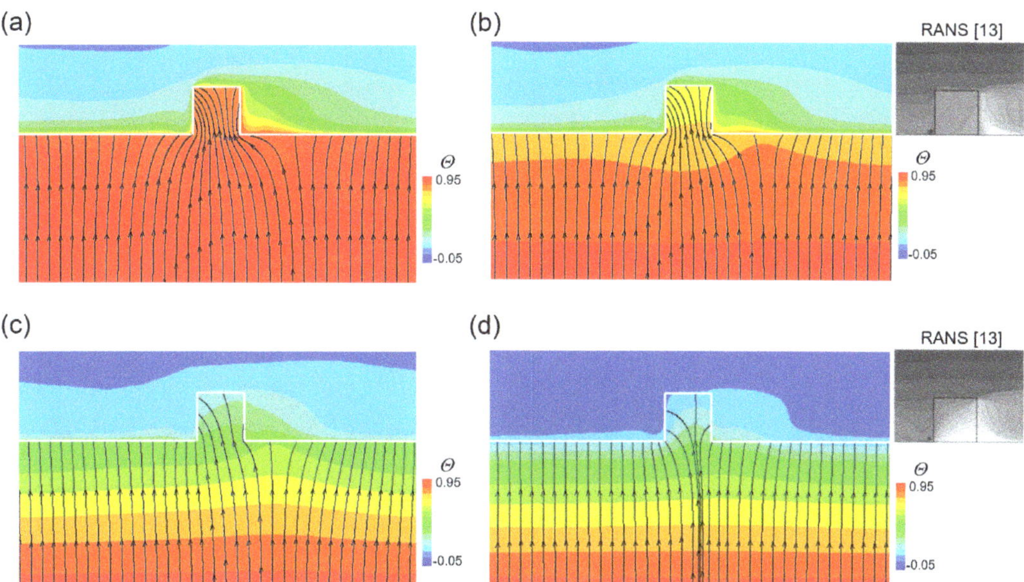

Figure 4. Heat flux vectors inside the solid wall: (**a**) K* = 566.26, (**b**) K* = 100.00, (**c**) K* = 10.00, and (**d**) K* = 1.00.

When the thermal conductivity ratio is small (K* ≤ 10), the thermal resistance inside the solid wall is dominant. Consequently, the heat flux vector flows toward the fluid in the wall normal direction (y), without being concentrated on the rib. In Figure 4c,d, the heat flux vector inside the rib is toward the upstream edge. However, the heat flux is smaller than that for a large thermal conductivity ratio. Furthermore, when the thermal conductivity ratio is 10 (Figure 4c), the fluid temperature is higher than that of the rib on the downstream side of the rib, and therefore, the heat flux cannot pass through the rib.

The cases of K* = 100 and 1 were compared with RANS data [13] by adopting the same thermal conductivity ratio. At K* = 100 (Figure 4b), the temperature inside the rib was almost uniform for both LES and RANS simulation. In the present LES, the isotherm inside the rib including the channel wall was close to the horizontal line, but in the RANS simulation in which only the conduction of the rib was considered, the isotherm inside the rib appeared in the diagonal direction. At K* = 1, both LES and RANS simulation increased the number of isotherms inside the rib. The number of isotherms of the LES including the channel wall was considerably smaller than that of the RANS simulation. In the RANS simulation, isotherms occurred in a diagonal direction, while in the LES, they occurred in the form of a parabola that was convex upwards. If K* = 100 or less, the conjugate effect can be accurately identified only when the channel wall is included in the computational domain.

Figure 5 shows the dependence of local temperature distributions at the solid–fluid interface on the thermal conductivity ratio. In the case of high conductivity ratios (K* ≥ 100), the temperature at the interface between the ribs (Figure 5a) is considerably close to the outer wall temperature and uniform in space since the solid wall tends to maintain a uniform temperature distribution. Similar behavior is observed on the rib surface (Figure 5b). However, the upstream edge of the rib has the lowest temperature because of the high heat transfer rate. This is clearly observed for K* = 100.

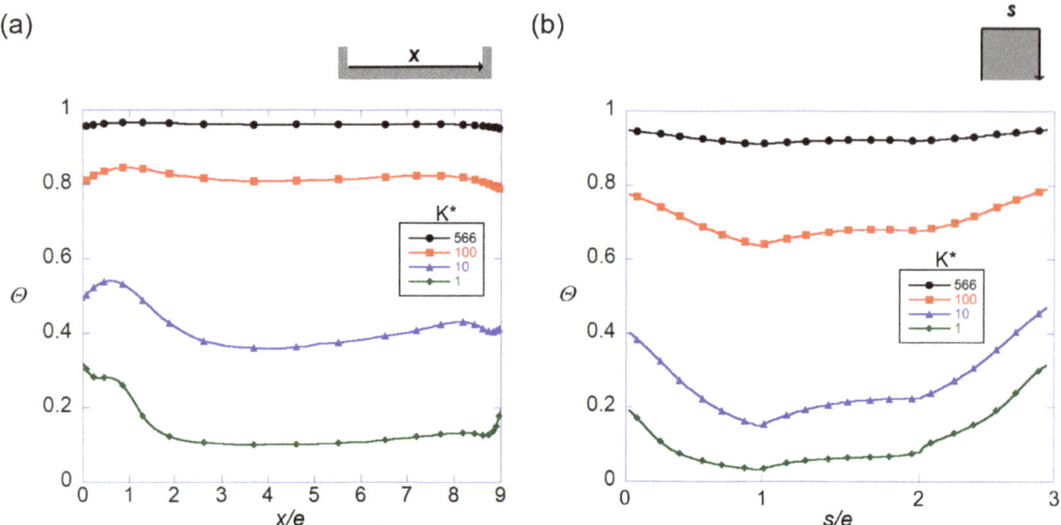

Figure 5. Local temperature distribution at the solid–fluid interface: (**a**) the temperature along the interface between the ribs and (**b**) the temperature on the rib.

In the case of low conductivity ratios ($K^* \leq 10$), the temperature was considerably lower than that in the case of high conductivity ratios, and a relatively large temperature variation was observed. The reason is that the spatially nonuniform thermal resistance of the flow field mainly influenced the temperature at the interface. In particular, the temperature near the upstream corner of the rib increased (Figure 5a). For $K^* \leq 10$, the fluid impinging on the upstream face of the rib was not very cold compared with the solid wall. However, the corner vortex prevented heat transfer from the solid to the fluid, resulting in the temperature near the upstream corner of the rib increasing.

Figure 6 shows the effect of the thermal resistance of the solid wall on the local heat transfer. Nu_0 is the Nusselt number without ribs, given by Equation (7). The Nusselt number presented in Figure 6 was defined on the basis of D_h. In Figure 6a, the local heat transfer coefficient evidently increases ($8 \leq x/e \leq 9$) near the upstream corner for all thermal conductivity ratios as the cold core fluid collides with the rib. However, as the thermal conductivity ratio increases, the local heat transfer coefficient decreases noticeably and becomes spatially uniform. Quantitatively, the local heat transfer coefficient depends strongly on the thermal conductivity ratio, but its qualitative distribution does not significantly depend on the local thermal conductivity ratio. Except in the vicinity of the rib, heat conduction in the flow direction is not significant since the heat flux vector inside the solid is directed in the $+y$ direction (Figure 4).

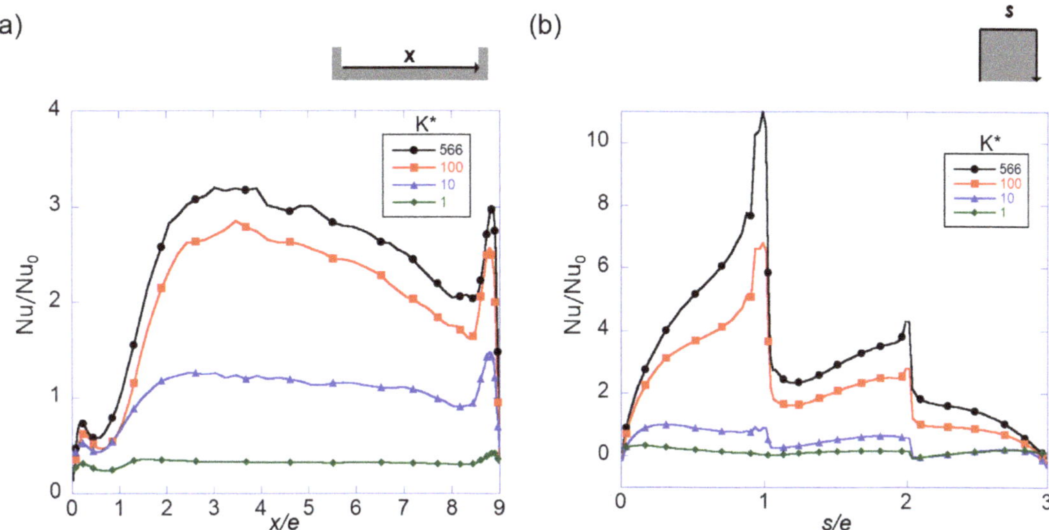

Figure 6. Effect of thermal resistance on local convective heat transfer along the solid–fluid interface: (**a**) heat transfer along the interface between the ribs and (**b**) heat transfer on the rib.

On the other hand, the local heat transfer distribution over the rib was strongly dependent on the thermal conductivity ratio (Figure 6b). Quantitatively, even at $K^* = 566.26$, the heat transfer rate at both edges of the rib was considerably lower than that in the isothermal case, and the local difference was quite small. Nevertheless, for large thermal conductivity ratios ($K^* \geq 100$), the distribution of local heat transfer was qualitatively similar to that of the isothermal wall. However, when the thermal conductivity ratio became small ($K^* \leq 10$), the heat transfer distribution changed qualitatively. For large thermal conductivity ratios, the maximum value of heat transfer occurred at the upstream edge of the rib, while for small thermal conductivity ratios, the maximum value of heat transfer occurred near the upstream edge of the rib ($s/e \approx 0.2$). Furthermore, in the vicinity of the downstream edge ($s/e \approx 2.1$), as the temperature of the solid decreased below that of the fluid, a region with negative heat transfer was formed.

3.3. Turbulent Heat Transfer

Figure 7 shows turbulent heat flux contours. It is well known that turbulent heat flux is an important factor in determining heat transfer [18,19]. The turbulent heat flux was large near the two edges of the rib and around $x/e = 4$ near the wall. For all heat flux ratios, the distribution of turbulent heat flux was similar, but when the thermal conductivity ratio decreased, the value of turbulent heat flux decreased and the distribution of turbulent heat flux became spatially uniform. As shown in Figure 3, when the heat conduction ratio decreased, the temperature distribution on the fluid side became uniform and the turbulent heat transfer decreased. In particular, the peak of the turbulent heat flux near the wall disappeared for $K^* < 10$.

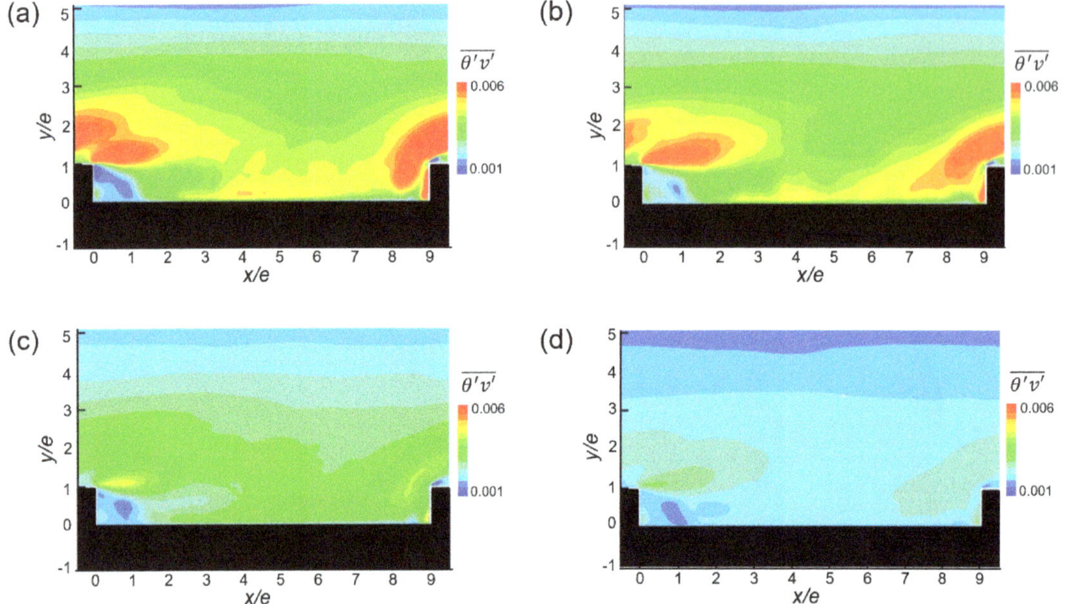

Figure 7. Turbulent heat flux $\overline{\theta'v'}$ contours for (**a**) K* = 566.26, (**b**) K* = 100.00, (**c**) K* = 10.00, and (**d**) K* = 1.00.

Locally, the turbulent heat flux near the rib varied with the thermal conductivity ratio. As evident in Figure 7a,b, when the thermal conductivity ratio was large, the turbulent heat flux at the front face of the rib was almost uniform. On the other hand, when the thermal conductivity ratio was small, the effect of the corner vortex relatively increased, resulting in the turbulent heat flux at the corner exceeding that at the edge. Consequently, the heat transfer distribution near the rib varied with the thermal conductivity ratio (Figure 6).

Figure 8 shows temperature fluctuations inside the solid wall. As explained in the introduction, the thermal condition of the solid wall is an important factor in determining convective heat transfer. In particular, in the turbulent flow section, temperature fluctuations in the solid are important along with thermal resistance. When the thermal conductivity ratio exceeds 1, most temperature fluctuations are caused by turbulent flows in the fluid region. At K* = 100, the temperature fluctuation at the upstream edge of the rib reaches 2%. When K* is 10 or higher, the temperature fluctuation inside the solid increases with the thermal resistance of the solid. When the thermal conductivity ratio is 1, the effect of the flow on convective heat transfer decreases and the temperature fluctuation in the fluid region weakens. Consequently, the temperature fluctuation in the solid region becomes similar to that in the fluid region. Furthermore, the depth at which temperature fluctuations occur inside the solid becomes smaller than that for K* = 10 or 100. When the thermal conductivity ratio is 1, the influence of the corner vortices becomes significant, and therefore, the maximum temperature fluctuation occurs at the corners of the rib.

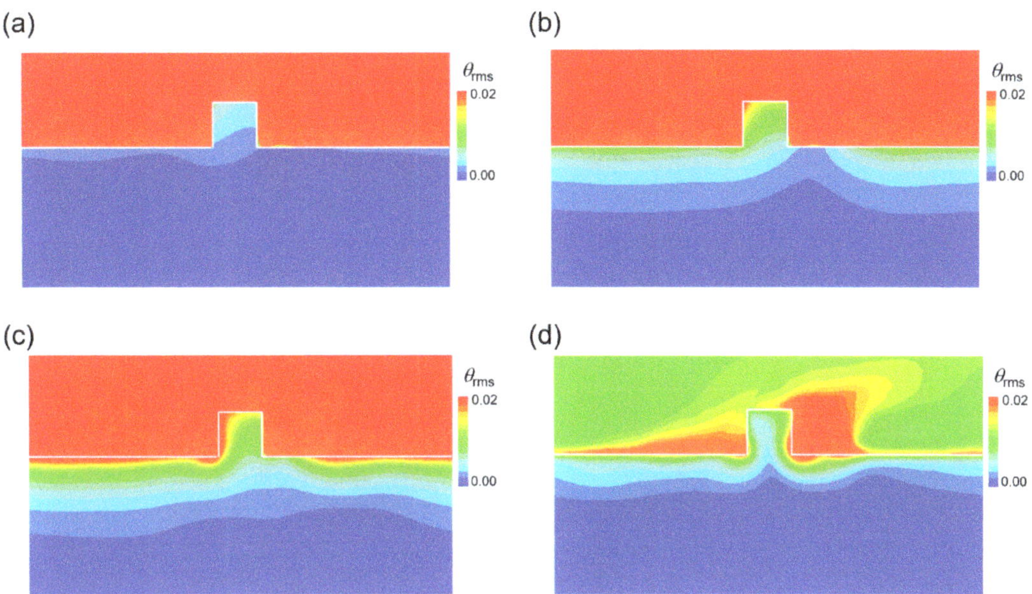

Figure 8. Temperature fluctuations (θ_{rms}) for (**a**) K* = 566.26, (**b**) K* = 100.00, (**c**) K* = 10.00, and (**d**) K* = 1.00.

3.4. Instantaneous Thermal Fields

Figure 9 shows instantaneous temperature fields near the rib. When the thermal conductivity ratio was large, the temperature inside the solid wall exceeded that of the fluid at all interfaces (Figure 5). In those cases, the difference between the bulk temperature of the fluid and the temperature on the solid surface was large (Figure 5), and therefore, the cold core fluid impinging became an important heat transfer mechanism.

However, as the thermal resistance of the solid increased (K* ≤ 10), the temperature difference between the fluid and the solid decreased. Consequently, the impingement of the fluid did not significantly influence the convective heat transfer in the near-upstream region of the rib. In this case, the corner vortex at the upstream corner of the rib induced a temperature change and reduced the local thermal resistance, resulting in the convective heat transfer increasing near both corners of the rib. In particular, the effect of the secondary vortex generated after the rib was very different when the thermal conductivity ratio was small. As shown in Figure 9c, the fluid heated by the channel wall was supplied to the downstream face of the rib by the recirculation flow and the secondary vortex. Consequently, the heat transfer rate at the downstream face of the rib was negative. In the temperature field of Figure 9d, compared with the temperature field of Figure 9c, it is evident that the isotherms are more concentrated near the surface as the relative thermal resistance of the solid increases. Consequently, the top surface of the rib is not heated and remains cold, and the cold flow over the rib takes heat from the rib and the hot fluid behind the rib.

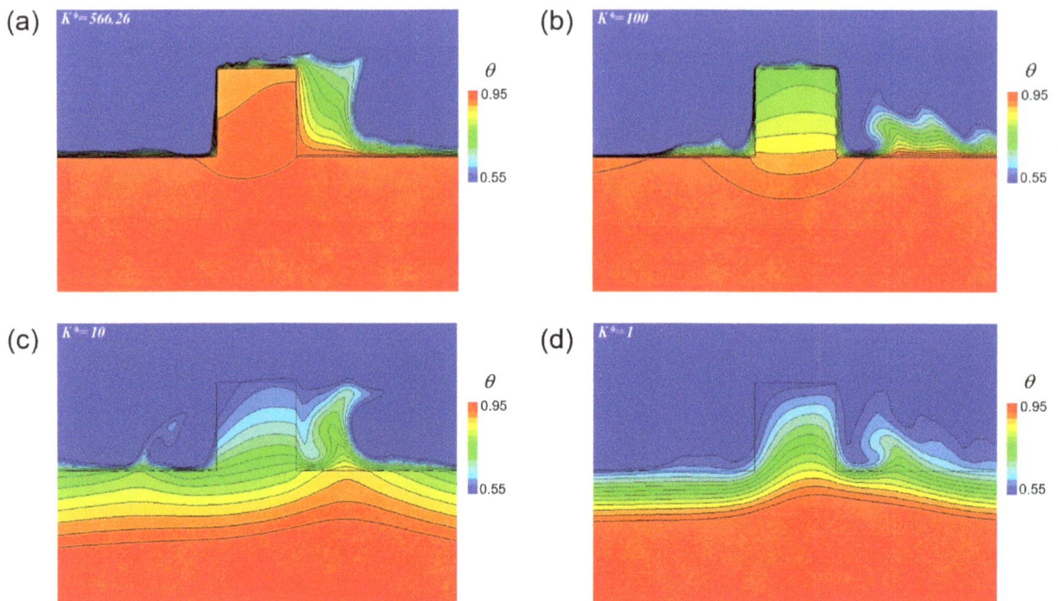

Figure 9. Instantaneous thermal fields near the rib for (**a**) K* = 566.26, (**b**) K* = 100.00, (**c**) K* = 10.00, and (**d**) K* = 1.00.

Figure 10 shows the local heat transfer distribution obtained for the instantaneous temperature field. Here, A, B, C and D represent the upstream corner, upstream edge, downstream edge, and downstream corner of the rib, respectively. At K* = 566.26, the cold fluid flowing along the wall collides with the rib, greatly promoting heat transfer to the upstream face of the rib. This can be confirmed by the repeated occurrence of high heat transfer after the corner at the location of the streak with a high heat transfer coefficient on the channel wall based on A, which corresponds to the upstream corner of the rib. At edge B, as the flow separates and recombines, spots with high heat transfer coefficient occur on the downstream side of the top of the rib. The downstream face of the rib is partially supplied with recombined cold fluid and has a high heat transfer coefficient, but the overall heat transfer is greatly reduced compared with the upstream face. In the downstream corner D, a locally negative heat transfer coefficient (white area in the contour) is observed. For K* = 100, the above-mentioned trend is maintained, but the heat transfer coefficient is reduced overall.

As the thermal conductivity ratio decreases (K* ≤ 10), the heat transfer rate decreases and becomes uniform overall. As the heat transfer rate decreases, the area of negative heat transfer (the white area in Figure 10) also widens. In the case of K* = 10 (Figure 10c), a negative heat transfer coefficient appears on the downstream side and along the downstream corner (D). For K* = 1, it appears along the downstream corner (C) but not at the downstream corner. This appears to be because the cold fluid flowing over the top of the rib cools the rib, as discussed in the context of Figure 9.

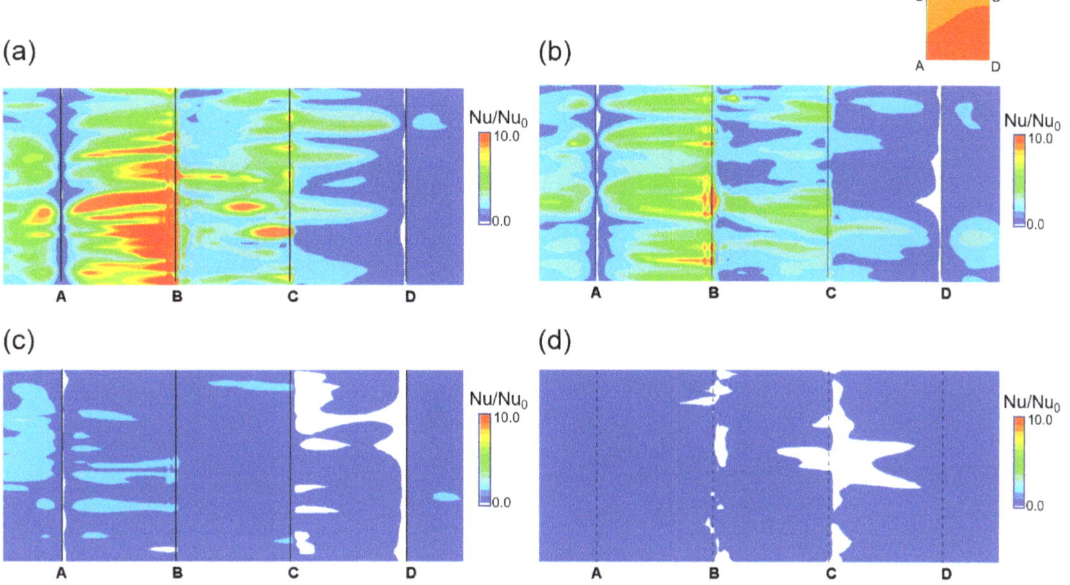

Figure 10. Instantaneous local heat transfer: (**a**) K* = 566.26; (**b**) K* = 100.0; (**c**) K* = 10.0; and (**d**) K* = 1.0.

3.5. Thermal Performance and Biot Number

Since ribs can be considered to be extended surfaces or fins, the performance of fins was analyzed. Fin performance is evaluated by examining the fin effectiveness and fin efficiency [41]. The fin effectiveness is defined as follows:

$$\varepsilon_f = \frac{q_{rib,conj}}{h_{conv}A_{c,b}(T_w - T_b)}.$$ (8)

Fins are judged to be effective when their effectiveness is 2 or more [41], and this condition is satisfied when K* is 100 or higher (see Figure 11a).

The fin efficiency is defined as follows [41]:

$$\eta_f = \frac{q_f}{q_{max}} = \frac{q_{rib,conj}}{h_{conv}A_{rib}(T_w - T_b)}.$$ (9)

Since numerator is identical to $q_{rib,conj}$, it shows the same trend as the fin effectiveness. In actual gas turbine materials (K* = 566), the fin efficiency is close to 100%, but at K* = 100 it decreases to 78%. At K* = 10, the fin efficiency is less than 20%, and the fin does not perform its intended function properly.

The total heat transfer rate (q) is shown in Figure 11b; q_0 is the heat transfer rate in the smooth channel for pure convection, and it is obtained from the Dittus–Boelter equation (Equation (7)). Figure 11b shows that the overall heat transfer rate decreases significantly as K* decreases. For K* = 566.26, there is no significant difference from the isothermal conditions, but for K* = 100, the overall heat transfer rate decreases by about 17%. At K* = 10, it is less than 1/3, and at K* = 1 it is considerably smaller than that in the isothermal smooth channel.

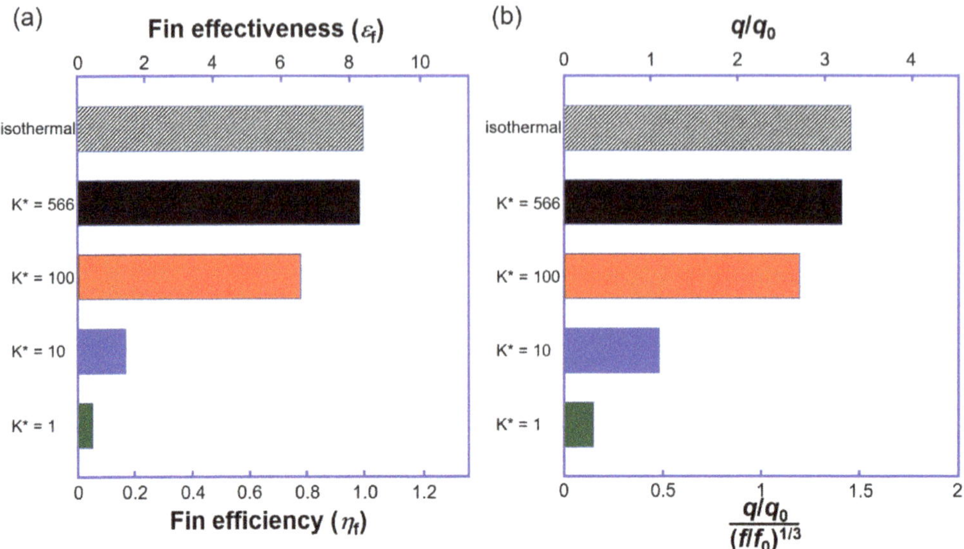

Figure 11. Thermal performance: (**a**) fin performance of the rib and (**b**) the total heat transfer rate.

Thermal performance considering both heat transfer and pressure drop is defined by Equation (10), where f stands for frication factor [10].

$$\text{Thermal performance} = \frac{q/q_0}{\left(f/f_0\right)^{1/3}}. \tag{10}$$

Thermal performance is proportional to the total heat transfer rate as the flow field (i.e., friction factor) does not change even when the conductivity changes. For the heat transfer enhancement to be greater than the pressure drop penalty, the value of thermal performance should exceed 1, but this criterion is satisfied only when K* = 100 or higher. If K* = 10 or lower, the rib is not effective in promoting heat transfer.

Figure 12 shows the variation of the local Biot number with the conductivity ratio. The Biot number on the coolant side, which is within the scope of this study, is defined by Equation (11), using the thickness of the solid wall (d) as the characteristic length [42]:

$$\text{Bi} = \frac{h_{conj}d}{k_s} = \frac{\text{Nu}}{K*}\frac{d}{D_h}. \tag{11}$$

In a cooled gas turbine blade, Bi is typically around 0.3 [43]. If the Biot number is less than 0.1, the fluid domain accounts for most of the thermal resistance, similar to the isothermal case [43]. Ahn et al. [27] reported that under typical gas turbine blade conditions, the Biot number is less than 0.1 (black circles in Figure 12) and reflects heat transfer characteristics close to pure convection.

At K* = 100 (red squares), Bi exceeds 0.1 at 3 < x/e < 6 on the channel wall, making the conduction thermal resistance non-negligible (Figure 12a), and at the rib surface, Bi goes up to 0.6 at the upstream edge (Figure 12b). At K* = 10 (blue triangles), Bi is close to 1, and the thermal resistance of the solid is similar to that of the fluid. About half of the temperature drop is expected in the solid region. The data for K* = 10 in Figure 5 shows that θ is around 0.4, confirming that the Biot number is an appropriate indicator. At K* = 1 (green diamonds), Bi exceeds 1 in most regions, resulting in a higher temperature drop in solids than in fluids, consistent with the results presented in Figures 3–5.

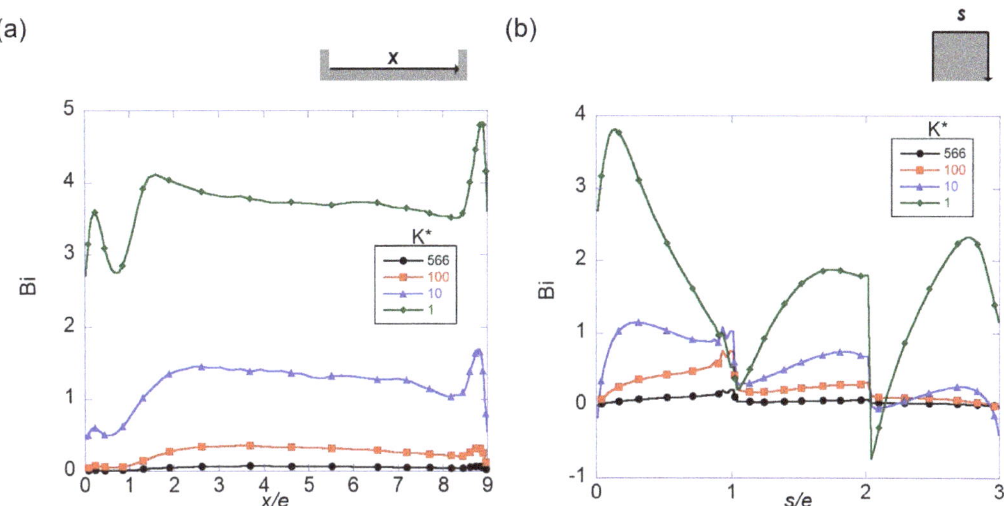

Figure 12. Local Biot number variations: (**a**) on the channel wall and (**b**) on the rib.

4. Conclusions

In this study, a CHT analysis including heat conduction in a ribbed channel was performed, and the effect of the thermal resistance of a solid wall was discussed on the basis of simulation data. The main results can be summarized as follows.

1. When the thermal conductivity ratio was large ($K^* \geq 100$), the heat transfer characteristics were similar to those in isothermal conditions. In this case, the impingement of the cold core fluid into the rib and recirculation of the flow mainly affected the convective heat transfer. The heat flux from the solid wall was concentrated on the rib and directed toward both edges of the rib.

2. When the thermal resistance of the solid increased ($K^* \leq 10$), the effect of cold core fluid impingement on the rib decreased, and the two vortices located at the corners played an important role in heat transfer. In this case, the temperature distribution of the solid wall that was affected by the two corner vortices determined the convective heat transfer. In particular, on the downstream face of the rib, a region with negative heat transfer appeared.

3. For $K^* \leq 10$, the turbulent heat flux on the front face of the rib was concentrated at a corner, and the turbulent heat flux whose peak occurred near the channel wall disappeared.

4. At $K^* = 100$, the temperature fluctuation at the upstream edge of the rib reached 2%, and at $K^* = 1$, the temperature fluctuation in the solid region was at a level similar to that in the fluid region.

5. Below $K^* = 100$, heat transfer enhancement was significantly reduced by conduction. Up to $K^* = 100$, the rib promoted heat transfer, but below $K^* = 10$, it did not promote heat transfer.

6. Compared with the thermal resistance of the solid and fluid for the CHT of the ribbed channel, the Biot number that was defined on the basis of the thickness of the channel wall appropriately represented the heat transfer characteristics. In other words, for $K^* = 100$ or higher, the Biot number at the channel wall was considerably smaller than 0.1, but at $K^* = 1$, it was considerably larger than 1, which was consistent with the thermal performance of the rib.

Author Contributions: Conceptualization, J.A. and J.S.L.; methodology, J.A. and J.C.S.; software, J.C.S.; validation, J.A. and J.S.L.; formal analysis, J.A.; investigation, J.A. and J.C.S.; resources, J.S.L.; data curation, J.A. and J.C.S.; writing—original draft preparation, J.A. and J.C.S.; writing—review and editing, J.A.; supervision, J.S.L. All authors have read and agreed to the published version of the manuscript.

Funding: This work was supported by the Korea Institute of Energy Technology Evaluation and Planning (KETEP) and the Ministry of Trade, Industry & Energy (MOTIE) of the Republic of Korea (No. 1415167167).

Institutional Review Board Statement: Not applicable.

Informed Consent Statement: Not applicable.

Data Availability Statement: Data is contained within this article.

Conflicts of Interest: The authors declare no conflict of interest.

Nomenclature

$A_{c,b}$	cross-sectional area at the base [m^2]
A_{rib}	rib surface area [m^2]
Bi	Biot number $(=hd/k_s)$
C^*	heat capacity ratio $(=(\rho\,c_p)_f/(\rho\,c_p)_s)$
d	thickness of the channel wall [m]
D_h	hydraulic diameter of the channel [m]
e	rib height [m]
f	friction factor
f_i	momentum forcing
h	heat transfer coefficient [W/m^{2}K]
H	channel height [m]
k_f	thermal conductivity of the fluid [W/mK]
k_s	thermal conductivity of the solid [W/mK]
K^*	thermal conductivity ratio $(=k_s/k_f)$
ms	mass source/sink
Nu	Nusselt number $(=hD_h/k_f)$
p	rib-to-rib pitch [m]
Pr	Prandtl number $(=v/\alpha)$
Q''	heat flux {W/m^2}
q	heat transfer rate [W]
q_f	heat transfer rate through a fin [W]
Re	bulk Reynolds number $(=U_b D_h/v)$
t	time [sec]
T	temperature [K]
T_b	bulk temperature [K]
T_w	wall temperature [K]
U_b	bulk velocity [m/s]
V'	wall-normal velocity fluctuation [m/s]
W	channel width [m]
Greek symbols	
α	thermal diffusivity [m^2/s]
β	mean pressure gradient [Pa/m]
γ	mean temperature gradient [K/m]
ε_ϕ	fin effectiveness
η_ϕ	fin efficiency
v	kinematic viscosity [m^2/s]
θ	dimensionless temperature $(=(T-T_b)/(T_w-T_b))$
Θ	time-averaged dimensionless temperature
ω	index function between the solid and the fluid

Subscripts
rms root-mean-square value
0 fully developed value in a smooth pipe
Abbreviations
IBM immersed boundary method
LES large eddy simulation
RANS Reynolds averaged Navier–Stokes simulation

References

1. Srinivasan, V.; Simon, T.W.; Goldstein, R.J. Synopsis. In *Heat Transfer in Gas Turbine Systems*; The New York Academy of Science: New York, NY, USA, 2001; pp. 1–10.
2. Han, J.C. Fundamental Gas Turbine Heat Transfer. *ASME J. Therm. Sci. Eng. Appl.* **2013**, *5*, 021007. [CrossRef]
3. Han, J.C.; Park, J.S. Developing Heat Transfer in Rectangular Channels with Rib Turbulators. *Int. J. Heat Mass Transf.* **1988**, *31*, 183–195. [CrossRef]
4. Park, J.S.; Han, J.C.; Huang, S.; Ou, S.; Boyle, R.J. Heat Transfer Performance Comparisons of Five Different Rectangular Channels with Parallel Angled Ribs. *Int. J. Heat Mass Transf.* **1992**, *35*, 2891–2903. [CrossRef]
5. Ruck, S.; Arbeiter, F. Detached Eddy Simulation of Turbulent Flow and Heat Transfer in Cooling Channels Roughened by Variously Shaped Ribs on One Wall. *Int. J. Heat Mass Transf.* **2018**, *118*, 388–401. [CrossRef]
6. Liou, T.M.; Hwang, J.J. Effect of Ridge Shapes on Turbulent Heat Transfer and Friction in a Rectangular Channel. *Int. J. Heat Mass Transf.* **1993**, *36*, 931–940. [CrossRef]
7. Han, J.C.; Huh, M. Recent Studies in Turbine Blade Internal Cooling. *Heat Transf. Res.* **2010**, *41*, 803–828.
8. Kim, H.M.; Kim, K.Y. Design Optimization of Rib-Roughened Channel to Enhance Turbulent Heat Transfer. *Int. J. Heat Mass Transf.* **2004**, *47*, 5159–5168. [CrossRef]
9. Taslim, M.E.; Wadsworth, C.M. An Experimental Investigation of the Rib Surface-Averaged Heat Transfer Coefficient in a Rib-Roughened Square Passage. *ASME J. Turbomach.* **1997**, *119*, 381–389. [CrossRef]
10. Coletti, F.; Scialanga, M.; Arts, T. Experimental Investigation of Conjugate Heat Transfer in a Rib-Roughened Trailing Edge Channel with Corssing Jets. *ASME J. Turbomach.* **2012**, *134*, 041016. [CrossRef]
11. Ju, Y.; Feng, Y.; Zhang, C. Conjugate Heat Transfer Simulation and Entropy Generation Analysis of Gas Turbine Blades. *ASME J. Eng. Gas Turbine Power* **2021**, *143*, 081012. [CrossRef]
12. Yusefi, A.; Nejat, A.; Sabour, H. Ribbed Channel Heat Transfer Enhancement of an Internally Cooled Turbine Vane Using Cooling Conjugate Heat Transfer Simulation. *Therm. Sci. Eng. Progress* **2020**, *19*, 100641. [CrossRef]
13. Iaccarino, G.; Ooi, A.; Durbin, P.A.; Behnia, M. Conjugate Heat Transfer Predictions in Two-dimensional Ribbed Passages. *Int. J. Heat Fluid Flow* **2002**, *23*, 340–345. [CrossRef]
14. Cukurel, B.; Arts, T.; Selcan, C. Conjugate Heat Transfer Characterization in Cooling Channels. *J. Therm. Sci.* **2012**, *21*, 286–294. [CrossRef]
15. Cukurel, B.; Arts, T. Local Heat Transfer Dependency on Thermal Boundary Condition in Ribbed Cooling Channel Geometries. *ASME J. Heat Transf.* **2013**, *135*, 101001. [CrossRef]
16. Ooi, A.; Iaccarino, G.; Durbin, P.A.; Behnia, M. Reynolds Averaged Simulation of Flow and Heat Transfer in Ribbed Ducts. *Int. J. Heat Fluid Flow* **2002**, *23*, 750–757. [CrossRef]
17. Duchaine, F.; Gicquel, L.; Grosnickel, T.; Koupper, C. Large-Eddy Simulation of the Flow Developing in Static and Rotating Ribbed Channels. *ASME J. Turbomach.* **2020**, *142*, 041003. [CrossRef]
18. Ahn, J.; Choi, H.; Lee, J.S. Large Eddy Simulation of Flow and Heat Transfer in a Channel Roughened by Square or Semicircle Ribs. *ASME J. Turbomach.* **2005**, *127*, 263–269. [CrossRef]
19. Ahn, J.; Choi, H.; Lee, J.S. Large Eddy Simulation of Flow and Heat Transfer in a Rotating Ribbed Channel. *Int. J. Heat Mass Transf.* **2007**, *50*, 4937–4947. [CrossRef]
20. Scholl, S.; Verstraete, T.; Duchaine, F.; Gicquel, L. Influence of the Thermal Boundary Conditions on the Heat Transfer of a Rib-Roughened Cooling Channel Using LES. *J. Power Energy* **2015**, *229*, 498–507. [CrossRef]
21. Scholl, S.; Verstraete, T.; Duchaine, F.; Gicquel, L. Conjugate Heat Transfer of a Rib-Roughened Internal Turbine Blade Cooling Channel Using Large Eddy Simulation. *Int. J. Heat Fluid Flow* **2016**, *61*, 650–664. [CrossRef]
22. Fedrizzi, R.; Arts, T. Determination of the Conjugate Heat Transfer Performance of a Turbine Blade Cooling Channel. *Quant. InfraRed Thermogr. J.* **2004**, *1*, 71–88. [CrossRef]
23. Patil, S.; Tafti, D.K. Large-Eddy Simulation with Zonal Near Wall Treatment of Flow and Heat Transfer in a Ribbed Duct for the Internal Cooling of Turbine Blades. *ASME J. Turbomach.* **2013**, *135*, 031006. [CrossRef]
24. Oh, T.K.; Tafti, D.K.; Nagendra, K. Fully Coupled Large Eddy Simulation-Conjugate Heat Transfer Analysis of a Ribbed Cooling Passage Using the Immersed Boundary Method. *ASME J. Turbomach.* **2021**, *143*, 041012. [CrossRef]
25. Halila, E.E.; Lenahan, D.T.; Thomas, T.T. *High Pressure Turbine Test Hardware Detailed Design Report*; NASA CR-167955; NASA: Washington, DC, USA, 1982; pp. 18–68.
26. Romeyn, A. *High Pressure Turbine Blade Fracture CFM56-3C1 Engine Test Cell, 7 July 2004*; ATSB Transport Safety Investigation Report; Australian Transport Safety Bureau: Canberra, Australia, 2006; pp. 7–8.

27. Ahn, J.; Song, J.C.; Lee, J.S. Fully Coupled Large Eddy Simulation of Conjugate Heat Transfer in a Ribbed Channel with a 0.1 Blockage Ratio. *Energies* **2021**, *14*, 2096. [CrossRef]
28. Neuberger, H.; Hernandez, F.; Ruck, S.; Arbeiter, F.; Bonk, S.; Rieth, M.; Stratil, L.; Muller, O.; Volker, K.-U. Advances in Additive Manufacturing of Fusion Materials. *Fusion Eng. Des.* **2021**, *167*, 112309. [CrossRef]
29. Bang, M.; Kim, S.; Park, H.S.; Kim, T.; Rhee, D.-H.; Cho, H.H. Impingement/Effusion Cooling with a Hollow Cylinder Structure for Additive Manufacturing: Effect of Channel Gap Height. *Int. J. Heat Mass Transf.* **2021**, *175*, 121420. [CrossRef]
30. Ditaranto, M.; Heggset, T.; Berstad, D. Concept of Hydrogen Fired Gas Turbine Cycle with Exhaust Gas Recirculation: Assessment of Process Performance. *Energy* **2020**, *192*, 116646. [CrossRef]
31. Choi, H.; Moin, P. Effects of the Computational Time Step on Numerical Solutions on Turbulent Flow. *J. Comp. Phys.* **1994**, *113*, 1–4. [CrossRef]
32. Kim, J.; Kim, D.; Choi, H. An Immersed Boundary Finite Volume Method for Simulations of Flow in Complex Geometries. *J. Comp. Phys.* **2001**, *171*, 132–150. [CrossRef]
33. Kim, J.; Choi, H. An Immersed Boundary Finite Volume Method for Simulation of Heat Transfer in Complex Geometries. *KSME Int. J.* **2004**, *18*, 1026–1035. [CrossRef]
34. Song, J.C.; Ahn, J.; Lee, J.S. An Immersed-Boundary Method for Conjugate Heat Transfer Analysis. *J. Mech. Sci. Technol.* **2017**, *31*, 2287–2294. [CrossRef]
35. Tafti, D.K. Evaluating the Role of Subgrid Stress Modeling in a Ribbed Duct for the Internal Cooling of Turbine Blades. *Int. J. Heat Fluid Flow* **2005**, *26*, 92–104. [CrossRef]
36. Patankar, S.V.; Liu, C.H.; Sparrow, E.M. Fully Developed Flow and Heat Transfer in Ducts Having Streamwise-Periodic Variations of Cross-Sectional Area. *ASME J. Heat Transf.* **1977**, *99*, 180–186. [CrossRef]
37. Germano, M.; Piomelli, P.; Moin, P.; Cabot, W.H. A Dynamic Sub-grid Scale Eddy Viscosity Model. *Phys. Fluids* **1991**, *A3*, 1760–1765. [CrossRef]
38. Lilly, D.K. A Proposed Modification of the Germano Sub-grid Scale Closure Model. *Phys. Fluids* **1992**, *A4*, 633–635. [CrossRef]
39. Kong, H.; Choi, H.; Lee, J.S. Dissimilarity Between the Velocity and Temperature Fields in a Perturbed Turbulent Thermal Boundary Layer. *Phys. Fluids* **2001**, *13*, 1466–1479. [CrossRef]
40. Casarsa, L.; Arts, T. Experimental Investigation of the Aerothermal Performance of a High Blockage Rib-Roughened Cooling Channel. *ASME J. Turbomach.* **2005**, *127*, 580–588. [CrossRef]
41. Incropera, F.P.; Dewitt, D.P.; Bergman, T.L.; Lavine, A.S. *Principles of Heat and Mass Transfer*, 1st ed.; Wiley: Hoboken, NJ, USA, 2017; pp. 153–158.
42. Ramachandran, S.G.; Shih, T.I.-P. Biot Number Analogy for Design of Experiments in Turbine Cooling. *ASME J. Turbomach.* **2015**, *137*, 061002. [CrossRef]
43. Jung, E.Y.; Chung, H.; Choi, S.M.; Woo, T.K.; Cho, H.H. Conjugate Heat Transfer on Full-coverage Film Cooling with Array Jet Impingement with Various Biot numbers. *Exp. Therm. Fluid Sci.* **2017**, *83*, 1–8. [CrossRef]

Article

Modeling Transient Pipe Flow in Plastic Pipes with Modified Discrete Bubble Cavitation Model

Kamil Urbanowicz [1,*], Anton Bergant [2,3], Apoloniusz Kodura [4], Michał Kubrak [4], Agnieszka Malesińska [4], Paweł Bury [5] and Michał Stosiak [5]

[1] Faculty of Mechanical Engineering and Mechatronics, West Pomeranian University of Technology in Szczecin, 70-310 Szczecin, Poland

[2] Litostroj Power d.o.o., 1000 Ljubljana, Slovenia; Anton.Bergant@litostrojpower.eu

[3] Faculty of Mechanical Engineering, University of Ljubljana, 1000 Ljubljana, Slovenia

[4] Faculty of Building Services, Hydro and Environmental Engineering, Warsaw University of Technology, 00-661 Warsaw, Poland; apoloniusz.kodura@pw.edu.pl (A.K.); michal.kubrak@pw.edu.pl (M.K.); agnieszka.malesinska@pw.edu.pl (A.M.)

[5] Faculty of Mechanical Engineering, Wrocław University of Science and Technology, 50-370 Wrocław, Poland; pawel.bury@pwr.edu.pl (P.B.); michal.stosiak@pwr.edu.pl (M.S.)

* Correspondence: kamil.urbanowicz@zut.edu.pl

Citation: Urbanowicz, K.; Bergant, A.; Kodura, A.; Kubrak, M.; Malesińska, A.; Bury, P.; Stosiak, M. Modeling Transient Pipe Flow in Plastic Pipes with Modified Discrete Bubble Cavitation Model. *Energies* **2021**, *14*, 6756. https://doi.org/10.3390/en14206756

Academic Editor: Dmitry Eskin

Received: 11 September 2021
Accepted: 15 October 2021
Published: 17 October 2021

Abstract: Most of today's water supply systems are based on plastic pipes. They are characterized by the retarded strain (RS) that takes place in the walls of these pipes. The occurrence of RS increases energy losses and leads to a different form of the basic equations describing the transient pipe flow. In this paper, the RS is calculated with the use of convolution integral of the local derivative of pressure and creep function that describes the viscoelastic behavior of the pipe-wall material. The main equations of a discrete bubble cavity model (DBCM) are based on a momentum equation of two-phase vaporous cavitating flow and continuity equations written initially separately for the gas and liquid phase. In transient flows, another important source of pressure damping is skin friction. Accordingly, the wall shear stress model also required necessary modifications. The final partial derivative set of equations was solved with the use of the method of characteristics (MOC), which transforms the original set of partial differential equations (PDE) into a set of ordinary differential equations (ODE). The developed numerical solutions along with the appropriate boundary conditions formed a basis to write a computer program that was used in comparison analysis. The comparisons between computed and measured results showed that the novel modified DBCM predicts pressure and velocity waveforms including cavitation and retarded strain effects with an acceptable accuracy. It was noticed that the influence of unsteady friction on damping of pressure waves was much smaller than the influence of retarded strain.

Keywords: retarded strain; cavitation; water hammer; unsteady friction; method of characteristics

1. Introduction

1.1. Gaseous and Vaporous Cavitation

Cavitation is one of the natural phenomena whose thorough understanding should be a scientific priority. Among others, it takes place when gas is released from the liquid. It occurs in hydraulic systems (water supply, hydropower, heating, cooling, etc.) in which the flow (forced by the pressure gradient) takes place through pressurized pipes. There are two types of cavitation: gaseous and vaporous [1–3].

More dangerous is vaporous cavitation, which occurs when the pressure drops to the saturated vapor pressure. This type of cavitation is rapidly changing, as it only takes place during the duration of the reduced pressure. In the literature, there is a group of mathematical models based on this type of cavitation, the so-called discrete vapor cavity models (DVCM) [1–3]. In the event of a water hammer, the reduction of pressure to the

131

saturated vapor pressure takes a relatively short time, which is followed by an implosion of the resulting vapor regions. The implosion is accompanied by large local increases in the velocity of the liquid, because the cavitation space must rapidly be filled with liquid at the time of pressure increase (above the vapor pressure). The impact of the liquid against the walls of the pipe (as well as the walls of other elements of the systems: valves, turbines, pumps, flow meters, etc.) results in cavitation erosion in the long term. Irreversible losses appear in the material of the walls of the pipes and other elements of the system. Sections in which such erosion takes place are systematically weakened in terms of strength, and it is in these places that leaks or, in extreme cases, complete damage of the structure can occur. Cavitation also leads to a reduction of the efficiency of hydraulic systems, contributing to the deterioration of the operation of energy-saving systems in hydraulic drives [4].

The second type of cavitation, namely gaseous cavitation, is a slowly changing phenomenon occurring in systems with unsteady flows (dynamic, rapid changes of velocity and pressure) or large pressure drops along the length of the system. Each liquid dissolves a certain amount of air (possibly a different gas). In water systems (water supply networks), the average amount of dissolved air is about 2%. In oils, on the other hand, the amount of dissolved air can reach up to about 10%. Hence, the influence of this type of cavitation is much more noticeable in oil-hydraulic systems than in water supply systems. Interestingly, during water hammer, such cavitation areas, due to the large time necessary for desorption and absorption, are beneficial. Their presence causes a faster damping of dynamic waveforms, as the "air bags" emitted by their action resemble local air–liquid shock absorbers. The influence of this type of cavitation is still poorly understood both experimentally and theoretically. There is a group of models called discrete gas cavity models (DGCM) [1,2], which take into account the influence of free gas in a simplified way.

The type of pipe material also significantly affects the intensity and timing of transient phenomena [5]. The flows in metal pipes with vapor–gas cavitation areas are well recognized and described. However, if we look at plastic pipes, which are now starting to displace metal pipes (especially in water supply systems), the researchers have mostly used the two basic cavitation models, i.e., the DVCM and the DGCM. Apart from these two models, alternative models have been developed, including a revised version of the DVCM model proposed by Adamkowski [6,7] as well as a model based on two-phase flow equations that can be called a discrete bubble cavity model (DBCM), which was developed by Shu [8]. Shu's model does not generate the unrealistic pressure spikes due to flow discontinuity at each computational section [8] that have been found in DVCM simulations. In DGCM, it is difficult to assign the physical amount of free air at computational sections along the pipeline. The model that is based on two-phase pipe flow equations is in principle more realistic than the model that is based on single-phase pipe flow equations with cavities lumped at computational sections. However, the aforementioned discrete Adamkowski cavity (DACM) and DBCM models have not been previously used to model transient cavitating flow in plastic pipes. The main objective of this paper is to present a novel DBCM that will enable the simulation of transient cavitating flows in plastic pipes.

1.2. Transient Cavitating Flow in Plastic Pipes

Among the phenomena accompanying transient flows, the most important are (1) unsteady friction (UF, other name: skin friction), (2) cavitation (CAV), (3) viscoelastic property of pipe deformation (flow in plastic conduits) associated with the retarded strain (RS), and (4) mutual fluid–structure interaction (FSI) of the flowing liquid with the vibrations of the pipe walls. In this work, we will focus in detail on the first three phenomena. They will be implemented in the revised DBCM model. The continuity equation with the retarded strain term was originally proposed by Rieutord and Blanchard [9]. Cavitation was modeled by Güney [10] using the column-separation modeling assumption proposed by Swaffield [11] and Safwat [12]. Another scientist examining the effect of cavitation occurring during transient flow in PE and PVC pipes was Mitosek [13,14], who showed that an increased pressure reduction is accompanied with gas desorption (reduced pressure oscillations with the increase time period of their existence). Hadj-Taïeb with Taïeb [15] proposed initially a numerical model based on the conservative finite difference method to solve the nonlinear system of hyperbolic partial differential equations describing the transient flow in which the degasification takes place (according to Henry's law). Their study showed that the degasification area is strongly connected with the wall elasticity. The same two authors [16] proposed an alternative modified mathematical model that includes retarded strain and cavitation, which was solved with the second-order finite difference scheme. The mixture density in this model was expressed by means of a non-linear expression of the liquid volume fraction. Borga et al. [17,18] conducted several transient tests with localized gas cavities in around 200 m long HDPE pipe and concluded that the presence of the leak (or air valves) in cavitating flow induces a greater damping and dispersion of transient pressure waves. Soares et al. [19,20] continued the research of Borga (which was done under the supervision of H. Ramos) and compared the effect of used cavitation models DVCM and DGCM for the prediction of transient flows with cavitation in HDPE pipes. The results indicated that the assumption of the ideal gas law (DGCM) is more appropriate than a simple adoption of vapor pressure when the pressure reaches vapor pressure (DVCM) and induces more attenuation and dispersion of transient pressures. For flows with cavitation, a new set of pipe-wall viscoelastic parameters was determined (calibration technique). The unsteady friction losses, pipe-wall viscoelasticity, and wave speed variation due to the formation of localized gas cavities were described only by the creep function. Such an approach lumped all these important phenomena in the coefficients of the creep function. Keramat et al. [21] utilized DVCM and modeled RS uses a modified Kelvin–Voigt model to study the unsteady flow with cavitation in plastic pipes. His model did not include at that time the unsteady friction effects. The main conclusion from the presented simulations (compared to simulated results with Covas [22] and Soares data [19]) is that viscoelastic pipes strongly diminish the dangers of column separation: "First, cavity opening and collapse occur only one or two times instead of tens of times (as inelastic pipes)". Two years later, Keramat and Tijsseling [23] were first to present a numerical model that included all four important phenomena that take place in transient pipe flows: UF, CAV, RS, and FSI. Unfortunately, to date, there are no experimental results that are conducted to check this interesting model in the full extent. In 2018, Urbanowicz and Firkowski developed the foundations of the model presented in this paper [24]. A year later, Urbanowicz et al. [25] compared the DACM and DBCM models, which had not been used before, for the analysis of cavitation flows in plastic pipes, indicating that they both model the phenomena in a similar way, despite the fact that they are characterized by a significantly different mathematical notation.

1.3. Recent Progress in Cavitation Modeling in Metal Pipes

Liu et al. [26] analyzed cavitation that can take place in long-distance transport pipelines. The water hammer due to the collapse of air cavities in the pipeline was discussed when the pump unit is shut down due to an incident. The theoretical and numerical analysis pointed out that it is very important to prevent the occurrence of large water hammer loads due to the collapse of flow interruption in such a system. Santoro et al. [27] tested the DVCM model, writing the continuity equation in terms of mass balance instead of volume balance. Such an assumption allowed calculations with appropriate computational fine grids. Additionally, the flow field was assumed to be two-dimensional (2D axial-symmetric flow), in order to evaluate unsteady friction without the need of parameters calibration. This research pointed out that one-dimensional (1D) models are weakly sensitive to grid size, whereas 2D model results are practically grid-independent, and in the opinion of the authors, the 2D model performs better than the 1D ones. Shankar et al. [28] studied the optimal operation of centrifugal pumps to avoid the major harmful issues as cavitation and water hammering. These authors built a system with a cascade parallel pumping setup. The extensive experimental study reveals that the preferable operating region enhances reliability as well as reduces the occurrence of faults. This paper can serve as a reference to VFD pumping systems and paves the way for sensor-less control. Zhao et al. [29] built an experimental test stand to realize a water hammer event with multipoint collapse. The influencing factors and laws of the cavity length and water hammer pressure have been summarized using the experimental data. They also reveal that the initial flow rate and valve-closing speed greatly affect the water hammer pressure rise and cavity length. In their next work [30], the authors presented a new water hammer velocity formula, a new cavity model, and introduced a floating grid method. An in-house program written in C++ confirmed that the simulation results of the new model matched the measured values.

Sun et al. [31] proposed a quasi-two-dimensional transient model coupled with DVCM which, according to the authors' analysis, can provide a better fit than classic 1D solutions. Warda et al. [32] performed three-dimensional computer fluid dynamics (CFD) simulations based on the finite volume numerical approach. The cavitation was modeled with the use of two models: the Volume of Fluid (VOF) and Schnerr–Sauer. They concluded that the 3D model that was adopted is "deemed physically superior to the existing 1D models as it removes the restriction of the 1D models that vapor cavities, when formed, fill the whole cross-section of the pipe without radial variation". Sanín-Villa et al. [33] considered the influence of the convective terms in the momentum and continuity equations (which standardly are neglected). The cavitation problem has been evaluated by use of the DVCM model. In conclusion, they stated that the influence of the convective term is small compared with a simple model where those terms are neglected. Tang et al. [34] used Fluent software to investigate the cavitation flow in the pipeline. A density–pressure model has been implemented into the continuity equation by using the further development of a user-defined function, which gives the possibility of studying the effects of the variable wave speed on the transient cavitation flow. The weakly compressible fluid RANS model (CFD) results agree well with the measured results. Saidani et al. [35] analyzed the temperature effect (in a range from 4 to 95 °C) on unsteady flow with cavitation. These authors simulated single-phase and two-phase transient flows in a hydraulic copper pipe system. The DVCM and DGCM models were used. From the performed simulations, it was evident that the water hammer is considerably sensitive to the temperature, and its proper value needs to be considered at the design stage of hydraulic systems. Yang et al. [36] used a uniform cavitation distribution model in which the critical flow velocity gradients are calculated both in front and at the back of the section and are the sufficient condition to define water column separation. Dynamic meshes were applied for tracking the change of vaporous cavitation. However, multidimensional models are computationally expensive.

2. Mathematical Model Derivation

The original bubble cavitation flow model was introduced by Shu more than fifteen years ago [8]. Its applicability was limited to systems made of metal pipes. Today, when plastic pipes are replacing traditional pipes (this trend is especially visible in water supply systems), there is a strong need to modify this interesting numerical model. In the model discussed and presented below, a single-phase flow is treated as its special solution.

2.1. Momentum and Continuity Equations

The derivation starts from the equation of momentum of two-phase vapor–liquid flow in a freely oriented conduit:

$$\frac{\partial}{\partial t}\left(\rho_m v_m A\right) + \frac{\partial}{\partial x}\left(\rho_m v_m^2 A\right) + A\frac{\partial p}{\partial x} + \pi D \tau_m + g\rho_m A \sin\theta = 0 \tag{1}$$

in which ρ_m being the mixture density is calculated:

$$\rho_m = \alpha\rho_l + (1-\alpha)\rho_v. \tag{2}$$

Please note that this starting momentum equation, as well as the set of the continuity equations (Equation (4), is identical to the one discussed in the IAHR Synthetic Report [37]. After the differentiation and ordering, one gets the following form:

$$\frac{\rho_m v_m}{A}\frac{dA}{dt} + \rho_m\frac{dv_m}{dt} + \rho_m v_m\frac{\partial v_m}{\partial x} + v_m\frac{d\rho_m}{dt} + \frac{\partial p}{\partial x} + \frac{2}{R}\tau_m + g\rho_m\sin\theta = 0. \tag{3}$$

For the sake of simplicity, let us assume that the pipe is horizontal, i.e., $\theta = 0$. Thus, the last term on the left-hand side of the above equation (Equation (3) is zero. Now, let us write the continuity equations written separately for the gas and liquid phase, respectively:

$$\left.\begin{array}{c}\frac{\partial}{\partial t}(\rho_v(1-\alpha)A) + \frac{\partial}{\partial x}(\rho_v(1-\alpha)Av_v) = 0 \\[6pt] \frac{\partial}{\partial t}(\rho_l\alpha A) + \frac{\partial}{\partial x}(\rho_l\alpha Av_l) = 0 \end{array}\right\} \tag{4}$$

where α—volumetric fraction of liquid.

Adding the continuity equations (Equation (4) together for the respective phases gives:

$$\frac{\partial}{\partial t}(\rho_v(1-\alpha)A + \rho_l\alpha A) + \frac{\partial}{\partial x}(\rho_v(1-\alpha)Av_v + \rho_l\alpha Av_l) = 0. \tag{5}$$

Next, assuming that a homogeneous bubbly flow takes place, then the dispersed vapor phase does have the same velocity as the surrounding continuous liquid phase $v_v = v_l = v_m$:

$$\frac{\partial}{\partial t}A\rho_m + \frac{\partial}{\partial x}(Av_m\rho_m) = 0. \tag{6}$$

By making differentiation and ordering in Equation (6), the following result is obtained:

$$A\frac{d\rho_m}{dt} + \rho_m\frac{dA}{dt} + A\rho_m\frac{\partial v_m}{\partial x} = 0. \tag{7}$$

Dividing Equation (7) by $A\rho_m$, the first useful form of this equation is derived:

$$\frac{1}{\rho_m}\frac{d\rho_m}{dt} + \frac{1}{A}\frac{dA}{dt} + \frac{\partial v_m}{\partial x} = 0. \tag{8}$$

However, when one multiplies Equation (7) by v_m and then divides by A, we get the second useful form of Equation (7):

$$v_m\frac{d\rho_m}{dt} + \frac{\rho_m v_m}{A}\frac{dA}{dt} + v_m\rho_m\frac{\partial v_m}{\partial x} = 0. \tag{9}$$

Using the above Equation (9) and the fact that the analyzed system is horizontal, the momentum Equation (3) can be reduced to a simpler form:

$$\rho_m\frac{dv_m}{dt} + \frac{\partial p}{\partial x} + \frac{2}{R}\tau_m = 0. \tag{10}$$

Please note that Equation (10) reduces to the equation of a single-phase flow (continuous liquid phase) when no cavitation occurs (mean values of pressure in an analyzed cross-section are larger than the vapor pressure).

The next step is to derive the continuity equation. From the works [22,37,38], it follows that in bubble flow and plastic pipes, the fluid elasticity (separately defined for liquid and vapor) and the pipe deformation can be defined as follows:

$$\frac{1}{\rho_l}\frac{d\rho_l}{dt} = \frac{1}{K_l}\frac{dp}{dt}; \ \frac{1}{\rho_v}\frac{d\rho_v}{dt} = \frac{1}{K_v}\frac{dp}{dt} \ \text{and} \ \frac{1}{A}\frac{dA}{dt} = \frac{\Xi}{E}\frac{dp}{dt} + 2\frac{d\varepsilon_r}{dt} \tag{11}$$

where $\Xi = \frac{D}{e}\zeta$—enhanced pipe restraint factor. The first two equations represent liquid and vapor elasticity, respectively, whereas the third one defines pipe deformation (the right-hand one). The derivation of the first two ones is straightforward. This is not the case for the third one—its derivation is presented in Appendix A.

The total derivative of Equation (2) mixture density ρ_m is:

$$\frac{d\rho_m}{dt} = \alpha\frac{d\rho_l}{dt} + (\rho_l - \rho_v)\frac{d\alpha}{dt} + (1-\alpha)\frac{d\rho_v}{dt}. \tag{12}$$

When Equations (11) and (12) are used in the continuity equation, Equation (8), one gets:

$$\left[\rho_m\frac{\Xi}{E} + \frac{\alpha\rho_l}{K_l} + \frac{(1-\alpha)\rho_v}{K_v}\right]\frac{dp}{dt} + (\rho_l - \rho_v)\frac{d\alpha}{dt} + 2\rho_m\frac{d\varepsilon_r}{dt} + \rho_m\frac{\partial v_m}{\partial x} = 0. \tag{13}$$

In addition, a constant pressure wave speed is assumed. In the proposed model, the value of the speed will be assumed for the steady flow occurring before the water hammer event. Then, there is only pure liquid phase ($\alpha = 1$). The last term of the square bracket in Equation (13) vanishes, and the formula under square bracket reduces to:

$$c^{-2} = \left[\rho_l\left(\frac{\Xi}{E} + \frac{1}{K_l}\right)\right] \tag{14}$$

which is the pressure wave speed of the pure liquid phase. The above wave speed equation includes elastic effects of the fluid (K_l) and of the pipe wall (E). Enhanced pipe restraint factor Ξ is calculated in a different way in thin $((D/e) < 25)$ and thick-walled pipelines [2].

Equation (14) governs the final form of the continuity equation for unsteady flows in plastic pipes:

$$\frac{1}{c^2}\frac{dp}{dt} + (\rho_l - \rho_v)\frac{d\alpha}{dt} + 2\rho_m\frac{d\varepsilon_r}{dt} + \rho_m\frac{\partial v_m}{\partial x} = 0. \tag{15}$$

In non-slip flow conditions, the proportion of the dispersed phase is of a statistical nature; i.e., the volumetric concentration and the mass are equal to the corresponding dynamic shares—the transport concentration and the degree of dryness [39,40]. Then, the following relationship applies in non-slip flows: $v_m = v/\alpha$, and the final set of fundamental equations (appropriately momentum and continuity) is as follows:

$$\begin{cases} \rho_m\frac{d}{dt}\left(\frac{v}{\alpha}\right) + \frac{\partial p}{\partial x} + \frac{2}{R}\tau_m = 0 \\ \frac{1}{c^2}\frac{dp}{dt} + (\rho_l - \rho_v)\frac{d\alpha}{dt} + 2\rho_m\frac{d\varepsilon_r}{dt} + \rho_m\frac{\partial}{\partial x}\left(\frac{v}{\alpha}\right) = 0 \end{cases} \tag{16}$$

The term v/α indicates the difference between the velocities of the liquid and vapor phase.

2.2. Wall Shear Stress and Retarded Strain

The wall shear stress in transient pipe flow can be calculated with the help of convolutional theory. Zielke [41] for laminar flow and later Vardy-Brown [42] for turbulent flow presented an initial version of this equation. For homogeneous bubble flow, the mixture density should be taken into account:

$$\tau_m = \left(\frac{fv|v|\rho_m}{8\alpha^2} + \frac{2\mu_m}{R}\int_0^t \frac{\partial}{\partial t}\left(\frac{v}{\alpha}(u)\right)\cdot w_{UF}(t-u)du\right). \tag{17}$$

The function $w_{UF}(t-u)$ [-] is a so-called *weighting function*. In our work, this function is identical to the functions used in other cavitational and single-phase flow models. For detailed form and more information about weighting functions, please refer to paper [43].

The numerical modified effective solution of the above convolution integral, which is used in this work, is based on the improved solution for single-phase flows [44,45]:

$$\tau(t+\Delta t) \approx \frac{\rho_m f v_{(t)} \left| v_{(t)} \right|}{8 \alpha_{(t)}^2} + \frac{2\mu_m}{R} \sum_{i=1}^{3} \underbrace{\left[A_i y_i(t) + \eta B_i \left[\frac{v_{(t)}}{\alpha_{(t)}} - \frac{v_{(t-\Delta t)}}{\alpha_{(t-\Delta t)}} \right] + [1-\eta]C_i \left[\frac{v_{(t-\Delta t)}}{\alpha_{(t-\Delta t)}} - \frac{v_{(t-2\Delta t)}}{\alpha_{(t-2\Delta t)}} \right] \right]}_{y_i(t+\Delta t)}. \tag{18}$$

where Δt [s] is a constant time step in the method of characteristics; μ_m is Dukler's [46] two-phase mixture dynamic viscosity [Pa·s], and

$$\eta = \frac{\int_0^{\Delta \hat{t}} w_{class.}(u)du}{\int_0^{\Delta \hat{t}} w_{eff.}(u)du}; \quad A_i = e^{-n_i \Delta \hat{t}}; \quad B_i = \frac{m_i}{\Delta \hat{t} n_i}[1 - A_i]; \quad C_i = A_i B_i. \tag{19}$$

The m_i and n_i coefficients are determined from the analytical formulas presented in a recent paper [47], and $\Delta \hat{t} = \frac{\mu_m}{\rho_m R^2}\Delta t$ is dimensionless time. In the case of turbulent flow, these coefficients must be scaled in accordance with the guidelines presented in paper [48].

The next step is to evaluate the partial derivative of retarded strain using a convolution integral. According to [49,50], the retarded strain can be written in a simpler form than the original one [38]:

$$\frac{\partial \varepsilon_r(t)}{\partial t} = \frac{D}{2e}\xi \int_0^t \frac{\partial}{\partial t}(p(u) - p(0)) \cdot \left(\sum_{i=1}^{k} \frac{J_i}{T_i} e^{-\frac{t-u}{T_i}} \right) du = \frac{\Xi}{2} \int_0^t \frac{\partial p(u)}{\partial t} \cdot w_J(t-u)du \tag{20}$$

where $w_J(t-u)$ is the creep weighting function [Pa^{-1}·s^{-1}]; J_i is the creep compliance of the i spring of the Kelvin–Voigt element [Pa^{-1}]; T_i is the retardation time of the dashpot of i-element [s].

Worth noting is the analogy of the above convolution integral with the convolutional integral representing the wall shear stress $\sum_{i=1}^{k} \frac{J_i}{T_i} e^{-\frac{t-u}{T_i}} = w_J(t-u)$. This analogy made it possible to solve numerically the above convolution integral in an effective manner using Schohl's effective scheme [51]:

$$\frac{\partial \varepsilon_r}{\partial t}(t+\Delta t) \approx \frac{\Xi}{2} \sum_{i=1}^{k} \underbrace{\left(z_i(t) \cdot e^{-\frac{\Delta t}{T_i}} + \frac{J_i}{\Delta t}\left[1 - e^{-\frac{\Delta t}{T_i}} \right]\left(p_{(t+\Delta t)} - p_{(t)} \right) \right)}_{z_i(t+\Delta t)}. \tag{21}$$

The above equation (Equation (21) may be written in a simpler form:

$$\frac{\partial \varepsilon_r}{\partial t}(t+\Delta t) = \left(p_{(t+\Delta t)}F - G(t) \right)\frac{\Xi}{2} \tag{22}$$

where:

$$F = \sum_{i=1}^{k} M_i \,; \quad G(t) = \sum_{i=1}^{k} \left(M_i p_{(t)} - z_i(t) \cdot N_i \right); \quad M_i = \frac{J_i}{\Delta t}\left[1 - e^{-\frac{\Delta t}{T_i}} \right]; \quad N_i = e^{-\frac{\Delta t}{T_i}}. \tag{23}$$

The use of convolution integrals (Equations (17) and (20)) in the basic system of Equation (16) results in:

$$\begin{cases} \rho_m \frac{d}{dt}\left(\frac{v}{\alpha} \right) + \frac{\partial p}{\partial x} + \frac{2}{R}\left(\frac{f v |v| \rho_m}{8\alpha^2} + \frac{2\mu_m}{R}\int_0^t \frac{\partial}{\partial t}\left(\frac{v}{\alpha}(u) \right) \cdot w_{UF}(t-u)du \right) = 0 \\[2ex] \frac{1}{\rho_m c^2}\frac{dp}{dt} + \frac{(\rho_l - \rho_v)}{\rho_m}\frac{d\alpha}{dt} + \frac{\partial}{\partial x}\left(\frac{v}{\alpha} \right) + \Xi \int_0^t \frac{\partial p(u)}{\partial t} \cdot w_J(t-u)du = 0 \end{cases} \tag{24}$$

2.3. Numerical Solution for Inner Nodes

The convective terms are less important and are omitted from the main analyzed set of equations. This procedure will ensure that the use of interpolation, which greatly affects the numerical solution [2], is excluded. The simplified equations of continuity and momentum are identified as L_1 and L_2, respectively:

$$L_1 = \frac{\partial p}{\partial t} + c^2(\rho_l - \rho_v)\frac{\partial \alpha}{\partial t} + 2\rho_m c^2 \frac{\partial \varepsilon_r}{\partial t} + \rho_m c^2 \frac{\partial}{\partial x}\left(\frac{v}{\alpha}\right) = 0 \tag{25}$$

$$L_2 = \frac{\partial}{\partial t}\left(\frac{v}{\alpha}\right) + \frac{1}{\rho_m}\frac{\partial p}{\partial x} + \frac{2}{\rho_m R}\tau_m = 0. \tag{26}$$

Combining linearly $L = \psi L_1 + L_2$, these equations by the unknown multiplier ψ gives:

$$\psi\left[\frac{\partial p}{\partial t} + \frac{1}{\rho_m \psi}\frac{\partial p}{\partial x}\right] + \left[\frac{\partial}{\partial t}\left(\frac{v}{\alpha}\right) + \psi \rho_m c^2 \frac{\partial}{\partial x}\left(\frac{v}{\alpha}\right)\right] + \psi c^2(\rho_l - \rho_v)\frac{\partial \alpha}{\partial t} + 2\psi \rho_m c^2 \frac{\partial \varepsilon_r}{\partial t} + \frac{2}{\rho_m R}\tau_m = 0. \tag{27}$$

By examination of the above Equation (27) with the definition of total derivatives, it can be noted that with:

$$\frac{dx}{dt} = \frac{1}{\rho_m \psi} = \psi \rho_m c^2 \tag{28}$$

it becomes the ordinary differential equation:

$$\psi \frac{dp}{dt} + \frac{d}{dt}\left(\frac{v}{\alpha}\right) + \psi c^2(\rho_l - \rho_v)\frac{\partial \alpha}{\partial t} + 2\psi \rho_m c^2 \frac{\partial \varepsilon_r}{\partial t} + \frac{2}{\rho_m R}\tau_m = 0. \tag{29}$$

The solution of Equation (28) yields two particular values of ψ,

$$\psi = \pm \frac{1}{c\rho_m}. \tag{30}$$

By inserting the above values into Equation (28) the particular relation between x and t is given as follows:

$$\frac{dx}{dt} = \pm c. \tag{31}$$

This leads to a set of two equations on positive C^+ and negative C^- characteristic lines:

$$C^+ : \frac{1}{c\rho_m}\frac{dp}{dt} + \frac{d}{dt}\left(\frac{v}{\alpha}\right) + \frac{c}{\rho_m}(\rho_l - \rho_v)\frac{\partial \alpha}{\partial t} + 2c\frac{\partial \varepsilon_r}{\partial t} + \frac{2}{\rho_m R}\tau_m = 0 \tag{32}$$

$$C^- : -\frac{1}{c\rho_m}\frac{dp}{dt} + \frac{d}{dt}\left(\frac{v}{\alpha}\right) - \frac{c}{\rho_m}(\rho_l - \rho_v)\frac{\partial \alpha}{\partial t} - 2c\frac{\partial \varepsilon_r}{\partial t} + \frac{2}{\rho_m R}\tau_m = 0. \tag{33}$$

From Equation (2), it follows that:

$$\alpha = \frac{\rho_m - \rho_v}{\rho_l - \rho_v}. \tag{34}$$

Considering the above Equation (34), the third term on the left-hand side in Equations (32) and (33) can take the form:

$$\frac{c}{\rho_m}(\rho_l - \rho_v)\frac{\partial \alpha}{\partial t} = \frac{c}{\rho_m}\frac{\partial}{\partial t}\left(\frac{\rho_m - \rho_v}{\rho_l - \rho_v}(\rho_l - \rho_v)\right) = \frac{c}{\rho_m}\frac{\partial}{\partial t}(\rho_m - \rho_v) = \frac{c}{\rho_m}\frac{\partial}{\partial t}\rho_m. \tag{35}$$

Shu [8] writes the partial derivative of the density of the mixture in a logarithmic form:

$$\frac{c}{\rho_m}\frac{\partial}{\partial t}\rho_m = c\frac{\partial}{\partial t}\ln\left(\frac{\rho_m}{\rho_l}\right). \tag{36}$$

Note that under the logarithm, we have the division of ρ_m by ρ_l, but the ρ_l could be exchanged to any other constant, and the above equation would be satisfied anyway.

$$\begin{cases} \dfrac{1}{c\rho_l}\dfrac{\Delta(p-\kappa p_v)_{A-D}}{\Delta t} + \dfrac{\Delta}{\Delta t}\left(\dfrac{v}{\alpha}\right)_{A-D} + c\dfrac{\Delta}{\Delta t}ln\left(\dfrac{\rho_m}{\rho_l}\right)_{\substack{E-D \\ F-A}} + 2c\left(\dfrac{\partial\varepsilon_r}{\partial t}\right)_{E-D} + \dfrac{2}{\rho_m R}(\tau_m)_{F-A} = 0 \\[2em] for\ \dfrac{dx}{dt} = c \end{cases} \tag{37}$$

$$\begin{cases} -\dfrac{1}{c\rho_l}\dfrac{\Delta(p-\kappa p_v)_{B-D}}{\Delta t} + \dfrac{\Delta}{\Delta t}\left(\dfrac{v}{\alpha}\right)_{B-D} - c\dfrac{\Delta}{\Delta t}\left(\dfrac{\rho_m}{\rho_l}\right)_{\substack{E-D \\ G-B}} - 2c\left(\dfrac{\partial\varepsilon_r}{\partial t}\right)_{E-D} + \dfrac{2}{\rho_m R}(\tau_m)_{G-B} = 0 \\[2em] for\ \dfrac{dx}{dt} = -c \end{cases} \tag{38}$$

where $\kappa = 1 + c^2\rho_l\Xi F\Delta t$.

From both characteristics at the inner node, the following explicit system of equations is obtained:

$$\begin{cases} \dfrac{1}{c\rho_l}\dfrac{((p_D-\kappa p_v)-(p_A-\kappa p_v))}{\Delta t} + \dfrac{\left(\frac{v_D}{\alpha_D}-\frac{v_A}{\alpha_A}\right)}{\Delta t} + \dfrac{c}{2\Delta t}\left(ln\dfrac{\rho_{mD}}{\rho_l} - ln\dfrac{\rho_{mE}\rho_{mF}}{\rho_l\rho_{mA}}\right) + 2c\left(\dfrac{\partial\varepsilon_r}{\partial t}\right)_{E-D} + \dfrac{2}{\rho_m R}(\tau_m)_{F-A} = 0 \\[1.5em] -\dfrac{1}{c\rho_l}\dfrac{((p_D-\kappa p_v)-(p_B-\kappa p_v))}{\Delta t} + \dfrac{\left(\frac{v_D}{\alpha_D}-\frac{v_B}{\alpha_B}\right)}{\Delta t} - \dfrac{c}{2\Delta t}\left(ln\dfrac{\rho_{mD}}{\rho_l} - ln\dfrac{\rho_{mE}\rho_{mG}}{\rho_l\rho_{mB}}\right) - 2c\left(\dfrac{\partial\varepsilon_r}{\partial t}\right)_{E-D} + \dfrac{2}{\rho_m R}(\tau_m)_{G-B} = 0 \end{cases} \tag{39}$$

The system can be further rewritten after introducing:

$$2c\left(\dfrac{\partial\varepsilon_r}{\partial t}\right)_{E-D} = p_D\Xi cF - \Xi cG_E(t) \tag{40}$$

$$\dfrac{2}{\rho_m R}(\tau_m)_{F-A} = \dfrac{f_A v_A|v_A|}{4R\alpha_A^2} + \dfrac{4v_{mA}}{R^2}\underbrace{\sum_{i=1}^{3}\left[A_i y_{iF} + \eta B_i\left[\dfrac{v_A}{\alpha_A} - \dfrac{v_F}{\alpha_F}\right] + [1-\eta]C_i\left[\dfrac{v_F}{\alpha_F} - \dfrac{v_{F'}}{\alpha_{F'}}\right]\right]}_{y_{iA}} = \dfrac{\lambda_{F-A}v_A|v_A|}{4R\alpha_A^2} \tag{41}$$

$$\dfrac{2}{\rho_m R}(\tau_m)_{G-B} = \dfrac{f_B v_B|v_B|}{4R\alpha_B^2} + \dfrac{4v_{mB}}{R^2}\underbrace{\sum_{i=1}^{3}\left[A_i y_{iG} + \eta B_i\left[\dfrac{v_B}{\alpha_B} - \dfrac{v_G}{\alpha_G}\right] + [1-\eta]C_i\left[\dfrac{v_G}{\alpha_G} - \dfrac{v_{G'}}{\alpha_{G'}}\right]\right]}_{y_{iB}} = \dfrac{\lambda_{G-B}v_B|v_B|}{4R\alpha_B^2} \tag{42}$$

where:

$$\begin{cases} \lambda_{F-A} = f_A + \dfrac{16v_{mA}\alpha_A^2}{Rv_A|v_A|}\underbrace{\sum_{i=1}^{3}\left[A_i y_{iF} + \eta B_i\left[\dfrac{v_A}{\alpha_A} - \dfrac{v_F}{\alpha_F}\right] + [1-\eta]C_i\left[\dfrac{v_F}{\alpha_F} - \dfrac{v_{F'}}{\alpha_{F'}}\right]\right]}_{y_{iA}} \\[1.5em] \lambda_{G-B} = f_B + \dfrac{16v_{mB}\alpha_B^2}{Rv_B|v_B|}\underbrace{\sum_{i=1}^{3}\left[A_i y_{iG} + \eta B_i\left[\dfrac{v_B}{\alpha_B} - \dfrac{v_G}{\alpha_G}\right] + [1-\eta]C_i\left[\dfrac{v_G}{\alpha_G} - \dfrac{v_{G'}}{\alpha_{G'}}\right]\right]}_{y_{iB}} \end{cases} \tag{43}$$

By transforming the system of Equation (39) in a way that the parameters searched for a given inner node D of the characteristics grid (Figure 1) remain on the left-hand side, one obtains:

$$\begin{cases} \dfrac{p_D}{c\rho_l} - \dfrac{\kappa p_v}{c\rho_l} + \dfrac{v_D}{\alpha_D} + \dfrac{c}{2}ln\dfrac{\rho_{mD}}{\rho_l} + p_D c\Xi F\Delta t = C_A \\[1em] -\dfrac{p_D}{c\rho_l} + \dfrac{\kappa p_v}{c\rho_l} + \dfrac{v_D}{\alpha_D} - \dfrac{c}{2}ln\dfrac{\rho_{mD}}{\rho_l} - p_D c\Xi F\Delta t = C_B \end{cases} \tag{44}$$

where C_A as well C_B are time-dependent functions that are iteratively calculated using the values known from the previous numerical time step:

$$\begin{cases} C_A = \dfrac{v_A}{\alpha_A} + \dfrac{p_A - \kappa p_v}{c\rho_l} - \dfrac{f_A \Delta t v_A |v_A|}{4R\alpha_A^2} + \dfrac{c}{2}ln\dfrac{\rho_{mE}\rho_{mF}}{\rho_l\rho_{mA}} + G_E(t)\Xi c\Delta t \\[3mm] C_B = \dfrac{v_B}{\alpha_B} - \dfrac{p_B - \kappa p_v}{c\rho_l} - \dfrac{f_B \Delta t v_B |v_B|}{4R\alpha_B^2} - \dfrac{c}{2}ln\dfrac{\rho_{mE}\rho_{mG}}{\rho_l\rho_{mB}} - G_E(t)\Xi c\Delta t \end{cases}. \tag{45}$$

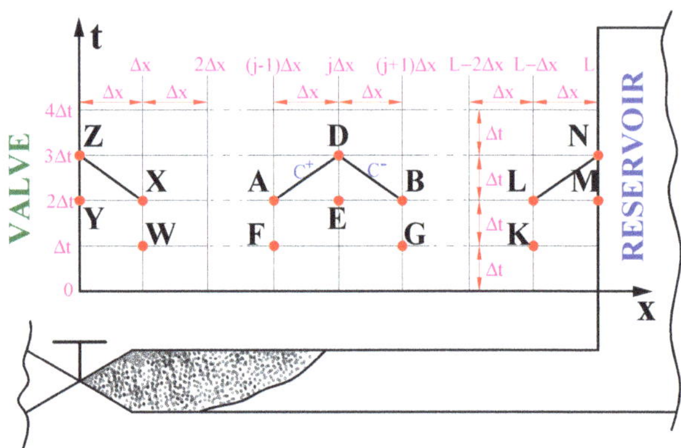

Figure 1. Rectangular grid in the method of characteristics.

From the above equations, the final solutions for the inner node D (Figure 1) of the grid of characteristics are obtained. The solution for the mean velocity at cross-section is:

$$v_D = \frac{\alpha_D(C_A + C_B)}{2} \tag{46}$$

and the solution for the pressure is:

$$p_D = \frac{(C_A - C_B)c\rho_l}{2\kappa} + p_v - \frac{c^2\rho_l}{2\kappa}ln\frac{\rho_{mD}}{\rho_l}. \tag{47}$$

The analysis of the above formulas shows that in order for $p_D > p_v$ (note that then $\rho_{mD} = \rho_l$), the condition that $C_A \geq C_B$ must be met. When $C_A \geq C_B$, there is no cavitation and $\alpha_D = 1$; $p_D = \frac{(C_A - C_B)c\rho_l}{2\kappa} + p_v$. Otherwise, when $C_A < C_B$ cavitation occurs, then $p_D = p_v$ and $\rho_{mD} = \rho_l e^{(\frac{C_A - C_B}{c})}$.

Having the instantaneous value of the mixture density ρ_{mD}, the vapor density ρ_v, and the liquid density ρ_l, the instantaneous value of the liquid phase concentration α_D should be determined from the formula (Equation (2)):

$$\alpha_D = \frac{\rho_l e^{(\frac{C_A - C_B}{c})} - \rho_v}{\rho_l - \rho_v}. \tag{48}$$

2.4. Boundary Conditions

The next step is to solve the boundary conditions. According to Figure 1, the instantaneous closing valve of an RVP system is at the left-hand side of the system ($x = 0$). The valve boundary condition is derived from the negative C^- characteristic:

$$\begin{cases} -\dfrac{p_Z}{c\rho_l} + \dfrac{\kappa p_v}{c\rho_l} + \dfrac{v_Z}{\alpha_Z} - \dfrac{c}{2}ln\dfrac{\rho_{mZ}}{\rho_l} - p_Z c\Xi F\Delta t = C_X \\[3mm] C_X = \dfrac{v_X}{\alpha_X} - \dfrac{p_X - \kappa p_v}{c\rho_l} - \dfrac{f_X \Delta t v_X |v_X|}{4R\alpha_X^2} - \dfrac{c}{2}ln\dfrac{\rho_{mY}\rho_{mW}}{\rho_l\rho_{mX}} - G_Y(t)\Xi c\Delta t \end{cases}. \tag{49}$$

Please note that the value of C_X is based only on known values from the previous time steps. The velocity at the valve section for time $t > 0$ has zero value, i.e., $v_Z = 0$ (closed valve). The above Equation (49) takes the form:

$$p_Z \left(\underbrace{1 + c^2 \rho_l \Xi F \Delta t}_{\kappa} \right) = -c\rho_l C_X + \kappa p_v - \frac{c^2 \rho_l}{2} ln \frac{\rho_{mZ}}{\rho_l} \tag{50}$$

which finally reduces to:

$$p_Z = p_v - \frac{\left(C_X + \frac{c}{2} ln \frac{\rho_{mZ}}{\rho_l} \right) c\rho_l}{\kappa}. \tag{51}$$

When the pressure p_Z at this boundary is higher than the vapor pressure p_v, then the natural logarithm is equal to 0 as $\rho_{mZ} = \rho_l$; this means that there is no cavitation when $C_X < 0$, and the pressure at the valve section can be calculated from the following equation:

$$p_Z = p_v - \frac{C_X c\rho_l}{\kappa}. \tag{52}$$

Otherwise, when $C_X \geq 0$ and $p_Z = p_v$, then the bubble mixture density and volumetric fraction of the liquid, respectively, should be calculated as follows:

$$\rho_{mZ} = \rho_l e^{-\frac{2C_X}{c}} \text{ and } \alpha_Z = \frac{\rho_{mZ} - \rho_v}{\rho_l - \rho_v}. \tag{53}$$

Next, the dynamic viscosity of the homogeneous bubble mixture using Dukler's formula [46] should be calculated:

$$\mu_{mZ} = \alpha_Z \mu_l + (1 - \alpha_Z)\mu_v. \tag{54}$$

At the opposite end ($x = L$) of the RPV system, Figure 1 is the reservoir. At the cross-section connecting the pipe with the reservoir, the pressure is assumed to be of constant value, i.e., $p_N = p_R$ during the complete transient event associated with the analyzed water hammer phenomenon. As the pressure does not pulsate at this cross-section, the retarded strain is neglected here. The final equation for the velocity pulsation at this section in which the pressure is always higher than the vapor pressure $p_N > p_v$ is:

$$v_N = \frac{v_L}{\alpha_L} + \frac{p_L - p_N}{c\rho_l} + \frac{c}{2} ln \frac{\rho_{mM}\rho_{mK}}{\rho_l \rho_{mL}} - \frac{2\Delta t}{R\rho_l} \tau_L. \tag{55}$$

3. Experimental Verification of New Model

In order to demonstrate the effectiveness of the newly presented model, in this section, the results of simulation tests will be compared with the experimental results presented by Güney [10]. The Güney experimental test stand located at the INSA research center in Lyon (France) was a simple system consisting of three main components: reservoir–pipe–valve (Figure 2).

In the analyzed RPV system, in steady flow, water flowed directly into the atmosphere. The pipe had a total length of $L = 43.1$ m and an internal diameter $D = 0.0416$ m (the wall thickness of the pipe was $e = 0.0042$ m). The test pipe was made of low-density polyethylene (LDPE). The experimental tests of the water hammer forced by the sudden (momentary) closure of the valve (shutting off the flow) have been carried out for five different temperatures of the flowing liquid (water). In Table 1, the parameters required to simulate the analyzed unsteady flows with cavitation are tabulated. It can be seen that although the initial flow velocity was similar, due to the change in viscosity, the value of the Reynolds number increased with the temperature increase (almost twice as high for *Case* 05 : $Re_{05} \approx 82,000$ than for *Case* 01 : $Re_{01} \approx 45,500$). After the temperature change, not only do the parameters related to the flowing liquid change (Table 1) but also the values of the parameters representing the mechanical properties of the pipe; thus, it is necessary to compare their values (J creep compliances and τ retardation time coefficients values are presented in Table 2).

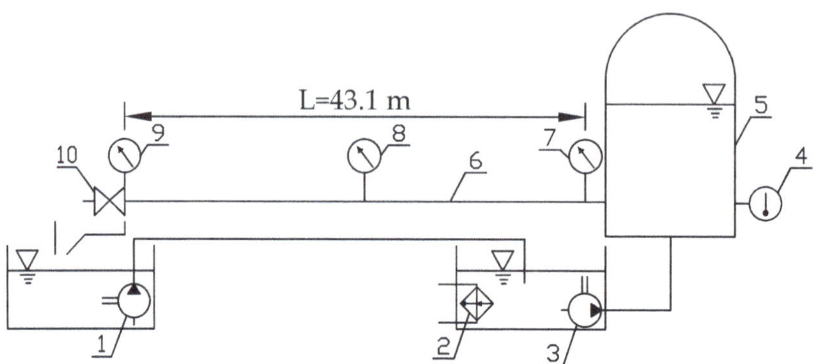

Figure 2. Schematic diagram of Güney's experimental test stand: 1—booster pump, 2—temperature stabilization system, 3—test stand supply pump, 4—thermometer, 5—reservoir, 6—LDPE pipe, 7, 8, and 9—pressure transducers, 10—quick-closing valve.

Güney used the time–temperature superposition principle (also known as time–temperature reducibility) to derive his creep compliance functions for different temperatures. During initial simulations, complete Güney creep compliance functions were used (three exponential terms) that can be found in the works [10,52]. As the initially obtained simulation results indicated that this creep function is a source of simulation error, we had a detailed look at the original coefficients. We noticed that the corresponding creep compliance values of small retardation times ($\tau < 1.5\Delta10^{-4}$ [s]–original J_1 and τ_1 coefficients) are out of the frequency range of the used dynamic viscoelastometer RHEOVIBRON. Filtering out this coefficient for small retardation times (rejecting from the analysis original J_1 and τ_1 coefficients without changing all other experimentally defined creep coefficients) helped to receive corrected comparisons results.

The creep functions for LDPE have different characteristics (Figure 3) than those for the typical currently used plastic material, namely HDPE. The LDPE material has higher values of creep compliance than the HDPE material. Additionally, we may see (Figure 3) that an increase in temperature increases the creep compliance values. The HDPE traces which are presented for comparison in Figure 3 were obtained experimentally by Covas et al. [22].

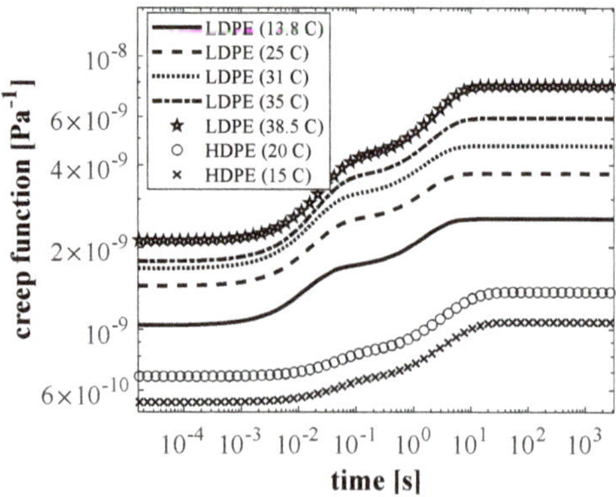

Figure 3. Creep functions for two different PE pipes.

Table 1. Güney cases details.

Case	T [°C]	v_0 [m/s]	Re_0 [−]	c [m/s]	p_R [Pa]	p_v [Pa]	K_l [Pa]	ρ_l [kg/m^3]	μ_l [Pa·s]	ρ_v [kg/m^3]	μ_v [Pa·s]
01	13.8	1.28	45,511	305	1.2955×10^5	1570	2.14×10^9	999.3	0.0012	0.012	9.6×10^{-6}
02	25	1.37	63,892	265	1.3056×10^5	3160	2.24×10^9	997.1	8.9×10^{-4}	0.023	9.9×10^{-6}
03	31	1.34	71,102	247	1.3041×10^5	4480	2.27×10^9	995.3	7.8×10^{-4}	0.032	1×10^{-5}
04	35	1.37	78,827	235	1.3038×10^5	5610	2.285×10^9	994.1	7.2×10^{-4}	0.040	1.02×10^{-5}
05	38.5	1.33	81,967	215	1.2985×10^5	6790	2.295×10^9	992.6	6.7×10^{-4}	0.047	1.03×10^{-5}

Table 2. Creep compliance function coefficients.

Case	T [°C]	J_0 [Pa^{-1}]	J_1 [Pa^{-1}]	J_2 [Pa^{-1}]	τ_1 [s]	τ_2 [s]
01	13.8	1.071×10^{-9}	0.637×10^{-9}	0.871×10^{-9}	0.0166	1.747
02	25	1.438×10^{-9}	1.046×10^{-9}	1.237×10^{-9}	0.0222	1.864
03	31	1.665×10^{-9}	1.397×10^{-9}	1.628×10^{-9}	0.0221	1.822
04	35	1.847×10^{-9}	1.797×10^{-9}	2.349×10^{-9}	0.0265	2.392
05	38.5	2.219×10^{-9}	2.097×10^{-9}	3.570×10^{-9}	0.0347	3.077

J_i—creep-compliance coefficients; τ_i—retardation times.

The pressure wave speeds were estimated based on the empirically observed duration of the first pressure amplitudes. Their values summarized in Table 1 enabled the determination of J_0 (see Table 2) from the transformed formula of the pressure wave speed:

$$J_0 = \frac{1}{\rho \Xi c^2} - \frac{1}{K_l \Xi} \tag{56}$$

where $\xi = 0.97$; $\Xi = \frac{D}{e}\xi = 9.61$.

The method of characteristics was used with a constant number of reaches $N = 64$. The selected number of reaches meets the computational compliance criteria discussed in paper [53], i.e., $N > 10$. Extra simulation studies performed during the preparation of this paper whose purpose was to investigate the impact of the number of reaches showed that there are no significant differences between the results of $N = 16, 32$, and the selected 64. A finer grid is favorable in the case of instantaneous valve closure. The time steps are calculated on the basis of the Courant–Friedrichs–Lewy (CFL) stability condition $C_n = \frac{(c \cdot \Delta t)}{\Delta x} \leq 1$. In order to keep the value of the CFL number equal to one, appropriate values of the time steps should be determined from $\Delta t = \Delta x / c$ (wave speeds c are given in Table 1). In the MOC $\Delta x = L/N$ i.e., $\Delta x \approx 0.67$ m. Then, the following time steps are obtained for the five cases: $\Delta t_{G01} = 0.0022$ s; $\Delta t_{G02} = 0.0025$ s; $\Delta t_{G03} = 0.0027$ s; $\Delta t_{G04} = 0.0029$ s; and $\Delta t_{G05} = 0.0031$ s, respectively. The results of the simulation tests compared with the experimental data are presented in Figure 4.

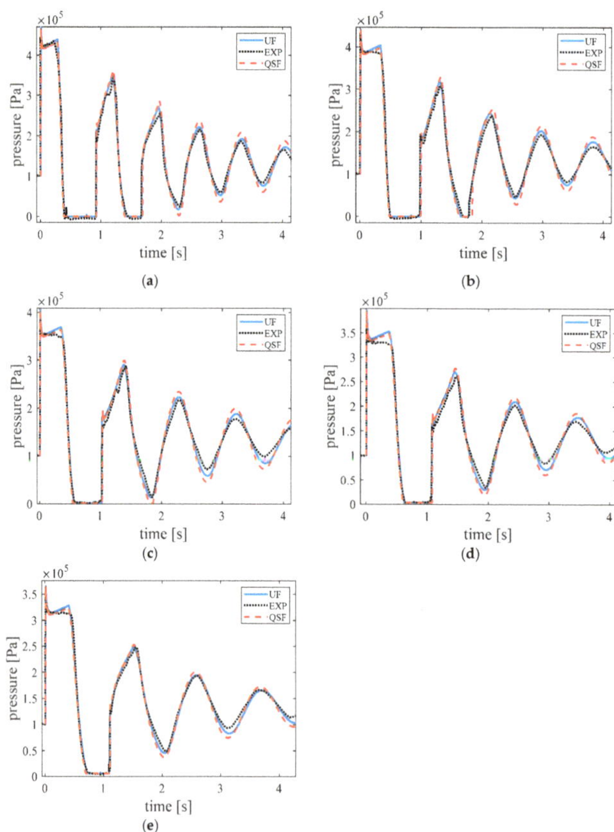

Figure 4. Computed and measured results for different cases: (**a**) Case 01 (13.8 °C); (**b**) Case 02 (25 °C); (**c**) Case 03 (31 °C); (**d**) Case 04 (35 °C); (**e**) Case 05 (38.5 °C).

The qualitative analysis of the obtained results (Figure 4) indicates the following:

- In systems based on plastic pipes, as the temperature of the flowing medium increases, the maximum value of the pressure at the first amplitudes decreases (assuming a similar value of the initial velocity in the steady flow just before the quick valve closure). The above is due to the decrease in pressure wave speed with increasing temperature;
- The decrease of pressure wave speed caused by the increase in temperature is also responsible for the change in the frequency of the water hammer itself. Based on the research carried out, one may notice that for a higher value of the pressure wave speed, the number of pressure amplitudes appearing in the same time interval increases (Figure 4a—five amplitudes), while for a small value, this amount is smaller (Figure 4d,e—four amplitudes);
- The omission of unsteady hydraulic resistances negatively affects the modeled waveforms, which are overestimated starting from the second amplitudes (Figure 4a–e). Large discrepancies are visible in the modeled pressures at the peaks and at the valleys of these pressure amplitudes;
- The modified proposed numerical solution modeled the first peak of pressure visible at the beginning of all tops of first amplitudes (Figure 5a) as well the small peak at the beginning of second amplitudes (Figure 5b). This proves the physics of the analyzed phenomena;
- In *Cases* 03, 04, and 05, where only single column separation takes place (after the first amplitude), it can be seen that the phase shift of the simulated pressure increased over time. This behavior can be explained by the fact that in a real situation, the pressure wave speed does not remain constant during the entire transient event but rather slightly changes. The change of the pressure wave speed is not included in the current version of the mathematical model, as it would force the use of interpolation, which would introduce additional numerical damping [2];
- Although qualitative studies indicate the advantage of the model taking into account unsteady friction, it is necessary to carry out quantitative studies to confirm the above hypothesis. Such research will be carried out in the next Chapter 4;
- The largest model discrepancies occurred in the runs carried out for *Cases* 02, 03, 04, and 05 at the top of the first amplitude. In the experimental studies in the final phase of the pressure increase (first amplitudes), no increase in pressure was observed as in *Case* 01 (Figure 4a), while from what we can see in Figure 4b–e, such an increase was modeled by the numerical model. This increase was not influenced by the way of taking into account skin friction (quasi-steady or unsteady resistances); hence, the applied creep functions were responsible for them.

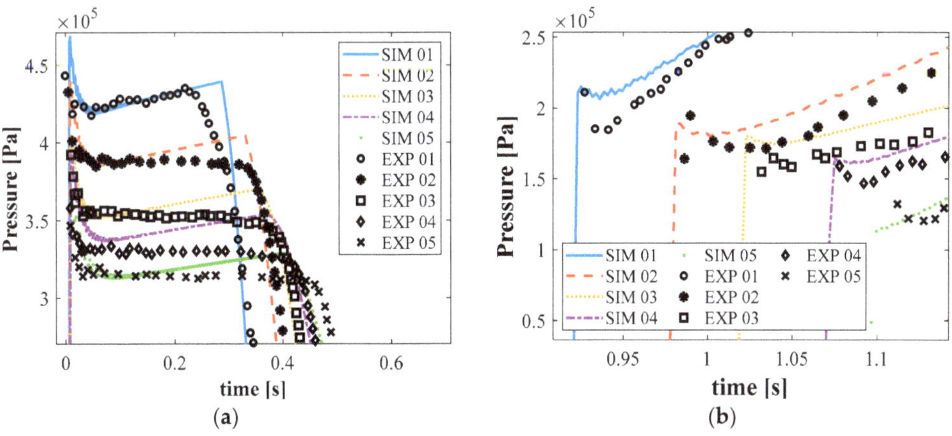

Figure 5. Enlargement of (**a**) top of first amplitude and (**b**) early stage of second amplitude.

4. Quantitative Analysis of Results

In this section, quantitative research is performed whose role is to define and determine important criteria parameters of the analyzed flow. It is difficult to find any favorable quantitative method in the literature on the subject of transient pipe flows. Here, we present a new methodology that results in two criteria parameters. The role of the final qualitative parameters is to determine the compliance of the simulated histories with respect to the experimental ones in a simplified mathematical way.

A MATLAB subprogram was written to search for maxima (peaks) and minimum values of pressure histories and their occurrence times (calculated from the beginning of the analyzed transient state). In a demonstrative way, Figure 6 illustrates the working idea of this proposed "collecting" subprogram. As can be seen, the pressure drops to the saturated vapor pressure were not taken into account, as the final results would be false. Additionally, when determining the time compliance, the time of the first pressure peak t_1 at the first amplitude was omitted, as it would also cause the final result to be distorted.

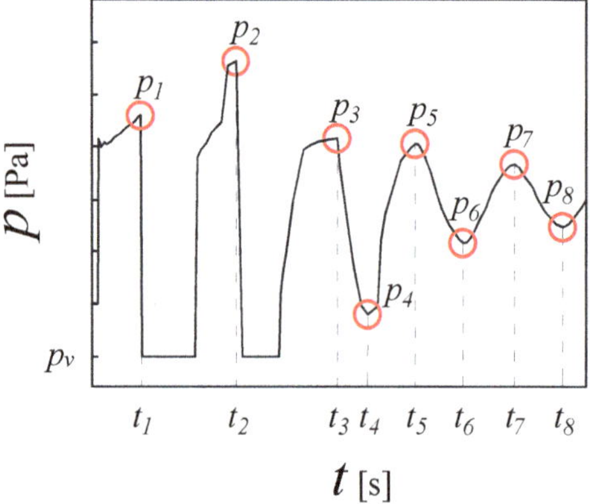

Figure 6. Selection of maximal and minimal pressures and their times of occurrence.

The pressure compliance parameter determining the compliance of the maximum and minimum simulated pressures is calculated by the following formula:

$$E_p = \frac{\sum_{i=1}^{k} \left| \frac{p_{s,i} - p_{e,i}}{p_{e,i}} \right|}{k} \cdot 100\% \tag{57}$$

where $p_{s,i}$—simulated maximal and minimal pressures and $p_{e,i}$—experimentally predicted maximal and minimal pressures

The time compliance parameter that determines the time fit of subsequent simulated amplitudes was calculated using the following formula:

$$E_t = \frac{\sum_{i=2}^{k} \left| \frac{t_{s,i} - t_{e,i}}{t_{e,i}} \right|}{k - 1} \cdot 100\% \tag{58}$$

where $t_{s,i}$—times of occurrence of maximal and minimal simulated pressures and $t_{e,i}$—experimentally observed times of maximal and minimal pressures (note that times $t_{s,1}$ and $t_{e,1}$ representing the maximum at first amplitude are not taken into account, because in some cases, their fit could distort the whole analysis).

The degree of simulation compatibility increases with decreasing values of the above coefficients. In Table 3, the complete quantitative results of the E_p and E_t parameters calculated for all comparative studies carried out in this work are summarized.

Table 3. Quantitative results.

Case	T [°C]	v_0 [m/s]	Re_0 [−]	Unsteady Friction (UF)		Quasi−Steady Friction (QSF)		ΔCav_{EXP} [s]	ΔCav_{UF} [s]	ΔCav_{QSF} [s]
				E_p [%]	E_t [%]	E_p [%]	E_t [%]			
01	13.8	1.28	45,511	8.39	0.70	23.51	0.39	0.74	0.77	0.83
02	25	1.37	63,892	5.32	0.49	15.15	0.31	0.59	0.61	0.72
03	31	1.34	71,102	6.76	1.02	23.17	0.88	0.45	0.46	0.53
04	35	1.37	78,827	8.57	0.90	18.04	0.62	0.43	0.44	0.47
05	38.5	1.33	81,967	4.38	1.62	10.46	1.36	0.38	0.38	0.42

The results of quantitative research indicate the following:

- Unsteady friction losses contribute to a significant reduction of pressure compliance errors (parameter E_p—values of simulated pressures). When only quasi-steady resistances were taken into account, the average error E_p from all the tests carried out was about 18%, while when the model of UF losses was taken into account, then the average error E_p was about 6.5%; i.e., almost three times smaller;
- The unsteady friction also influences the second analyzed parameter, i.e., the phase compatibility E_t. However, in this case, there was a slight increase in the value of the time fit error of the modeled waveforms. When the quasi-steady nature of the resistance was taken into account, the average error E_t was 0.71%, while when the unsteady nature of the resistance was taken into account, the average error E_t was 0.95%. The difference is very small and can be neglected; however, it is recommended to use models of unsteady friction when the experimentally obtained creep functions are used during modeling;
- The deterioration of the quantitative parameter E_t, which was noted in the previous paragraph, after taking into account the unsteady hydraulic resistances, prompted us to analyze another quantitative parameter, which is the duration of the cavitation phenomenon at the analyzed cross-section (cross-section at the valve). From the data presented in Table 3 (ΔCav_{EXP}), it can be seen that the duration of the cavitation phenomenon at the cross-section at the valve decreased with increasing temperature (decreasing the speed of pressure wave propagation). It can also be seen that the numerical model which takes into account unsteady friction predicts the duration of cavitation areas slightly longer than it was in the experiments. The quasi-steady resistance model overestimated the duration of cavitation quite significantly.

5. Conclusions

The paper presents a modified unsteady discrete bubble cavity model DBCM. The new model is developed in a very simple form, which makes it easy for implementation in commercial programs for unsteady pipe flow analysis. The new model takes into account three very important phenomena: unsteady wall shear stress, vaporous cavitation, and pipe-wall retarded strain. The latter mentioned phenomenon occurs in plastic pipes.

The conducted comparative studies have shown that with the help of the presented model, it is possible to simulate pressure and velocity waveforms in which vapor areas appear as a result of the cavitation phenomenon in plastic pipes. It was noticed that the influence of unsteady friction was much smaller than the influence of retarded strain. An innovative method of calculating the convolutional integral describing the retarded strain was applied by using the analogy to the convolutional integral defining the wall shear stress (Schohl's method). Taking into account the commonly known boundary conditions related to the method of characteristics enables the use of the novel model in complex networks: water supply, oil hydraulics, heating, etc.

The modified solution is an alternative to the two commonly used transient cavitating pipe flow models, namely the DGCM (discrete gas cavity model) and the DVCM (discrete vapor cavity model). In our future work, we are planning to execute broad comparisons of the presented new model with the existing ones.

The use of the experimentally determined creep functions (obtained by Güney) showed that such functions can be an alternative to the calibration methods commonly developed today. This indicates that the creation of the so-called "maps" of creep function curves for various polymeric materials currently used in the world for pressure pipes will significantly help designers at the design stage and will enable the study of the most dangerous unsteady cases "a priori". Presentation of such "maps" obtained for different temperatures should be a priority of the current scientific research.

Author Contributions: Conceptualization, K.U., A.B. and M.S.; methodology, K.U. and A.K.; software, K.U., A.M. and A.B.; validation, K.U., A.M. and M.K.; formal analysis, K.U., P.B. and M.S.; investigation, K.U., A.B, M.S. and M.K.; resources, K.U.; data curation, K.U.; writing—original draft preparation, K.U. and A.B.; writing—review and editing, K.U., A.B., A.K., M.S.; visualization, K.U., P.B.; supervision, K.U. and A.B.; project administration, K.U. and M.S.; funding acquisition, K.U. All authors have read and agreed to the published version of the manuscript.

Funding: A.B. gratefully acknowledges the support of Slovenian Research Agency (ARRS) conducted through the research project L2-1825 and the programme P2-0162.

Institutional Review Board Statement: Not applicable.

Informed Consent Statement: Not applicable.

Data Availability Statement: All codes generated during the study and experimental data are available from the corresponding author by request.

Conflicts of Interest: The authors declare no conflict of interest.

Nomenclature

A	pipe cross-sectional area (m^2)
A_i, B_i and C_i	unsteady friction coefficients (-)
D	pipe internal diameter (m)
E	pipe modulus of elasticity (Pa)
E_p and E_t	pressure and time compliance parameters (%)
$J_0 = 1/E$	instantaneous creep compliance (Pa^{-1})
J_i	creep compliance of the i-th Kelvin-Voigt element (Pa^{-1})
K_l	bulk modulus of liquid phase (Pa)
K_v	bulk modulus of vapor phase (Pa)
L	pipe length (m)
N	number of computational reaches (-)
R	pipe internal radius (m)
T	temperature in Celsius degrees (°C)
T_i	the retardation time of i-th Kelvin-Voigt element (s)
c	pressure wave speed (m/s)
e	pipe-wall thickness (m)
f	Darcy–Weisbach friction factor (-)
g	acceleration due to gravity (m/s^2)
m_i and n_i	frictional weighting function coefficients (-)
p	pressure (Pa)
p_v	saturated vapor pressure (Pa)
p_R	reservoir pressure (Pa)
Re_0	initial Reynolds number (-)
t	time (s)
u	dummy variable (s)

w_{UF}	weighting function of unsteady friction (-)
w_J	weighting function of creep ($\mathrm{Pa^{-1} \cdot s^{-1}}$)
v	average flow velocity (m/s)
v_m	mixture velocity (m/s)
v_0	initial liquid velocity (m/s)
x	space coordinate (m)
y_i	time-dependent velocity history effect (m/s)
z_i	time-dependent strain history effect ($\mathrm{s^{-1}}$)
α	volumetric fraction of liquid phase (-)
Δt	numerical time step (s)
$\Delta \hat{t}$	dimensionless time step (-)
Δx	numerical spatial step (m)
ε_r	retarded strain (-)
η	correction factor of unsteady friction (-)
θ	pipe slope angle ($^{\circ}$)
λ	transient friction factor (-)
μ_l	liquid dynamic viscosity (Pa·s)
μ_m	mixture dynamic viscosity (Pa·s)
μ_v	vapor dynamic viscosity (Pa·s)
ν_l	kinematic viscosity of liquid ($\mathrm{m^2/s}$)
ν_m	kinematic viscosity of liquid-vapor mixture ($\mathrm{m^2/s}$)
ν_v	kinematic viscosity of vapor ($\mathrm{m^2/s}$)
ξ	pipe restraint factor (-)
Ξ	enhanced pipe restraint factor (-)
ρ_l	density of liquid phase ($\mathrm{kg/m^3}$)
ρ_m	mixture density ($\mathrm{kg/m^3}$)
ρ_v	density of vapor phase ($\mathrm{kg/m^3}$)
σ_e	elastic component of the hoop stress (Pa)
τ_m	mixture wall shear stress (Pa)
ψ	MOC multiplier ($\mathrm{m \cdot Pa^{-1} \cdot s^{-1}}$)

Abbreviations

CAV	cavitation
CFL	Courant–Friedrichs–Lewy condition
DACM	discrete Adamkowski cavity model
DBCM	discrete bubble cavity model
DGCM	discrete gas cavity model
DVCM	discrete vapor cavity model
EXP	experimental
FSI	fluid structure interaction
HDPE	high-density polyethylene
LDPE	low-density polyethylene
MOC	method of characteristics
ODE	ordinary differential equation
PDE	partial differential equation
QSF	quasi-steady friction
RS	retarded strain
SIM	simulation
UF	unsteady friction
VOF	volume of fluid method

Appendix A

The total derivative of strain represents a derivative of pipe inner diameter changes [38]:

$$\frac{d\varepsilon}{dt} = \frac{d}{dt}\left(\frac{D - D_0}{D_0}\right) = \frac{1}{D}\frac{dD}{dt}. \tag{A1}$$

These changes in plastic pipes differ from those reported in elastic pipes. The relation between the cross-section derivative and strain derivative is:

$$\frac{dA}{dt} = \frac{d}{dt}\left(\frac{\pi D^2}{4}\right) = \frac{\pi D}{2}\frac{dD}{dt} = \frac{\pi D^2}{2}\frac{1}{D}\frac{dD}{dt} = \frac{\pi D^2}{2}\frac{d\varepsilon}{dt} = 2A\frac{d\varepsilon}{dt}. \tag{A2}$$

The circumferential strain can be decomposed into an instantaneous elastic strain ε_e and a retarded strain ε_r:

$$\varepsilon = \varepsilon_e + \varepsilon_r. \tag{A3}$$

Using the above decomposition (Equation (A3)) in Equation (A2) gives:

$$\frac{dA}{dt} = 2A\left(\frac{d\varepsilon_e}{dt} + \frac{d\varepsilon_r}{dt}\right). \tag{A4}$$

Typically, the instantaneous strain ε_e, which is assumed to be linear–elastic, can be related to the hoop stress as follows:

$$\varepsilon_e = \frac{\zeta \sigma_e}{E} \tag{A5}$$

where σ_e—elastic component of the hoop stress.

The hoop stress is also related to the fluid pressure and the ratio between the pipe inner diameter and wall thickness:

$$\sigma_e = \frac{pD}{2e}. \tag{A6}$$

Let us take the derivative of the above Equation (A6):

$$\frac{d\sigma_e}{dt} = \frac{p}{2e}\frac{dD}{dt} + \frac{D}{2e}\frac{dp}{dt}. \tag{A7}$$

The derivative of the elastic strain component (Equation (A5)) can be written with the help of Equation (A7):

$$\frac{d\varepsilon_e}{dt} = \frac{d}{dt}\left(\frac{\zeta \sigma_e}{E}\right) = \frac{\zeta}{2eE}\left(p\frac{dD}{dt} + D\frac{dp}{dt}\right) = \frac{\zeta}{2eE}\left(pD\frac{1}{D}\frac{dD}{dt} + D\frac{dp}{dt}\right). \tag{A8}$$

Finally, with help of Equation (A1), one gets:

$$\frac{d\varepsilon_e}{dt} = \frac{\Xi}{2E}\left(p\frac{d\varepsilon_e}{dt} + \frac{dp}{dt}\right). \tag{A9}$$

After rearrangement:

$$\frac{d\varepsilon_e}{dt} = \frac{\frac{\Xi}{2E}\frac{dp}{dt}}{1 - \frac{\Xi}{2E}p}. \tag{A10}$$

Introducing Equation (A10) into Equation (A4) results in:

$$\frac{dA}{dt} = \frac{A\frac{dp}{dt}}{\frac{E}{\Xi} - \frac{p}{2}} + 2A\frac{d\varepsilon_r}{dt}. \tag{A11}$$

Since in most practical applications $p/2 \ll E/\Xi$, then:

$$\frac{1}{A}\frac{dA}{dt} = \frac{\Xi}{E}\frac{dp}{dt} + 2\frac{d\varepsilon_r}{dt}. \tag{A12}$$

This equation has been used in the manuscript during the derivation of the continuity equation—see Equation (11).

References

1. Wylie, E.B. Simulation of Vaporous and Gaseous Cavitation. *J. Fluids Eng.* **1984**, *106*, 307–311. [CrossRef]
2. Wylie, E.B.; Streeter, V.L.; Suo, L. *Fluid Transients in Systems*; Prentice Hall: Englewood Cliffs, NJ, USA, 1993.
3. Bergant, A.; Simpson, A.; Tijsseling, A. Water hammer with column separation: A historical review. *J. Fluids Struct.* **2006**, *22*, 135–171. [CrossRef]
4. Karpenko, M.; Bogdevičius, M. Review of Energy-Saving Technologies in Modern Hydraulic Drives. Mokslas—Lietuvos Ateitis: Statyba, Transportas, Aviacinės Technologijos = Science − Future of Lithuania: Civil and Transport Engineering, Aviation Technologies. *Vilnius VGTU Press* **2017**, *9*, 553–558.
5. Bogdevičius, M.; Karpenko, M.; Bogdevičius, P. Determination of Rheological Model Coefficients of Pipeline Composite Material Layers Based on Spectrum Analysis and Optimization. *J. Theor. Appl. Mech.* **2021**, *59*, 265–278. [CrossRef]
6. Adamkowski, A. *Transient States in the Vortex Systems of Water Machines*; Scientific Papers; Institute of Fluid-Flow Machinery, Polish Academy of Sciences: Gdańsk, Polish, 2004.
7. Adamkowski, A.; Lewandowski, M. A New Method for Numerical Prediction of Liquid Column Separation Accompanying Hydraulic Transients in Pipelines. *ASME J. Fluids Eng.* **2009**, *131*, 071302. [CrossRef]
8. Shu, J.-J. Modelling vaporous cavitation on fluid transients. *Int. J. Press. Vessel. Pip.* **2003**, *80*, 187–195. [CrossRef]
9. Rieutord, E.; Blanchard, A. Influence d'un Comportement Viscoélastique de la Conduite dans le Phénomène du Coup de Bélier (Influence of a Viscoelastic Pipe Behavior in the Phenomenon of Water Hammer). *Rep. Acad. Sci.* **1972**, *274*, 1963–1966. (In French)
10. Güney, M.S. Contribution à l'étude du Phénomène de Coup de Bélier en Conduite Viscoélastique (Contribution to the Study of the Water Hammer Phenomenon in Viscoelastic Pipe). Ph.D. Thesis, Claude Bernard University, Lyon, French, 1977.
11. Swaffield, J.A. Column Separation in an Aircraft Fuel System. In Proceedings of the 1st International Conference: Pressure Surges, Canterbury, UK, 6–8 September 1972; pp. 13–28.
12. Safwat, H.H.; De Kluyver, J.P. Digital Computations for Waterhammer-Column Separation. In Proceedings of the 1st International Conference: Pressure Surges, Canterbury, UK, 6–8 September 1972; pp. 53–68.
13. Mitosek, M. Study of Cavitation due to Water Hammer in Plastic Pipes. *Plast. Rubber Compos.* **1997**, *26*, 324–329.
14. Mitosek, M. Study of Transient Vapor Cavitation in Series Pipe Systems. *J. Hydraul. Eng.* **2000**, *126*, 904–911. [CrossRef]
15. Hadj-Taïeb, E.; Taïeb, L. Écoulements Transitoires Dans les Conduites Déformables avec Dégazage de l'Air Dissous (Transient Flows in Plastic Pipes with Dissolved Air Degasification). *La Houille Blanche* **2001**, *87*, 99–107. (In French)
16. Hadj-Taïeb, L.; Hadj-Taïeb, E. Numerical Simulation of Transient Flows in Viscoelastic Pipes with Vapour Cavitation. *Int. J. Model. Simul.* **2009**, *29*, 206–213. [CrossRef]
17. Borga, A.; Ramos, H.; Covas, D.; Dudlik, A.; Neuhaus, T. Dynamic Effects of Transient Flows with Cavitation in Pipe Systems. In Proceedings of the 9th International Conference: Pressure Surges, Chester, UK, 24–26 March 2004; Volume 2, pp. 605–617.
18. Ramos, H.M.; Borga, A.; Zhang, C.; Tang, H. Surge Effects in Pressure Systems for Different Pipe Materials. In *Advances in Water Resources and Hydraulic Engineering*; Springer: Berlin/Heidelberg, Germany, 2009; pp. 2152–2156.
19. Soares, A.K.; Covas, D.I.C.; Ramos, H.M.; Reis, L.F.R. Unsteady Flow with Cavitation in Viscoelastic Pipes. *Int. J. Fluid Mach. Syst.* **2009**, *2*, 269–277. [CrossRef]
20. Soares, A.K.; Covas, D.I.; Carriço, N.J. Transient vaporous cavitation in viscoelastic pipes. *J. Hydraul. Res.* **2012**, *50*, 228–235. [CrossRef]
21. Keramat, A.; Tijsseling, A.S.; Ahmadi, A. Investigation of Transient Cavitating Flow in Viscoelastic Pipes. *IOP Conf. Ser. Earth Environ. Sci.* **2010**, *12*, 012081. [CrossRef]
22. Covas, D.; Stoianov, I.; Mano, J.F.; Ramos, H.M.; Graham, N.; Maksimovic, C. The dynamic effect of pipe-wall viscoelasticity in hydraulic transients. Part II—model development, calibration and verification. *J. Hydraul. Res.* **2005**, *43*, 56–70. [CrossRef]
23. Keramat, A.; Tijsseling, A.S. Waterhammer with Column Separation, Fluid-Structure Interaction and Unsteady Friction in a Viscoelastic Pipe. In Proceedings of the 11th International Conference: Pressure Surges, Lisbon, Portugal, 24–26 October 2012; pp. 443–460.
24. Urbanowicz, K.; Firkowski, M. Extended Bubble Cavitation Model to predict water hammer in viscoelastic pipelines. *J. Phys. Conf. Ser.* **2018**, *1101*, 012046. [CrossRef]
25. Urbanowicz, K.; Bergant, A.; Duan, H. Simulation of unsteady flow with cavitation in plastic pipes using the discrete bubble cavity and Adamkowski models. *IOP Conf. Ser. Mater. Sci. Eng.* **2019**, *710*, 012013. [CrossRef]
26. Liu, E.; Wen, D.; Peng, S.; Sun, H.; Yang, Y. A study of the numerical simulation of water hammer with column separation and cavity collapse in pipelines. *Adv. Mech. Eng.* **2017**, *9*, 168781401771812. [CrossRef]

27. Santoro, V.C.; Crimì, A.; Pezzinga, G. Developments and Limits of Discrete Vapor Cavity Models of Transient Cavitating Pipe Flow: 1D and 2D Flow Numerical Analysis. *J. Hydraul. Eng.* **2018**, *144*, 04018047. [CrossRef]
28. Shankar, V.A.; Subramaniyan, U.; Sanjeevikumar, P.; Holm-Nielsen, J.B.; Blaabjerg, F.; Paramasivam, S. Experimental Investigation of Power Signatures for Cavitation and Water Hammer in an Industrial Parallel Pumping System. *Energies* **2019**, *12*, 1351. [CrossRef]
29. Zhao, L.; Yang, Y.; Wang, T.; Han, W.; Wu, R.; Wang, P.; Wang, Q.; Zhou, L. An Experimental Study on the Water Hammer with Cavity Collapse under Multiple Interruptions. *Water* **2020**, *12*, 2566. [CrossRef]
30. Zhao, L.; Yang, Y.; Wang, T.; Zhou, L.; Li, Y.; Zhang, M. A Simulation Calculation Method of a Water Hammer with Multpoint Collapsing. *Energies* **2020**, *13*, 1103. [CrossRef]
31. Sun, Q.; Wu, Y.; Liang, H. Numerical Simulation of Transient Cavitating Flows in Pipeline Systems. In Proceedings of the 11th International Symposium on Heating, Ventilation and Air Conditioning (ISHVAC 2019), Harbin, China, 12–15 July 2019; Springer: Singapore, 2019. [CrossRef]
32. Warda, H.A.; Wahba, E.M.; Salah El-Din, M. Computational Fluid Dynamics (CFD) Simulation of Liquid Column Separation in Pipe Transients. *Alex. Eng. J.* **2020**, *59*, 3451–3462. [CrossRef]
33. Sanín-Villa, D.; Florez, D.S.; Del Rio, J.S. Numerical simulation of water hammer and cavitation phenomena including the convective term in pipeline problems. *J. Phys. Conf. Ser.* **2020**, *1708*, 012027. [CrossRef]
34. Tang, X.; Duan, X.; Gao, H.; Li, X.; Shi, X. CFD Investigations of Transient Cavitation Flows in Pipeline Based on Weakly-Compressible Model. *Water* **2020**, *12*, 448. [CrossRef]
35. Saidani, A.; Fourar, A.; Massouh, F. Influence of temperature on transient flow with cavitation in copper pipe-rig. *Model. Earth Syst. Environ.* **2021**, 1–11. [CrossRef]
36. Yang, J.B.; Guo, W.C.; Luo, J.J.; Fu, G.F. Investigation of water column separation induced by pressure pulsation based on critical cavitation rate model. *IOP Conf. Ser. Earth Environ. Sci.* **2021**, *774*, 012055. [CrossRef]
37. Fanelli, M. *Hydraulic Transients with Water Column Separation: IAHR Working Group 1971–1991 Synthesis Report*; Enel-Cris: Milan, Italy, 2000.
38. Covas, D. Inverse Transient Analysis for Leak Detection and Calibration of Water Pipe Systems—Modelling Special Dynamic Effects. Ph.D. Thesis, Imperial College of Science, Technology and Medicine, University of London, London, UK, 2003.
39. Orzechowski, Z. *Two-Phase, One-Dimensional, Steady, Adiabatic Flows*; PWN: Warsaw, Polish, 1990.
40. Hsu, Y.Y.; Graham, R. *Transport Processes in Boiling and Two-Phase Systems*; McGraw-Hill: New York, NY, USA, 1976.
41. Zielke, W. Frequency-Dependent Friction in Transient Pipe Flow. *J. Basic Eng.* **1968**, *90*, 109–115. [CrossRef]
42. Vardy, A.; Brown, J. Transient turbulent friction in smooth pipe flows. *J. Sound Vib.* **2003**, *259*, 1011–1036. [CrossRef]
43. Urbanowicz, K.; Zarzycki, Z. New Efficient Approximation of Weighting Functions for Simulations of Unsteady Friction Losses in Liquid Pipe Flow. *J. Theor. Appl. Mech.* **2012**, *50*, 487–508.
44. Urbanowicz, K. Simple Modelling of Unsteady Friction Factor. In Proceedings of the 12th International Conference on Pressure Surges, Dublin, Ireland, 18–20 November 2015; pp. 113–130.
45. Urbanowicz, K. Fast and accurate modelling of frictional transient pipe flow. *ZAMM* **2018**, *98*, 802–823. [CrossRef]
46. Dukler, A.E.; Wicks, M.; Cleveland, R.G. Frictional pressure drop in two-phase flow: B. An approach through similarity analysis. *AIChE J.* **1964**, *10*, 44–51. [CrossRef]
47. Urbanowicz, K. Analytical expressions for effective weighting functions used during simulations of water hammer. *J. Theor. Appl. Mech.* **2017**, *55*, 1029–1040. [CrossRef]
48. Urbanowicz, K.; Zarzycki, Z.; Kudźma, S. Universal Weighting Function in Modeling Transient Cavitating Pipe Flow. *J. Theor. Appl. Mech.* **2012**, *50*, 889–902.
49. Urbanowicz, K.; Duan, H.; Bergant, A. Transient Liquid Flow in Plastic Pipes. *Stroj. Vestn. J. Mech. Eng.* **2020**, *66*, 77–90. [CrossRef]
50. Urbanowicz, K.; Firkowski, M. Modelling Water Hammer with Quasi-Steady and Unsteady Friction in Viscoelastic Pipelines. In *Springer Proceedings in Mathematics & Statistics*; Springer: Cham, Switzerland, 2018; pp. 385–399.
51. Schohl, G.A. Improved Approximate Method for Simulating Frequency-dependent Friction in Transient Laminar Flow. *J. Fluids Eng.* **1993**, *115*, 420–424. [CrossRef]
52. Gally, M.; Güney, M.; Rieutord, E. An Investigation of Pressure Transients in Viscoelastic Pipes. *J. Fluids Eng.* **1979**, *101*, 495–499. [CrossRef]
53. Urbanowicz, K. Computational compliance criteria in water hammer modelling. *E3S Web Conf.* **2017**, *19*, 3021. [CrossRef]

Article

Numerical Study on the Influence of Vortex Generator Arrangement on Heat Transfer Enhancement of Oil-Cooled Motor

Junjie Zhao [1,†], **Bin Zhang** [1,†], **Xiaoli Fu** [1,*] and **Shenglin Yan** [2,*]

1   College of Civil Engineering, Tongji University, Shanghai 200092, China; zhaojj@tongji.edu.cn (J.Z.); zhangb@tongji.edu.cn (B.Z.)

2   School of Mechanical and Power Engineering, East China University of Science and Technology, Shanghai 200237, China

*   Correspondence: xlfu@tongji.edu.cn (X.F.); yanshenglin.1988@163.com (S.Y.)

†   These authors contributed equally to this work and should be considered co-first authors.

Abstract: At present, vortex generators have been extensively used in radiators to improve the overall heat transfer performance. However, there is no research on the effect of vortex generators on the ends of motor coils. Meanwhile, the current research mainly concentrates on the attack angle, shape and size, and lacks a detailed study on the transverse and longitudinal distance and arrangement of vortex generators. In this paper, the improved dimensionless number R is used as the key index to evaluate the overall performance of enhanced heat transfer. Firstly, the influence of the attack angle on heat transfer enhancement is discussed through a single pair of rectangular vortex generators, and the results demonstrate that the vortex generator with a 45° attack angle is superior. On this basis, we compare the effects of different longitudinal distances (2 h, 4 h, and 6 h, h meaning the height of vortex generator) on enhanced heat transfer under four distribution modes: *Flow-Up (FU)*, *Flow-Down (FU)*, *Flow-Up-Down (FUD)*, *Flow-Down-UP (FDU)*. Thereafter, the performances of different transverse distances (0.25 h, 0.5 h, and 0.75 h) of the vortex generators are numerically simulated. When comparing the longitudinal distances, *FD* with a longitudinal distance of 4 h (*FD-4 h*) performs well when the Reynolds number is less than 4000, and *FU* with a longitudinal distance of 4 h (*FU-4 h*) performs better when the Reynolds number is greater than 4000. Similarly, in the comparison of transverse distances, *FD-4 h* still performs well when the Reynolds number is less than 4000, and *FU* with a longitudinal distance of 4 h and transverse distance of 0.5 h (*FU-4 h–0.5 h*) is more prominent when the Reynolds number is greater than 4000.

Keywords: vortex generator; arrangement; heat transfer; numerical simulation

Citation: Zhao, J.; Zhang, B.; Fu, X.; Yan, S. Numerical Study on the Influence of Vortex Generator Arrangement on Heat Transfer Enhancement of Oil-Cooled Motor. *Energies* **2021**, *14*, 6870. https://doi.org/10.3390/en14216870

Academic Editor: Artur Bartosik

Received: 25 September 2021
Accepted: 17 October 2021
Published: 20 October 2021

1. Introduction

The motor is widely used in ship, municipal, electric power, port handling, and other fields, with broad development prospects and considerable market capacity. In recent years, with the upgrading of power system requirements, the traditional motor with low efficiency and low power to weight ratio has been unable to meet the market demand, which has prompted the need for the research and development of new oil-cooled motors with high efficiency, high power density, low vibration and noise, and strong overload capacity [1,2]. During operation, the motor will produce immense heat, which will reduce the operating efficiency of the motor. In order to ensure the efficient operation of the motor, a heat exchanger is often used to enhance heat transfer and cool the motor. Fluid mediums in heat exchangers are diverse, and the oil phase is widely used because of its superior heat exchange effect, low cost, and long service life [3]. However, when the motor is compact, there will still be areas with a high temperature in the motor after the oil phase heat exchange [4,5]. For instance, in an oil-cooled motor, the temperature at the

153

end of the coil outlet is higher than that of the iron core section, so further improving the heat transfer effect at the end of the coil outlet can enhance the performance of the whole motor [6–8]. Three ways are acknowledged to enhance heat transfer: increasing heat transfer area, increasing average temperature difference, and increasing the heat transfer coefficient [9]. According to the characteristics of coil oil cooling, the feasible heat transfer enhancement method of the coil end can be analyzed, and the vortex generator can be used to increase the heat transfer coefficient [10].

As a passive heat transfer enhancement technology, the vortex generator can produce vortexes to effectively improve the heat transfer rate of the heat transfer system. Because of its economy and convenience, it has attracted extensive attention in recent years. When it was initially proposed, it was mainly used in the field of aerodynamics [11]. Later, Johnson and Joubert [12] studied the heat transfer enhancement effect of the delta wing vortex generator on the air of the heat exchanger, which initiated the application of the vortex generator in the field of heat exchangers. Chai et al. [13] investigated the improvement of heat exchanger performance via the installation of vortex generators, based on the mechanism of the longitudinal vortex destroying the growth of the boundary layer, increasing the turbulence intensity, and producing secondary fluid flow on the heat transfer surface. There are various structures of vortex generators, such as rectangular wing, triangular wing, trapezoidal wing, cylindrical trapezoidal wing, cylindrical triangular wing, cylindrical rectangular wing, and so on [14–17]. Promvonge et al. [18] studied the influence of the vortex generator, combined with a fin and airfoil, on the heat transfer and drag characteristics of the flow passage under the condition of uniform heat flow boundary. The results showed that the heat transfer efficiency and friction loss of the fluid with the fin and airfoil vortex generator were higher than those with a smooth channel. Chen et al. [19] optimized the aspect ratio of the fluid channel and the height of the vortex generator. The results showed that, in a fluid channel with a large aspect ratio, the heat transfer performance could be enhanced while reducing the pressure loss. So far, many scholars have done a lot of research on the size and attack angle of vortex generators [20–25]. Wijayanta et al. [26] used the $k - \varepsilon$ turbulence model to explore the heat transfer and pressure drop characteristics of vortex generators with various attack angles, and found that the maximum increases in Nusselt number and friction coefficient are 269% and 10.1 times higher than those of smooth tubes, respectively. Zhang et al. [27,28] explored the best combination of length, width, and longitudinal distance of the vortex generator, and its total efficiency was 7.2% higher than that without the vortex generator. Ebrahimi et al. [29,30] studied the heat transfer and fluid characteristics in the laminar flow channel installed with the vortex generator. It was noted that the channel of the vortex generator had higher efficiency, the friction coefficient increased by 2–25%, and the Nusselt number increased by 4–30%.

So far, the application of the vortex generator to enhance heat transfer has mainly been used in the heat exchanger, and air is primarily used as the fluid medium. In addition, the current research on the heat transfer of the vortex generator mainly focuses on the attack angle, size, and shape. However, only a few studies roughly explore the influence of the longitudinal distribution mode of the vortex generator on heat transfer, while there is a lack of detailed and orderly analysis and research on the specific longitudinal and transverse distribution mode of the vortex generator. Therefore, it is of great significance to study the arrangement of the vortex generator on the coil in the motor with oil as the fluid medium. With the help of computational fluid dynamics (CFD) software, the effect of different distribution types of the vortex generator on the heat transfer effect can be simulated, and the temperature change at the end of the coil and the pressure loss before and after the installation of the vortex generator can be analyzed. At the same time, the best distribution type can be obtained, and the mechanism of heat transfer enhancement by turbulence at the end of the coil can be revealed. Thus, through these explorations, this paper can provide a feasible idea for the design and application of the motor in industrial manufacturing.

2. Materials and Methods

2.1. Physical Model

As shown in Figure 1a, a three-dimensional rectangular fluid channel is built on the basis of the rectangular coil. The length of the fluid channel is 888 mm, while the height and width of the inlet and outlet are 245 mm. Besides, the length, width, and height of the rectangular coil are 365 mm, 85 mm, and 20 mm respectively, and the distance from the rectangular coil to the inlet and outlet is 261.5 mm, and the distance to the upper and lower boundaries is 112.5 mm. Moreover, as illustrated in Figure 1b, the rectangular coil is composed of copper wire surrounded by a 0.5 mm thick insulating layer.

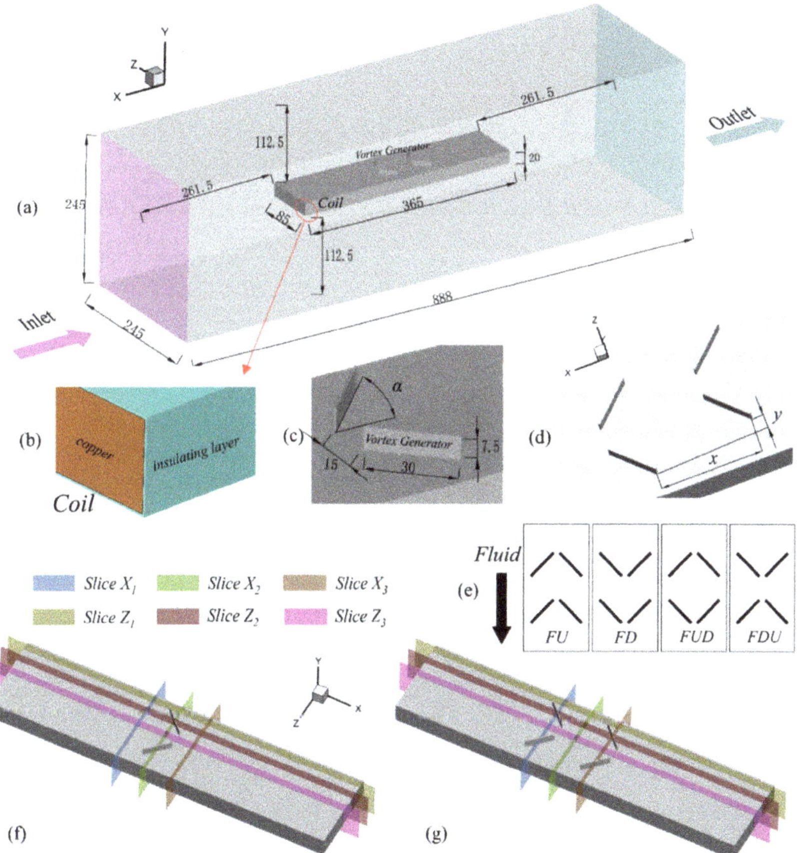

Figure 1. (**a**) Diagrams of a channel with coil and vortex generators, (**b**) specific graph of the coil, (**c,d**) specific graphs of vortex generator, (**e**) four different distribution types, (**f,g**) slice diagrams in X and Z directions.

Figure 1c shows the detailed data of the vortex generator. The length and width of vortex generators are 30 mm, 1 mm respectively, while the height, which is defined as h, of the vortex generator, is 7.5 mm. Furthermore, the attack angle of the vortex generator is defined as α, and the distance between each pair of wings is 15 mm. As presented in Figure 1d, when investigating the best arrangement of the vortex generator, the longitudinal distance between two adjacent vortex generators is x, while the transverse distance is defined as y. Additionally, besides the variety of the longitudinal and transverse distance, four different vortex generator distribution modes are shown in Figure 1e, which are specified as *Flow-Up (FU)*, *Flow-Down (FD)*, *Flow-Up-Down (FUD)*, *Flow-Down-UP (FDU)*. It is worth noting that when investigating the influence of vortex generator arrangement on heat transfer performance, this paper first explores $x = 2h, 4h, 6h$ and four different distribution modes to obtain a better longitudinal distance and distribution mode. On this basis, $y = 0.25h, 0.5h, 0.75h$ are discussed to explore the influence of the transverse distance of vortex generator on heat transfer. Based on the above results, a relatively better heat transfer arrangement can be obtained.

When discussing and analyzing the effect of vortex generator arrangement on heat transfer, there are six different slices in X and Z directions, specified as *Slice* X_1, *Slice* X_2, *Slice* X_3, *Slice* Z_1, *Slice* Z_2, *Slice* Z_3. The X coordinates of *Slice* X_1, X_2, and X_3 were -40 mm, -10 mm, and 20 mm separately, while the Z coordinates of *Slice* Z_1, Z_2, and Z_3 were 10 mm, 25 mm, and 42.5 mm respectively. On this foundation, the straight line 3.75 mm above the coil is selected on the slice to obtain the temperature data to explore the uniformity of temperature distribution. Since only one pair of vortex generators are used to discover the change of attack angle, while two pairs of vortex generators are used to explore the distance and arrangement, the position of the slice relative to the vortex generator changes slightly, which can be observed in Figure 1f,g.

2.2. Materials

Table 1 illustrates the materials and properties of all parts of the physical models [31,32].

Table 1. Materials and properties.

Models	Materials	ρ (kgm^{-3})	C_p (Jkg^{-1}k^{-1})	λ (wm^{-1}k^{-1})	μ (kgm^{-1}s^{-1})
Copper wire	Copper	8978	381	387.6	-
Insulating layer	Polymer	1190	1.05	0.2	-
Vortex generator	Epoxy	1200	550	2	-
Fluid	Oil	875	2093	0.135	0.008

2.3. Governing Equations

No. 25# transformer oil is used as the working fluid, due to its good viscosity temperature property. Considering that the temperature change in the whole fluid domain is very small, the viscosity of oil is assumed to be constant. As pointed out by Chai et al. [13], the standard $k - \varepsilon$ turbulence model is the most widely used and validated model, so the standard $k - \varepsilon$ turbulence model is used as the mathematical model for numerical simulation, the near-wall treatment is enhanced wall treatment, and the governing equations include continuity equation, momentum equation, and energy equation, which are expressed as follows:

Continuity equation:

$$\frac{\partial}{\partial x_i}(\rho_o u_i) = 0 \tag{1}$$

Momentum equation:

$$\frac{\partial}{\partial t}(\rho_o u_i) + \frac{\partial}{\partial x_w}(\rho_o u_i u_w) = \frac{\partial}{\partial x_w}(\mu \frac{\lambda_o \partial u_i}{\partial x_w} - \rho_o \overline{u_i' u_w'}) - \frac{\partial p}{\partial x_i} \tag{2}$$

Energy equation:

$$\frac{\partial x}{\partial t}(\rho_o T) + \frac{\partial}{\partial x_i}(\rho_o u_i T) = \frac{\partial}{\partial x_i}(\frac{\lambda_o}{C_p}\frac{\partial T}{\partial x_i} - \rho_o \overline{u_i' T'}) \tag{3}$$

k equation:

$$\frac{\partial}{\partial t}(\rho_o k) + \frac{\partial}{\partial x_i}(\rho_o k u_i) = \frac{\partial}{\partial x_w}\left[\left(\mu + \frac{\mu_t}{\sigma_k}\right)\frac{\partial k}{\partial x_w}\right] + G - \rho_o \varepsilon \tag{4}$$

ε equation:

$$\frac{\partial}{\partial t}(\rho_o \varepsilon) + \frac{\partial}{\partial x_i}(\rho_o \varepsilon u_i) = \frac{\partial}{\partial x_w}\left[\left(\mu + \frac{\mu_t}{\sigma_\varepsilon}\right)\frac{\partial \varepsilon}{\partial x_w}\right] + \frac{C_1 \varepsilon}{k}G - C_2 \rho_o \frac{\varepsilon^2}{k} \qquad (5)$$

2.4. Boundary Conditions

As shown in Table 2, the inlet boundary is the velocity inlet with a speed range of 0.04 to 0.2 m/s, the outlet is the pressure outlet, and the relative pressure is 0. For the no-slip wall conditions used for the wall, the specific material properties can be referred to in Table 1.

Table 2. Boundary conditions.

Boundary Conditions	Methods	Values
Inlet	Velocity inlet	0.04–0.2 m/s
Outlet	Pressure outlet	0 Pa
Wall	No slip	-

2.5. Simulation Method and Initial Conditions

The commercial software FLUENT 17.1 is used to simulate the heat transfer process, and the finite volume method is used as the governing equation. The simple algorithm is used to realize the coupling between pressure and velocity. The second order upwind scheme is used for the momentum and energy equations. The convergence residuals of the velocity dependent equation and the energy dependent equation are less than 10^{-6}. The initial conditions are heating capacity of 392,857 w/m^3, the initial temperature of 303 K.

3. Grid Independence Test

As illustrated in Figure 2a, all models calculated in this paper use hexahedral mesh. When figuring out the grid near the wall, it is necessary to take the thickness of the boundary layer into account, thus the use of the dimensionless number Y plus is required for characterization [33]. When calculating the grid spacing of the first layer, Y plus = 1 is taken to obtain the grid spacing of the first layer. After the grid spacing of the first layer is determined as 0.02 mm, it is extended outward in the proportion of 1.1 ratio. Before the numerical simulation, the grid independence is analyzed with five grid numbers, i.e., 262×10^4, 421×10^4, 596×10^4, 685×10^4, and 799×10^4. From Figure 2b, it can be observed that with the increase of the number of grids, the growth of Nu and f gradually decreases. Besides, it can be seen from Figure 2c that the maximum temperature of copper wire decreases and the ΔP increases with the increase in grid number. When the number of grids is greater than 685×10^4, the Nu, f, maximum temperature, ΔP change little, and when the number of grids changes from 685×10^4 to 799×10^4, the relative change rates are less than 0.3%. Therefore, considering the accuracy and calculation time, a grid number of 685×10^4 is adopted in the present study.

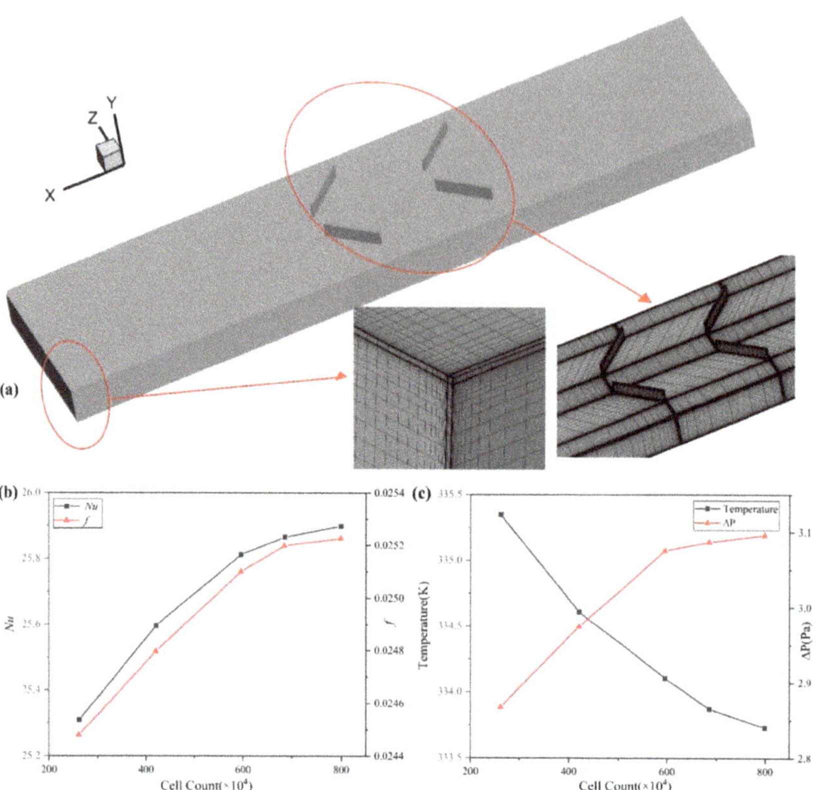

Figure 2. (**a**) Representative grids in numerical simulation, (**b**,**c**) grid independence test results.

4. Results and Discussions

4.1. Parameter Definition

The Nusselt number (Nu), Reynolds number (Re), Prandtl number (Pr), Colburn factor (j), and friction factor (f) are employed to describe the thermal and flow characteristics of the oil channel and are defined as

$$Nu = \frac{h_c D}{\lambda_o} \tag{6}$$

$$\mathrm{Re} = \frac{\rho_o u D}{\mu} \tag{7}$$

$$\mathrm{Pr} = \frac{C_p \mu}{\lambda_o} \tag{8}$$

$$j = \frac{Nu}{\mathrm{Re}\mathrm{Pr}^{\frac{1}{3}}} \tag{9}$$

$$f = \frac{2\Delta P D}{\rho_o u^2 L} \tag{10}$$

where ΔP is the pressure drop between the inlet and outlet of the test model.

In order to evaluate the overall performance considering both maximum temperature and fluid characteristics, the dimensionless number R, is improved in the study based on Zhou's [14] research. The R is defined as

$$R = \frac{T_0}{T_1} \times \frac{\frac{j}{j_0}}{\frac{f}{f_0}} \tag{11}$$

where T_0 represents the maximum temperature of the coil without vortex generators, while T_1 represents the maximum temperature of the coil with different dislocations of vortex generators. Based on the dimensionless number R, the larger R is, the better the comprehensive performance will be.

4.2. Verification of Numerical Results

Since the numerical model of the smooth channel proposed in this paper is similar to the physical models presented by Ma et al. [34] and Zhou et al. [14], the Nu, f of the numerical model and the experimental model are compared to verify the reliability of the numerical model. As is found in Figure 3a, the Nusselt numbers are in reasonable conformity with two correlations from Ma et al. and Zhou et al., with average deviations of 9.5% and 10.8%, respectively. Similarly, according to Figure 3b, the friction factors of the simulation results are in reasonable agreement with Ma and Zhou's results, with average differences of 26.9% and 16.9%, separately. Therefore, it is credible that the numerical model and solution method in this study are reliable.

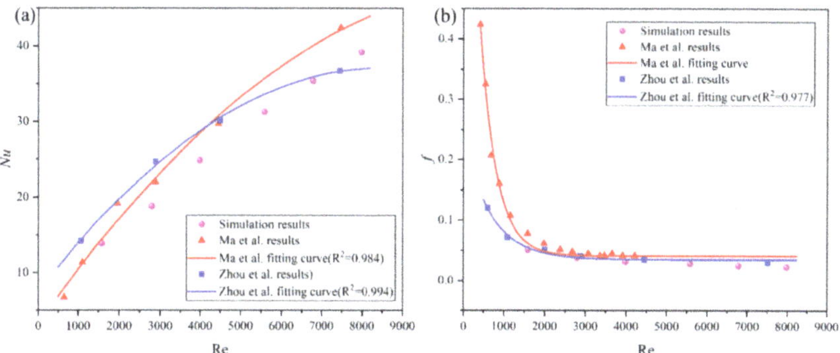

Figure 3. Validation of numerical results (**a**) Nu, (**b**) f.

4.3. Effect of Attack Angle of the Vortex Generator

Four different attack angles of vortex generators are used to explore the best choice, the four models differ only in the attack angle, and the other control conditions are consistent. By comparing the performance of the R of vortex generator under four different attack angles of 15°, 30°, 45°, and 60°, a better attack angle can be obtained. Figure 4a shows the change of the R of vortex generators with different attack angles in the process of Re increasing. It can be seen that the R is very close when the attack angles are 30° and 45°, and both of them are better than 15° and 60° attack angles, which is similar to the law obtained by Lotfi et al. [35] and Gholami et al. [36] when studying the attack angles of vortex generators in the past. Considering that the maximum temperature has a greater impact on oil-cooled motors, when the R of different types have the same performance, the vortex generator with a lower maximum temperature is used for calculation and discussion. In order to more widely explore the effect of the vortex generator on the coil end heat transfer, we introduce the thermal enhancement factor $\frac{\Delta T}{T_0}$, where ΔT represents the difference between the maximum temperature of the channel with vortex generator and the maximum temperature of the channel without vortex generator, and T_0 represents the maximum temperature of the channel without vortex generator. As presented in Figure 4b, the thermal enhancement factor $\frac{\Delta T}{T_0}$ of the 45° attack angle is higher than that of the 30° attack angle. Therefore, the 45° attack angle vortex generator is selected as the basis for subsequent exploration.

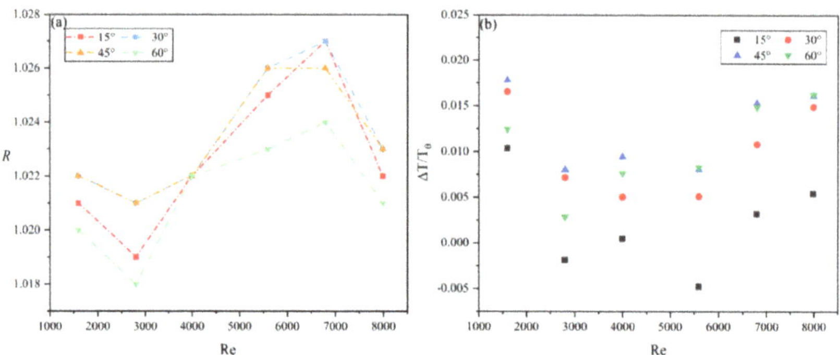

Figure 4. Performance of (**a**) R and (**b**) thermal enhancement factor $\frac{\Delta T}{T_0}$ at different Re.

When the iterative error is less than 10^{-5} and the maximum temperature does not change, the calculation is considered to be completed, and the data required for temperatures, velocities, and slices are extracted and analyzed. Under steady state conditions, Figure 5 shows pictures of a smooth channel without vortex generator at Re = 3993, Figure 5a shows the streamline diagram at 3.75 mm above the coil. Corresponding to this is the temperature contour diagram shown in Figure 5b. Moreover, Figure 5c,d shows the point line diagrams of the temperatures of different points extracted from the slices in the X and Z directions, respectively. Figure 6 shows pictures of the channel after installing a pair of vortex generators with an attack angle of 45 degrees. In Figure 6a,b, it can be seen that the region with low velocity in the streamline diagram shows higher temperature in the temperature contour diagram. Comparing with Figure 5a,b, due to the addition of vortex generator, two vortexes are formed behind the vortex generator, and the velocity at the two vortexes is higher than that in the adjacent region to a certain extent; therefore, this area shows lower temperature in the temperature contour diagram. However, by observing Figure 6c,d, since the thermal conductivity of the solid is significantly higher than that of the oil phase, the temperature of the vortex generator is higher than that of the oil phase on the same section, so the temperature of the vortex generator increases obviously at the place where the vortex generator is placed. In the middle of the two wings of the vortex generator and the area where the fluid flows through the vortex generator, the heat dissipation is accelerated due to the generation of the vortex, and the temperature is significantly lower than that in Figure 5c,d.

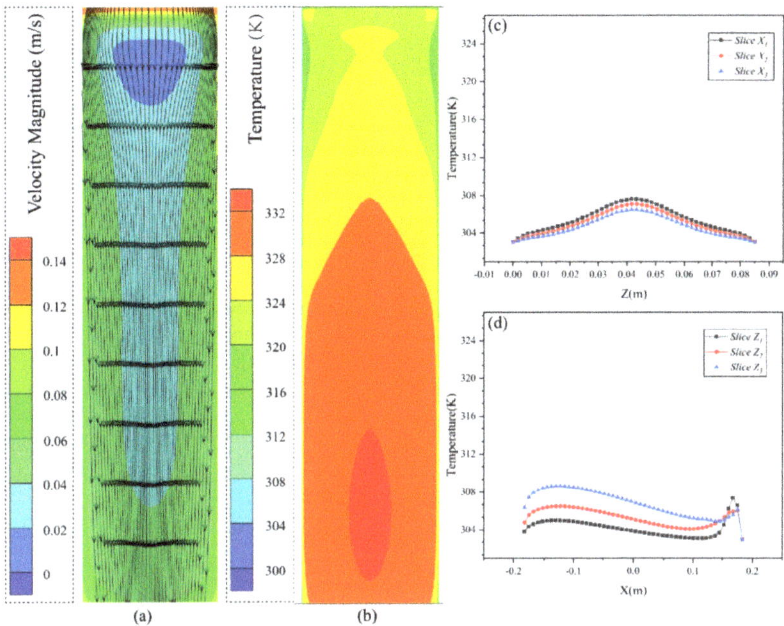

Figure 5. Diagrams of smooth channel under steady state conditions: (**a**) streamline diagram, (**b**) contour diagram of temperature, (**c**) slice temperature diagram in *X* direction, (**d**) slice temperature diagram in *Z* direction.

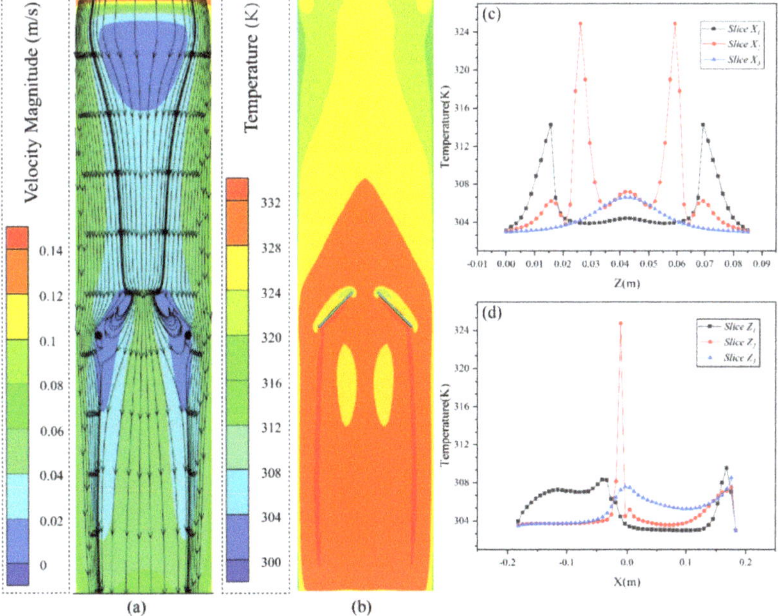

Figure 6. Diagrams of channel with 45° attack angle vortex generator under steady state conditions: (**a**) streamline diagram, (**b**) temperature contour diagram, (**c**) slice temperature diagram in *X* direction, (**d**) slice temperature diagram in *Z* direction.

4.4. Effect of Longitudinal Distance of the Vortex Generator

After selecting $45°$ as the better attack angle of the vortex generator, the best longitudinal distance and distribution mode are explored. In this part, two winglets of the vortex generator are symmetrically distributed, with a total of 12 different cases. Figure 7a–d shows the variation of the *R* of vortex generators with *FU, FD, FUD,* and *FDU* distribution modes at different longitudinal distances at different Re. Through comparing the performance of the *R*, it can be concluded that in the three distributions of *FU, FD,* and *FUD,* the performance of longitudinal distance of 4*h* is better than that of 2*h* and 6*h*. Especially in the two cases of *FU* and *FD,* the performance of the longitudinal distance of 4*h* is significantly better than that of 2*h* and 6*h* at small Re. However, in the case of *FDU* distribution, the performance of longitudinal distance of 4*h* is similar to that of 6*h*. Therefore, as shown in Figure 7e, comparing the five cases of *FU-4h, FD-4h, FUD-4h, FDU-4h,* and *FDU-6h,* the conclusion that when the Re is less than 4000, the comprehensive performance of *FD-4h* is the best, while when the Re is greater than 4000, *FU-4h* has a slight advantage over other cases can be summarized. Besides, according to Figure 8, it can be concluded that the thermal enhancement factor $\frac{\Delta T}{T_0}$ of *FD-4h* performs more prominent when the Re < 4000, while the *FU-4h* tends to be better when the Re > 4000, and the overall thermal performance can be improved by up to 10.2%.

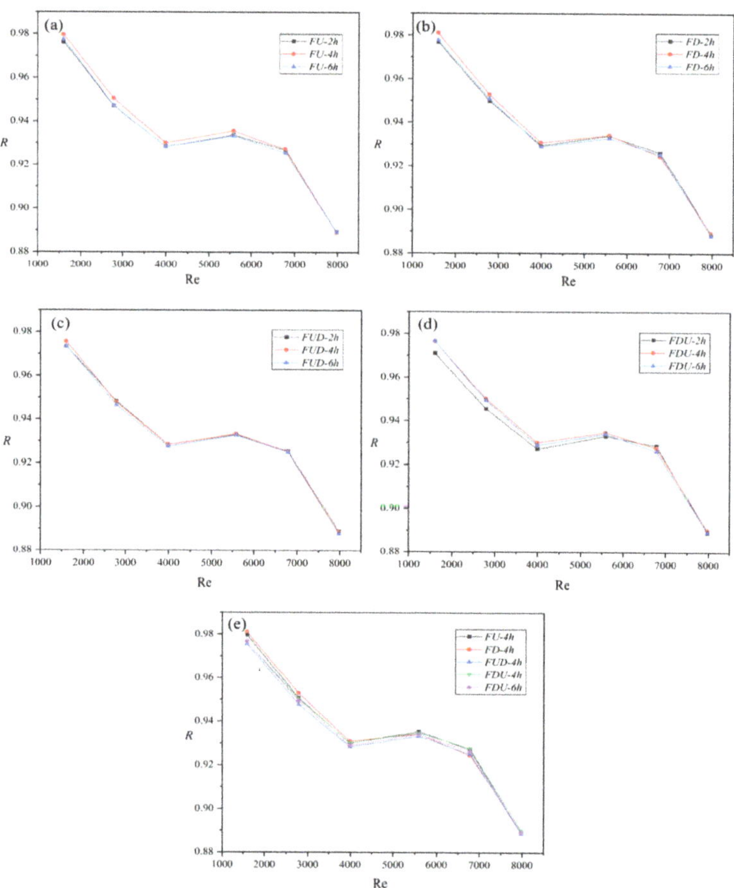

Figure 7. Performance of *R* at different longitudinal distances (2 *h*, 4 *h*, and 6 *h*) and distribution modes, (**a**) *FU*; (**b**)*FD*; (**c**) *FUD*; (**d**) *FDU* and (**e**) relatively better distribution mode.

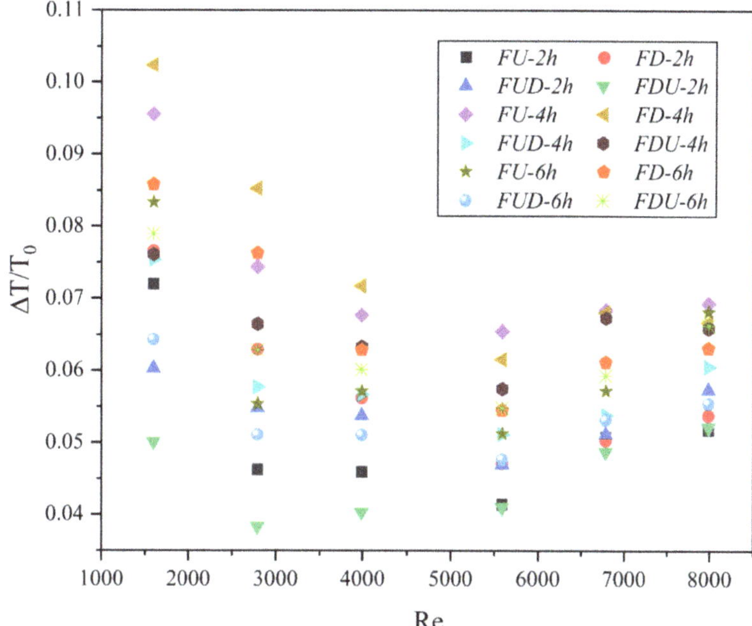

Figure 8. Performance of thermal enhancement factor $\frac{\Delta T}{T_0}$ at different Re.

Figures 9 and 10 show pictures of *FD-4h* and *FU-4h*. By observing Figures 9a and 10a, four vortexes are formed behind the vortex generator due to the installation of two pairs of vortex generators at this time. In *FD-4h*, the vortexes are created at the contracted two wings of the vortex generator and develop towards the center of the vortex generator, while in *FU-4h*, the vortexes are formed at the expanded two wings of the vortex generator and develop towards both sides of the vortex generator. Through analyzing Figure 9c,d and Figure 10c,d, it can be found that, in *FD-4h*, the flow velocity at the center of the vortex generator is relatively slow and the temperature is high, while in the vicinity of the center, due to the formation of the vortex, more energy loss and temperature drop are generated, whereas in *FU-4h*, the flow velocity at both sides of the vortex generator is relatively slow and the temperature is high, and the vortexes generated near both sides speeds up the heat dissipation and reduces the temperature. Moreover, due to the change of flow mode, the position of the highest temperature point on the section also changes. There are two symmetrical peaks in *Slice* Z_1 in Figure 10d, while the two peaks in Figure 9d are asymmetrical.

4.5. Effect of Transverse Distance of the Vortex Generator

Through the above research results, eight transverse distributions are explored based on *FD-4h* and *FU-4h*. Figure 11a,b illustrates the change of the *R* of *FD-4h* and *FU-4h* with different transverse distances under different Re. According to Figure 11a, compared with other transverse distances, *FD-4h* without transverse distance still performs best, while in Figure 11b, *FU-4h-0.5h* has slight advantages. Therefore, as shown in Figure 11c, by comparing *FD-4h* and *FU-4h-0.5h*, the results can be conducted that when the Re is less than 4000, the comprehensive performance of *FD-4h* is better, while when the Re is greater than 4000, the effect of *FU-4h-0.5h* is better than other situations. Furthermore, from Figure 12, the conclusion that when the Re < 4000, the thermal enhancement factor $\frac{\Delta T}{T_0}$ of *FD-4h* still performs better, while when the Re > 4000, the *FU-4h-0.5h* tends to have some gradual preponderance.

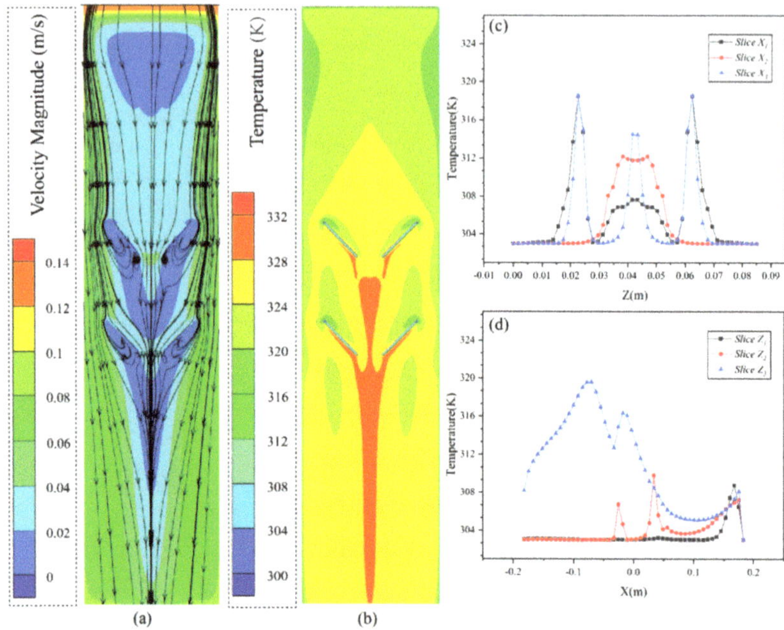

Figure 9. Diagrams of *FD*-4*h* under steady state conditions: (**a**) streamline diagram, (**b**) temperature contour diagram, (**c**) slice temperature diagram in *X* direction, (**d**) slice temperature diagram in *Z* direction.

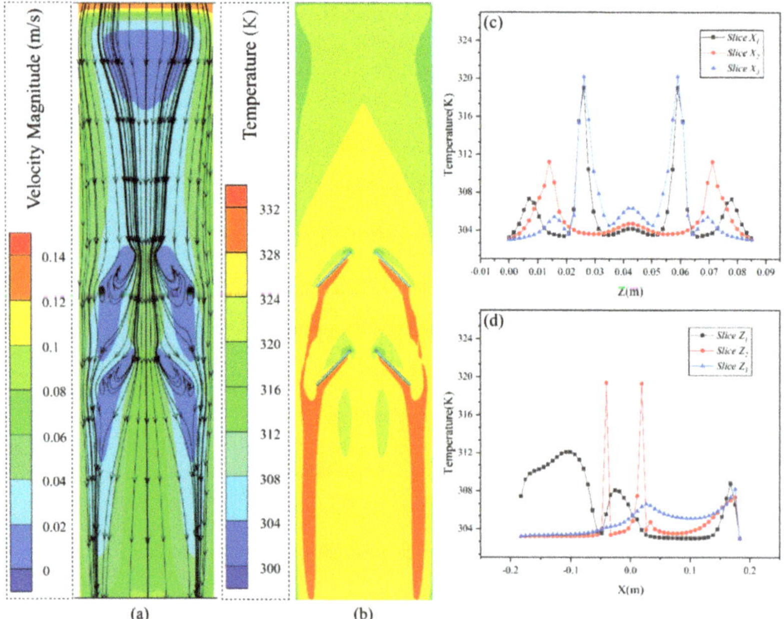

Figure 10. Diagrams of *FU*-4*h* under steady state conditions: (**a**) streamline diagram, (**b**) temperature contour diagram, (**c**) slice temperature diagram in *X* direction, (**d**) slice temperature diagram in *Z* direction.

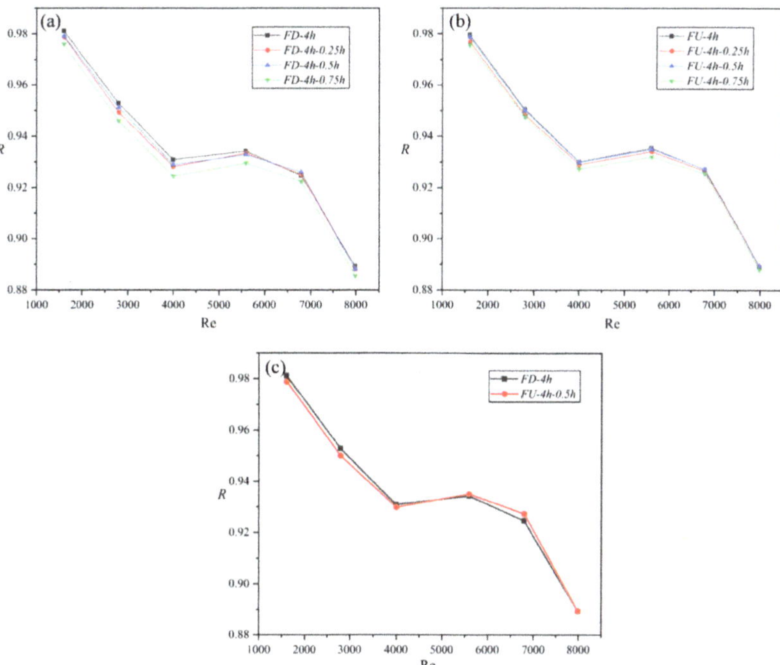

Figure 11. Performance of *R* at different transverse distances (0.25 *h*, 0.5 *h*, and 0.75 *h*) and distribution modes, (**a**) *FD-4h*; (**b**) *FU-4h*; (**c**) relatively better distribution mode.

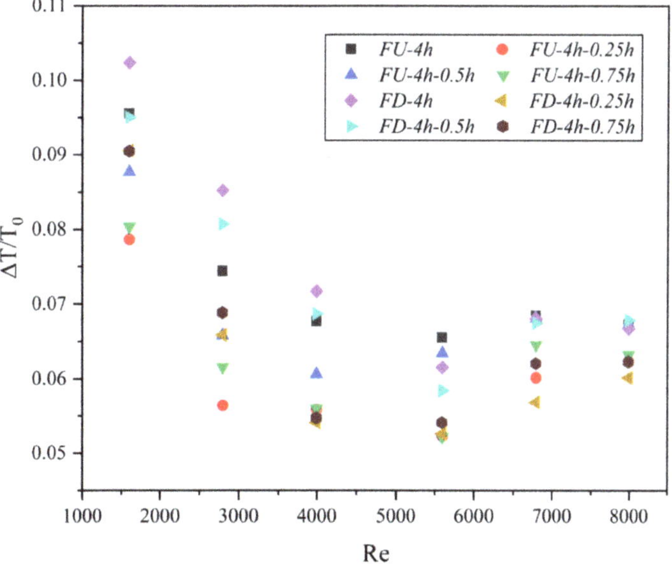

Figure 12. Performance of thermal enhancement factor $\frac{\Delta T}{T_0}$ at different Re.

Since the *FD-4h* situation has been analyzed above, there will be no more details here, and only the *FU-4h*-0.5*h* situation will be discussed. Figure 13a shows the flow line diagram of *FU-4h-0.5h*, from which it can be found that the flow line in the center of the vortex generator is no longer a straight line, but a curve with secondary disturbance on the left side of the second pair of vortex generators. In addition, the vortexes generated on the right side of the two pairs of vortex generators have a superposition effect, which impacts the heat transfer effect. The temperature contour diagram of Figure 13b and the slice diagrams of Figure 13c,d verify the analysis of the streamline in Figure 13a, the lower temperatures at the left wing of the second pair of vortex generators and the rear area of the second pair of vortex generators can be obtained. In addition, compared with Figure 10, due to the influence of transverse distance, the maximum temperature in Slice X_2 in Figure 13c is no longer symmetrically distributed, but the temperature is higher on one side and lower on the other.

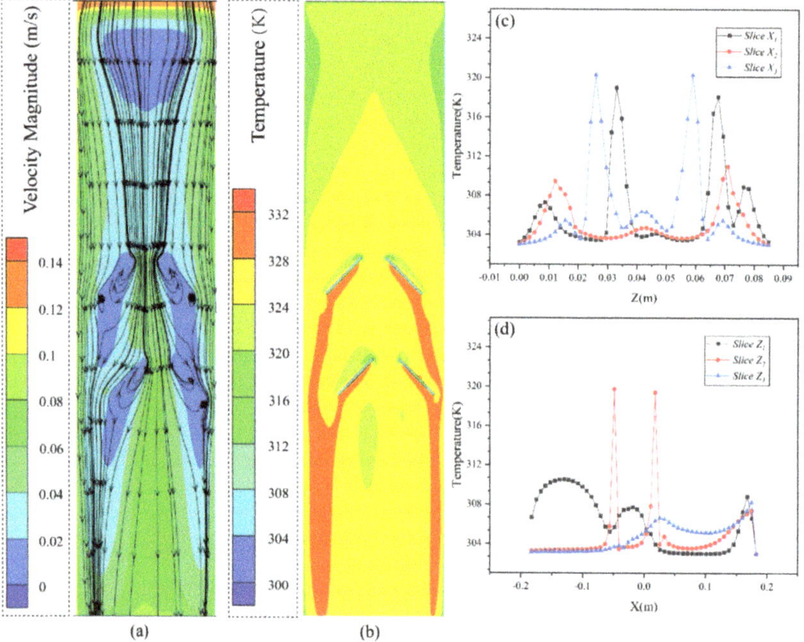

Figure 13. Diagrams of $FU\text{-}4h - 0.5h$ under steady state conditions, (**a**) streamline diagram, (**b**) temperature contour diagram, (**c**) slice temperature diagram in X direction, (**d**) slice temperature diagram in Z direction.

5. Conclusions

Equipped with vortex generators, the coil end aims to solve the local overheating of the oil-cooled motor, so as to increase the efficiency of the oil-cooled motor. The oil phase is used as the coolant in the fluid domain. Firstly, the numerical model is verified and compared with the experimental data in the references. On this basis, the improved dimensionless number R is used to optimize the attack angle, longitudinal distance, and transverse distance of the vortex generator. The main conclusions are shown below:

1. Compared with the vortex generators with attack angles of $15°$ and $60°$, the vortex generators with attack angles of $30°$ and $45°$ have better enhanced heat transfer effect. The performances of $30°$ and $45°$ attack angles are almost the same. Nevertheless, from the point of view of the maximum temperature, the effect of the $45°$ vortex generator is better than that of the $30°$ vortex generator.
2. The effect of vortex generator with the longitudinal distance of $4h$ is better than that of $2h$ and $6h$. When the Reynolds number is less than 4000, the enhanced heat transfer

performance of *FD* with a longitudinal distance of 4 h (*FD-4h*) is more prominent, and when the Reynolds number is greater than 4000, the performance of *FU* with a longitudinal distance of 4 h (*FU-4h*) is superior.

3. Through the numerical simulation of the transverse distance of *FU-4h* and *FD-4h*, it can be concluded that, when the Reynolds number is less than 4000, the enhanced heat transfer performance of *FD-4h* is extraordinary, and when the Reynolds number is greater than 4000, the improved heat transfer performance of *FU* with a longitudinal distance of 4 h and transverse distance of 0.5 h (*FU-4h* − *0.5h*) is better.

Author Contributions: Conceptualization, J.Z. and B.Z.; methodology, J.Z. and B.Z.; software, J.Z., S.Y. and B.Z.; validation, J.Z. and B.Z.; formal analysis, J.Z. and B.Z.; investigation, J.Z. and B.Z.; resources, B.Z., S.Y. and X.F.; data curation, J.Z. and B.Z.; writing—original draft preparation, J.Z.; writing—review and editing, J.Z., B.Z., S.Y. and X.F.; visualization, J.Z.; supervision, X.F.; project administration, X.F.; funding acquisition, X.F. All authors have read and agreed to the published version of the manuscript.

Funding: This research was funded by the National Natural Science Foundation of China, grant number 51879191.

Institutional Review Board Statement: Not applicable.

Informed Consent Statement: Not applicable.

Data Availability Statement: The datasets used and analyzed during the current study are available from the corresponding author on reasonable request.

Acknowledgments: Not applicable.

Conflicts of Interest: The authors declare no conflict of interest.

Nomenclature

Nu	Nusselt number
Re	Reynolds number
Pr	Prandtl number
j	Colburn factor
j_0	Colburn factor of smooth channel
f	Darcy friction factor
f_0	Darcy friction factor of smooth channel
h_c	Convective heat transfer coefficient $(\mathrm{Wm^{-2}K^{-1}})$
D	Hydraulic diameter (mm)
u	Fluid velocity (m/s)
C_p	Specific heat capacity $(\mathrm{Jkg^{-1}K^{-1}})$
P	Pressure (Pa)
t	Time (s)
T	Temperature (K)
X, Y, Z	Cartesian coordinates
G	Turbulence generation term
C_1, C_2	Constant number
k	Turbulent kinetic energy
Greek letters	
λ	Thermal conductivity $(\mathrm{Wm^{-1}K^{-1}})$
λ_o	Oil Thermal conductivity $(\mathrm{Wm^{-1}K^{-1}})$
ρ	Density $(\mathrm{kg/m^3})$
ρ_o	Oil density $(\mathrm{kg/m^3})$
μ	Dynamic viscosity
ε	Turbulent Dissipation Rate

μ_t	Turbulent viscosity coefficient
$\sigma_k, \sigma_\varepsilon$	Constant number
Subscripts	
i, w	i, w = 1, 2, and 3, respectively, represent the components along the X, Y and Z axes

References

1. Gundabattini, E.; Mystkowski, A.; Idzkowski, A.; Singh, R.R.; Solomon, D.G. Thermal Mapping of a High-Speed Electric Motor Used for Traction Applications and Analysis of Various Cooling Methods-A Review. *Energies* **2021**, *14*, 1472. [CrossRef]
2. Divakaran, A.M.; Hamilton, D.; Manjunatha, K.N.; Minakshi, M. Design, Development and Thermal Analysis of Reusable Li-Ion Battery Module for Future Mobile and Stationary Applications. *Energies* **2020**, *13*, 1477. [CrossRef]
3. Ha, T.; Han, N.G.; Kim, M.S.; Rho, K.H.; Kim, D.K. Experimental Study on Behavior of Coolants, Particularly the Oil-Cooling Method, in Electric Vehicle Motors Using Hairpin Winding. *Energies* **2021**, *14*, 956. [CrossRef]
4. Guo, F.L.; Zhang, C.N. Oil-Cooling Method of the Permanent Magnet Synchronous Motor for Electric Vehicle. *Energies* **2019**, *12*, 2984. [CrossRef]
5. Lehmann, R.; Petuchow, A.; Moullion, M.; Kunzler, M.; Windel, C.; Gauterin, F. Fluid Choice Based on Thermal Model and Performance Testing for Direct Cooled Electric Drive. *Energies* **2020**, *13*, 5867. [CrossRef]
6. Srinivasan, C.; Yang, X.; Schlautman, J.; Wang, D.; Gangaraj, S. Conjugate Heat Transfer CFD Analysis of an Oil Cooled Automotive Electrical Motor. *SAE Int. J. Adv. Curr. Pract. Mobil.* **2020**, *2*, 1741–1753. [CrossRef]
7. Davin, T.; Pelle, J.; Harmand, S.; Yu, R. Experimental study of oil cooling systems for electric motors. *Appl. Therm. Eng.* **2015**, *75*, 1–13. [CrossRef]
8. Saadi, M.S.; Ismail, M.; Fotowat, S.; Quaiyum, M.; Fartaj, A. Study of Motor Oil Cooling at Low Reynolds Number in Multi- Port Narrow Channels. *SAE Int. J. Engines* **2013**, *6*, 1287–1298. [CrossRef]
9. Liu, S.; Sakr, M. A comprehensive review on passive heat transfer enhancements in pipe exchangers. *Renew. Sust. Energy Rev.* **2013**, *19*, 64–81. [CrossRef]
10. He, Y.L.; Zhang, Y.W. Advances and Outlooks of Heat Transfer Enhancement by Longitudinal Vortex Generators. *Adv. Heat. Transfer.* **2012**, *44*, 119–185. [CrossRef]
11. Schubauer, G.B.; Spangenberg, W.G. Forced mixing in boundary layers. *J. Fluid Mech.* **1960**, *8*, 10–32. [CrossRef]
12. Johnson, T.; Joubert, P. The Influence of Vortex Generators on the Drag and Heat Transfer From a Circular Cylinder Normal to an Airstream. *J. Heat Transf.* **1969**, *91*, 1969. [CrossRef]
13. Chai, L.; Tassou, S.A. A Review of Airside Heat Transfer Augmentation with Vortex Generators on Heat Transfer Surface. *Energies* **2018**, *11*, 2737. [CrossRef]
14. Zhou, G.B.; Feng, Z.Z. Experimental investigations of heat transfer enhancement by plane and curved winglet type vortex generators with punched holes. *Int. J. Therm. Sci.* **2014**, *78*, 26–35. [CrossRef]
15. Wu, J.M.; Tao, W.Q. Effect of longitudinal vortex generator on heat transfer in rectangular channels. *Appl. Therm. Eng.* **2012**, *37*, 67–72. [CrossRef]
16. Aris, M.S.; McGlen, R.; Owen, I.; Sutcliffe, C.J. An experimental investigation into the deployment of 3-D, finned wing and shape memory alloy vortex generators in a forced air convection heat pipe fin stack. *Appl. Therm. Eng.* **2011**, *31*, 2230–2240. [CrossRef]
17. Wu, J.M.; Tao, W.Q. Numerical study on laminar convection heat transfer in a rectangular channel with longitudinal vortex generator. Part A: Verification of field synergy principle. *Int. J. Heat Mass. Trans.* **2008**, *51*, 1179–1191. [CrossRef]
18. Promvonge, P.; Chompookham, T.; Kwankaomeng, S.; Thianpong, C. Enhanced heat transfer in a triangular ribbed channel with longitudinal vortex generators. *Energy Convers. Manag.* **2010**, *51*, 1242–1249. [CrossRef]
19. Chen, C.; Teng, J.T.; Cheng, C.H.; Jin, S.P.; Huang, S.Y.; Liu, C.; Lee, M.T.; Pan, H.H.; Greif, R. A study on fluid flow and heat transfer in rectangular microchannels with various longitudinal vortex generators. *Int. J. Heat Mass. Trans.* **2014**, *69*, 203–214. [CrossRef]
20. Saha, P.; Biswas, G.; Sarkar, S. Comparison of winglet-type vortex generators periodically deployed in a plate-fin heat exchanger—A synergy based analysis. *Int. J. Heat Mass. Trans.* **2014**, *74*, 292–305. [CrossRef]
21. Sinha, A.; Raman, K.A.; Chattopadhyay, H.; Biswas, G. Effects of different orientations of winglet arrays on the performance of plate-fin heat exchangers. *Int. J. Heat Mass. Trans.* **2013**, *57*, 202–214. [CrossRef]
22. Min, C.H.; Qi, C.Y.; Kong, X.F.; Dong, J.F. Experimental study of rectangular channel with modified rectangular longitudinal vortex generators. *Int. J. Heat Mass. Trans.* **2010**, *53*, 3023–3029. [CrossRef]
23. Eiamsa-ard, S.; Wongcharee, K.; Eiamsa-ard, P.; Thianpong, C. Heat transfer enhancement in a tube using delta-winglet twisted tape inserts. *Appl. Therm. Eng.* **2010**, *30*, 310–318. [CrossRef]
24. Tian, L.T.; He, Y.L.; Lei, Y.G.; Tao, W.Q. Numerical study of fluid flow and heat transfer in a flat-plate channel with longitudinal vortex generators by applying field synergy principle analysis. *Int. Commun. Heat Mass.* **2009**, *36*, 111–120. [CrossRef]
25. Kim, E.; Yang, J.S. An experimental study of heat transfer characteristics of a pair of longitudinal vortices using color capturing technique. *Int. J. Heat Mass. Trans.* **2002**, *45*, 3349–3356. [CrossRef]
26. Wijayanta, A.T.; Aziz, M.; Kariya, K.; Miyara, A. Numerical Study of Heat Transfer Enhancement of Internal Flow Using Double-Sided Delta-Winglet Tape Insert. *Energies* **2018**, *11*, 3170. [CrossRef]

27. Zhang, J.F.; Jia, L.; Yang, W.W.; Taler, J.; Oclon, P. Numerical analysis and parametric optimization on flow and heat transfer of a microchannel with longitudinal vortex generators. *Int. J. Therm. Sci.* **2019**, *141*, 211–221. [CrossRef]
28. Zhang, J.F.; Wang, J.S.; Sun, J. Principles and Characteristics of Heat Transfer Enhancement on Small-scale Vortex Generators. *Energy Conserv. Technol.* **2006**, *24*, 399–401.
29. Ebrahimi, A.; Rikhtegar, F.; Sabaghan, A.; Roohi, E. Heat transfer and entropy generation in a microchannel with longitudinal vortex generators using nanofluids. *Energy* **2016**, *101*, 190–201. [CrossRef]
30. Ebrahimi, A.; Roohi, E.; Kheradmand, S. Numerical study of liquid flow and heat transfer in rectangular microchannel with longitudinal vortex generators. *Appl. Therm. Eng.* **2015**, *78*, 576–583. [CrossRef]
31. Minakshi, M.; Blackford, M.; Ionescu, M. Characterization of alkaline-earth oxide additions to the MnO_2 cathode in an aqueous secondary battery. *J. Alloys Compd.* **2011**, *509*, 5974–5980. [CrossRef]
32. Wickramaarachchi, K.; Sundaram, M.M.; Henry, D.J.; Gao, X.P. Alginate Biopolymer Effect on the Electrodeposition of Manganese Dioxide on Electrodes for Supercapacitors. *ACS Appl. Energy Mater.* **2021**, *4*, 7040–7051. [CrossRef]
33. Pugachev, A.O.; Ravikovich, Y.A.; Savin, L.A. Flow structure in a short chamber of a labyrinth seal with a backward-facing step. *Comput. Fluids* **2015**, *114*, 39–47. [CrossRef]
34. Ma, J.; Huang, Y.P.; Huang, J.; Wang, Y.L.; Wang, Q.W. Experimental investigations on single-phase heat transfer enhancement with longitudinal vortices in narrow rectangular channel. *Nucl. Eng. Des.* **2010**, *240*, 92–102. [CrossRef]
35. Lotfi, B.; Sunden, B.; Wang, Q.W. An investigation of the thermo-hydraulic performance of the smooth wavy fin-and-elliptical tube heat exchangers utilizing new type vortex generators. *Appl. Energ* **2016**, *162*, 1282–1302. [CrossRef]
36. Gholami, A.A.; Wahid, M.A.; Mohammed, H.A. Heat transfer enhancement and pressure drop for fin-and-tube compact heat exchangers with wavy rectangular winglet-type vortex generators. *Int. Commun. Heat Mass.* **2014**, *54*, 132–140. [CrossRef]

*energies*

Article

Thermal Calculations of Four-Row Plate-Fin and Tube Heat Exchanger Taking into Account Different Air-Side Correlations on Individual Rows of Tubes for Low Reynold Numbers

Mateusz Marcinkowski *, Dawid Taler, Jan Taler and Katarzyna Węglarz

Faculty of Environmental Engineering and Energy, Cracow University of Technology, 31-155 Cracow, Poland; dawid.taler@pk.edu.pl (D.T.); jan.taler@pk.edu.pl (J.T.); katarzyna.weglarz@doktorant.pk.edu.pl (K.W.)
* Correspondence: mateusz.marcinkowski@doktorant.pk.edu.pl

Abstract: Currently, when designing plate-fin and tube heat exchangers, only the average value of the heat transfer coefficient (HTC) is considered. However, each row of the heat exchanger (HEX) has different hydraulic–thermal characteristics. When the air velocity upstream of the HEX is lower than approximately 3 m/s, the exchanged heat flow rates at the first rows of tubes are higher than the average value for the entire HEX. The heat flow rate transferred in the first rows of tubes can reach up to 65% of the heat output of the entire exchanger. This article presents the method of determination of the individual correlations for the air-side Nusselt numbers on each row of tubes for a four-row finned HEX with continuous flat fins and round tubes in a staggered tube layout. The method was built based on CFD modelling using the numerical model of the designed HEX. Mass average temperatures for each row were simulated for over a dozen different airflow velocities from 0.3 m/s to 2.5 m/s. The correlations for the air-side Nusselt number on individual rows of tubes were determined using the least-squares method with a 95% confidence interval. The obtained correlations for the air-side Nusselt number on individual rows of tubes will enable the selection of the optimum number of tube rows for a given heat output of the HEX. The investment costs of the HEX can be reduced by decreasing the tube row number. Moreover, the operating costs of the HEX can also be lowered, as the air pressure losses on the HEX will be lower, which in turn enables the reduction in the air fan power.

Keywords: plate-fin and tube heat exchanger; air-side Nusselt number; different heat transfer coefficient in particular tube row; mathematical simulation; numerical simulation; CFD simulation

Citation: Marcinkowski, M.; Taler, D.; Taler, J.; Węglarz, K. Thermal Calculations of Four-Row Plate-Fin and Tube Heat Exchanger Taking into Account Different Air-Side Correlations on Individual Rows of Tubes for Low Reynold Numbers. *Energies* **2021**, *14*, 6978. https://doi.org/10.3390/en14216978

Academic Editor: Artur Bartosik

Received: 27 September 2021
Accepted: 20 October 2021
Published: 25 October 2021

Publisher's Note: MDPI stays neutral with regard to jurisdictional claims in published maps and institutional affiliations.

1. Introduction

Plate and fin heat exchangers (PFTHEs) are complicated devices. Their construction, which involves cross thin metal sheets and tubes with different numbers of rows, different fin pitches, and tube sizes, causes complicated phenomena on the air- and/or water-side [1]. Each row in PFTHEs operates in a different way. Differences are caused by variations in air and water temperatures, airflow turbulence, or even air velocity. Those elements cause different HTCs in each row [2].

Performance characteristics of tube heat exchangers are most often determined experimentally. This is usually due to the high complexity of the systems studied. An example is the experimental testing of photovoltaic (PV) cells cooled by phase-change materials (PCM) [3]. Fouling phenomena in exchangers are also investigated experimentally. Ali et al. [4] investigated the influence of dust deposited on the surface of two types of PVs, monocrystalline silicon, and polycrystalline silicon modules. Experimental studies are also widely used to determine the flow and heat characteristics of finned tube heat exchangers. Heat transfer correlations are determined on the side of the flowing gas, which is usually air, in a wide range of Reynolds number variations [5]. Additionally, the effectiveness of various types of improvements in the design of heat exchangers, such as oval tubes [6], new fin shapes, or guide vanes forming the air flow in the exchanger [5], is usually evaluated experimentally. Despite the high reliability of the obtained results, a disadvantage is the cost of experimental studies. To determine the experimental characteristics of an exchanger with a different number of rows, tubes, or different construction, it is necessary to build an experimental test facility equipped with a measuring apparatus and a computer data acquisition system. Moreover, the time and therefore costs of experimental investigations are high. Therefore, CFD modelling is increasingly used in the development of tubular cross-flow heat exchangers [7]. The influence of various innovations in the design of the exchanger is modelled in different ranges of Reynolds numbers on the air-side and on the fluid-side flowing inside the tube. Four round-convex strips were placed around the tube to enhance air-side heat transfer [8].

If the experimental testing cannot be fully replaced, then the experiment comes down to verification of a certain part of the results [9]. The CFD modelling gives us increased flexibility in the research and industry. This study shows the newly determined correlations of the air-side Nusselt number for the four-row PFTHE. Those correlations illustrate new functions to determine HTC in a particular row of PFTHE. Individual correlations can also change the way we look at the heat flow inside PFTHE.

Until recently, plenty of Nusselt number, Colburn factor, or HTC correlations were determined. However, the majority are referred to as average correlations. Some studies presented local HTCs and only a few of them showed results or determined Nusselt number correlations for a particular row in multi-row PFTHEs.

The first group of research overviews is about average HTCs in PFTHEs. Khan et al. [10] investigated twisted oval tube HEXs and designated the average Nusselt numbers and pressure drop correlations. Lindqvist et al. [11] conducted CFD research taking into account different tube bundle array angles. They also presented diagrams with the Colburn factors for low Reynold numbers. Łęcki et al. [12] presented a comparison between HTCs obtained using CFD simulation to the one calculated with VDI correlation for three-row inline PFTHE. Sadeghianjahromi et al. [13] determined HTCs and pressure drop factors of PFTHEs with different fin types and circular and flat tubes under dry and wet conditions. Elmekawy et al. [14] showed that attaching the splitter plates to the tubes could increase the Nusselt number and reduce the pressure drop. Okbaz et al. [15] presented different Colburn factors correlations for PFTHEs with different numbers of rows. However, those correlations were averaged for entire PFTHEs. Petrik and Szepesi [16] determined Nusselt number correlations for one and two-row U-shape HEXs. Additionally in another piece of research, Petrik et al. [17] presented Nusselt number correlations but for standard PFTHEs. González et al. [18] presented the average Nusselt numbers as a function of fin material and Reynolds number for two-row PFTHE with inline tube arrangement. Awais and Bhuiyan [19] collected plenty of Colburn factor correlations in their state-of-the art paper. It is possible to present many more average correlations. However, it is not necessary for this research. Adam et al. [20], in a state-of-the-art paper, shows studies of local HTCs and local heat/mass transfers coefficients.

All of the above research is related to the average Nusselt numbers, Colburn factors, or HTC correlations, or presents local HTCs to predict heat transfer correlations for the entire PFTHE. There are only a couple instances of research referring to heat transfer correlations for a particular row in PFTHEs. During the 1970s, Rich et al. [21] showed experimental research where a Colburn factor in a particular row for multi-row PFTHEs, with one-row to eight-rows, was presented. This was the first time when someone showed variation in the HTC inside PFTHEs. Taler et al. [2] determined individual Nusselt number correlations for each row in the case of two-row PFTHE. Despite its high practical importance and current marginal existence in the literature, there is a lack of heat transfer correlations on individual tube rows for heat exchangers with different number of tube rows. Additionally, the literature is full of average Nusselt number, Colburn factor, or HTC correlations for different fin geometries, tube geometries, fin pitches, tube sizes, air velocities, etc.

Only Rich et al. [21] showed that further tube rows in multi-row PFTHEs are inefficient when the air velocity is low and the airflow in the exchanger is laminar. However, correlations on Nusselt number for individual tube rows have not been developed. Taler et al. [22] showed that in a two-row automotive radiator, made with round or oval tubes, the first row of tubes is more efficient than the second row when the air velocity is less than 2.5 m/s. However, coolers with larger tube numbers have not been modelled by CFD or studied experimentally. There are many problems to solve, such as: whether a larger cross-section of a PFTHE, but with a smaller number of rows, would be more efficient than a three-, four-, or five-row PFTHE. Which row is the least effective? Should we consider individual correlations in the case of multi-row PFTHEs?

The present research refers to four-row PFTHE with the air velocity before the HEX changing in the range from 0.5 m/s to 2.5 m/s. This paper covers:

- Determination of individual air-side Nusselt number correlations by CFD simulation;
- Determination of average air-side Nusselt number correlations by CFD simulation;
- Comparison of determined correlations to correlations available in the literature.

2. Problem Statement

This paper presents a three-dimensional (3D) steady-state CDF simulation of PFTHE under the following assumptions:

- Constant air inlet temperature: 20 °C.
- Air outlet as an open condition with checking the average mass temperature in the air outlet cross-section.
- Constant air-side wall surface temperature (including tube and fins temperature): 70 °C.
- Variable air parameters dependent on the temperature.
- Shear Stress Transport (SST) turbulence model.
- Relative differences (e_{T_n}) between different mass average air temperatures in the last (the fourth) row of PFTHE for mesh with particular element numbers (T_{ni}) and the reference mass average temperature in the last (the fourth) row of PFTHE for mesh with 6,790,572 mesh elements ($T_{6,790,572}$) are shown in Equation (1). The above reference mesh has been selected for further calculations as the best ratio of calculation accuracy to calculation time.

$$e_{T_n} = \frac{(T_{ni} - T_{6,790,572})}{T_{6,790,572}} \tag{1}$$

- Relative differences (e_{T_w}) between the Nusselt number for a particular fin and tube surface temperature (Nu_{T_w}) and Nusselt number for the reference fin and tube surface temperature: 70 °C ($Nu_{70\,°C}$) as calculated according to Equation (2).

$$e_{T_w} = \frac{(Nu_{T_w} - Nu_{70\,°C})}{Nu_{70\,°C}} \tag{2}$$

3. Description of the Geometry of the Repeating Element of the Exchanger

Figure 1 presents a 3D view and geometry of analysed PFTHE. Directions of air and water flow are shown. Figure 1b illustrates a repetitive fragment of air between one fin pitch in PFTHE (grey colour). In this case, fin pitch equals 3 mm. However, the distance between surfaces of the adjacent plate fins is 2.84 mm. This is because the fin thickness is 0.14 mm. In the same picture, we can observe yellow colours in front and behind the PFTHE. Those fragments are free zones of air for the correct performance of CFD simulation. Each zone is 5 mm in length (Figure 1b). The diameter of the tubes is 12mm. Air Inlet in CFX software is set as a constant temperature: 20 °C. Air outlet is set as an opening condition with checking mass average temperature in the outlet cross section of air. The rest of the boundary conditions of the CFX software are shown in Figure 2a.

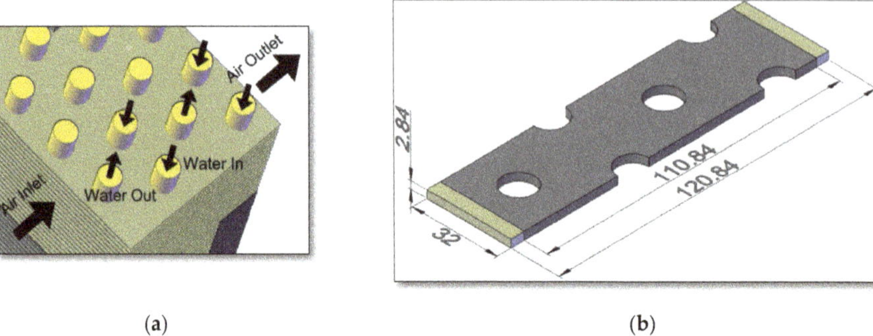

(a) (b)

Figure 1. (**a**) The analysed PFTHE: 3D view; (**b**) Repetitive fragment of the PFTHE with run-up zone using in ANSYS-CFX 2020 R2: geometry, mm.

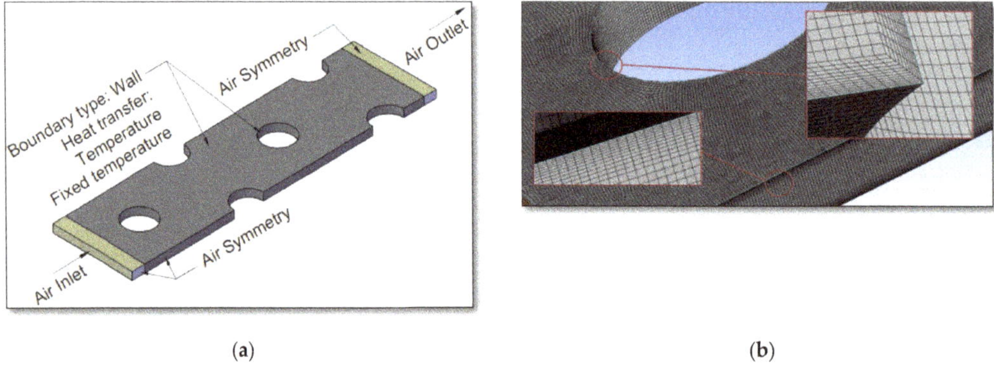

(a) (b)

Figure 2. Repetitive fragment of the PFTHE with run-up zone using in ANSYS-CFX 2020 R2: (**a**) boundary conditions; (**b**) finite element mesh.

4. Mesh Parameters and Mesh Independent Study

Mesh characteristics are presented in Table 1. The division of the modelled domain into finite elements considers mesh compaction at the junction of the air and fin wall surface and air and tube wall surface (Figure 2b).

Table 1. Specification of the finite element mesh for the model presented in Figure 2b.

Number of Finite Elements	6,790,572
Number of nodes	6,989,646
The dimension of the element, mm	0.15
The maximal dimension of the element, mm	0.25
Boundary layer	First layer thickness: 0.023 mm $(y^+ = 1)$ Growth rate: 1.1 Number of layers: 12
Minimum orthogonal angle	Air: 36°
Mesh expansion factor	Air: 12
Maximum aspect ratio	Air: 20

All simulations for the entire range of air velocity were preceded by mesh independent studies, separately for all considered air velocities. The final mesh was selected only for the most favourable simulation. Temperature stabilisation for air velocity—4 m/s for the fourth row of PFTHE—is shown in Figure 3. It was observed that, from air velocity in front of PFTHE, more than 2.5 m/s temperature stabilisation is not much different in comparison with temperature stabilisation in the fourth (last) row of PFTHE. Temperature stabilisation showed that, between chosen mesh element size and mesh with almost 30 million elements, relative differences were less than 1% for the fourth row of PFTHE (Figure 3). Moreover, for the first row, even for 10 m/s relative differences between chosen mesh and mesh with over 30 million, elements are less than 0.1% (Equation (1). This shows that, in the considered scenario, the greatest irregularity of flow and turbulence exists in the last row of PFTHE.

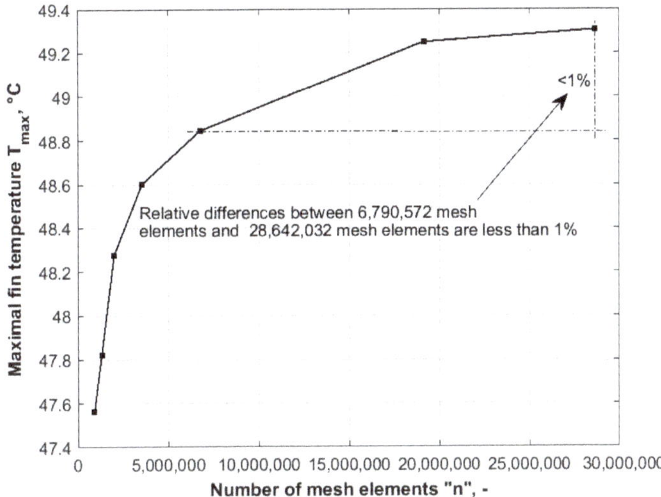

Figure 3. Temperature stabilisation for mesh independent study from the fourth row of PFTHE for 4 m/s air velocity in front of the heat exchanger.

A mesh with 6,790,572 elements was selected. Mesh elements' size belongs to the following CFX mesh sizes: mesh element size: 0.15 mm; mesh max size: 0.25 mm (Table 1). The whole mesh contains cuboidal elements only (Figure 1b).

5. Method of Determining HTC on the Individual Row Based on CFD Simulations

A method for determining HTCs for the individual rows is illustrated with an example of four-row PFTHE (Figure 4). The HTC is appointed, factoring constant fin and tube temperatures, which are equal to 70 °C Further simulations have also been performed for two different temperatures: 60 °C and 80 °C. It has been shown that, for three different surface temperatures of the tubes and fins, the determined HTCs do not differ more than 1.5% (Figure 5), and that proposed procedure for determining the air-side HTC can be successfully used in practice. The method consists of the following steps.

- Export the average mass flow temperature $\overline{T}^i_a$ from CFX simulations of considered air velocity ranging for each row independently. Cross-section to export temperatures in a half distance between the next two tubes in rows.

- Calculate the mean logarithmic temperature ($\Delta \overline{T}_{m,a}$) difference between fin and tube surface temperature (T_w) and average mass flow temperature ($\overline{T}^i_a$) (Equation (3).

$$\Delta T^i_{m,a} = \frac{\left(T_w - \overline{T}^{i+1}_a \right)\left(T_w - \overline{T}^i_a \right)}{ln\left(\frac{T_w - \overline{T}^{i+1}_a}{T_w - \overline{T}^i_a} \right)} \tag{3}$$

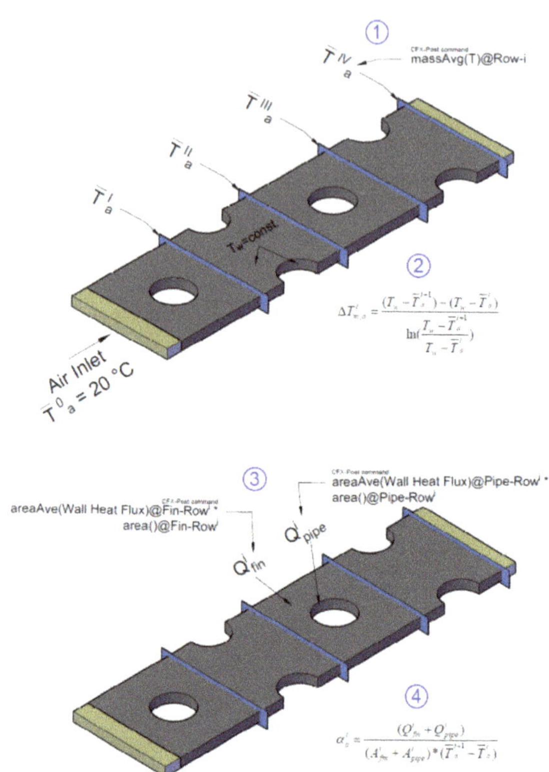

Figure 4. Graphical representation of determining HTC on the individual row based on a CFD simulation. The numbers on the picture show the order of calculation to determine HTC in the case of PFTHE.

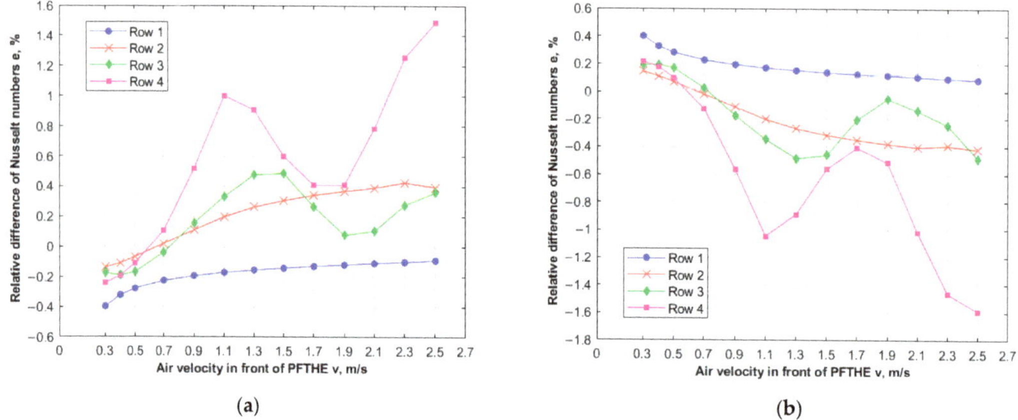

Figure 5. Relative differences of Nusselt numbers for each row of PFTHE for different air wall temperatures (constant temperature of fin and tube wall), where reference temperature is 70 °C: (**a**) 70 °C and 80 °C; (**b**) 70 °C and 60 °C.

The symbol T_w denotes a constant fin and tube surface temperature. The symbols $\overline{T}_a^i$ and $\overline{T}_a^{i+1}$ designate the inlet and outlet mass average air temperature for the *i*-th row of PFTHE. Then:

- Export total heat flow rate transferred from fin and tube wall surface to air (Figure 4).
- Calculate individual HTC for each row (Equation (4) (Figure 4)).

$$\alpha_a^i = Q_i / \left(A_w \Delta T_{m,a}^i \right) \tag{4}$$

The following designations are used in Equation (4): Q_i the total heat flow rate for the entire row of the repetitive fragment of PFTHE; α_a^i the air-side HTC in a particular row.

- Calculate Nusselt number $Nu_a^i = (\alpha_a^i\, d_h) / \lambda_a$ and Colburn factor $j_a^i = Nu_a^i / \left(Re_a^i\, Pr_a^{i\,1/3} \right)$.

6. CFD Simulation Results and Nusselt Number Correlations

The CFD simulation results of average mass flow temperatures behind the tube rows are shown in Table 2.

Table 2. CFD simulation data.

i	$w_{o,i}$, **m/s**	$T^I_{a,i}$, °C	$T^{II}_{a,i}$, °C	$T^{III}_{a,i}$, °C	$T^{IV}_{a,i}$, °C
1	0.3	59.95	67.67	69.42	69.85
2	0.4	55.55	65.13	68.23	69.34
3	0.5	52.06	62.51	66.62	68.44
4	0.7	47.07	57.91	63.10	66.03
5	0.9	43.72	54.37	59.97	63.59
6	1.1	41.33	51.69	57.45	61.63
7	1.3	39.53	49.60	55.46	60.08
8	1.5	38.11	47.94	53.87	58.71
9	1.7	36.96	46.57	52.51	57.43
10	1.9	36.01	45.41	51.27	56.25
11	2.1	35.21	44.41	50.17	55.21
12	2.3	34.52	43.54	49.21	54.34
13	2.5	33.92	42.76	48.35	53.58

Presented correlations for air-side Nusselt numbers relate to Reynolds numbers. The Reynolds number (Equation (5) is related to the hydraulic diameter, which is calculated using the definition proposed by Kays and London [23], where w_{max} is given by Equation (6). Parameter w_{max} has been calculated for the minimum airflow cross-section between tubes and it can exist in a different place in case of different PFTHE construction. The hydraulic diameter has been calculated by assumption dividing the volume through which air flows in one row by the surface area in contact with the air. Most often, this hydraulic diameter in the case of PFTHE is a bit smaller than double the fin pitch, and using geometry from Figure 1, equals 5.35 mm.

Reynolds numbers ($Re_{d_h,a}$) (Equation (5) and maximal velocities (w_{max}) (Equation (6)) have been calculated for each row separately, due to a different maximum air velocity caused by higher air temperature causing increased air volume. The calculated parameters such as Reynolds numbers, HTCs, Nusselt numbers, and Colburn factors are presented in Tables 3–6.

$$Re_{d_h,a} = \frac{w_{max} \cdot d_h}{\nu_a} \tag{5}$$

Table 3. Calculated thermal parameters for the first row of PFTHE.

i	$w_{o,i}$, m/s	$w_{max,i}$, m/s	$Re^{I}_{a,i}$	$\alpha^{I}_{a,i}$	$Nu^{I}_{a,i}$	$j^{I}_{a,i}$
1	0.3	0.54	169.6	17.07	3.33	0.022647
2	0.4	0.71	227.4	17.57	3.45	0.017474
3	0.5	0.89	285.4	18.14	3.58	0.014413
4	0.7	1.23	402.0	19.32	3.84	0.010949
5	0.9	1.57	519.1	20.53	4.09	0.009035
6	1.1	1.91	636.3	21.72	4.35	0.007813
7	1.3	2.25	753.7	22.89	4.59	0.006959
8	1.5	2.60	871.2	24.02	4.83	0.006325
9	1.7	2.94	988.7	25.12	5.06	0.005832
10	1.9	3.28	1106.4	26.19	5.28	0.005437
11	2.1	3.62	1224.1	27.22	5.49	0.005112
12	2.3	3.96	1341.8	28.24	5.70	0.004839
13	2.5	4.30	1459.6	29.22	5.91	0.004606

Table 4. Calculated thermal parameters for the second row of PFTHE.

i	$w_{o,i}$, m/s	$w_{max,i}$, m/s	$Re^{II}_{a,i}$	$\alpha^{II}_{a,i}$	$Nu^{II}_{a,i}$	$j^{II}_{a,i}$
1	0.3	0.54	150.9	15.44	2.83	0.021261
2	0.4	0.72	204.1	15.31	2.83	0.015750
3	0.5	0.90	258.2	15.35	2.86	0.012582
4	0.7	1.25	368.2	15.74	2.97	0.009166
5	0.9	1.60	479.6	16.44	3.13	0.007413
6	1.1	1.95	591.8	17.34	3.32	0.006379
7	1.3	2.29	704.6	18.38	3.54	0.005708
8	1.5	2.64	817.7	19.49	3.77	0.005234
9	1.7	2.98	931.2	20.61	4.00	0.004876
10	1.9	3.33	1044.9	21.73	4.23	0.004593
11	2.1	3.67	1158.9	22.83	4.45	0.004360
12	2.3	4.02	1273.0	23.90	4.67	0.004165
13	2.5	4.36	1387.4	24.93	4.88	0.003992

Table 5. Calculated thermal parameters for the third row of PFTHE.

i	$w_{o,i}$, m/s	$w_{max,i}$, m/s	$Re^{III}_{a,i}$	$\alpha^{III}_{a,i}$	$Nu^{III}_{a,i}$	$j^{III}_{a,i}$
1	0.3	0.55	147.6	14.62	2.65	0.020270
2	0.4	0.73	198.3	14.23	2.59	0.014768
3	0.5	0.91	250.0	13.97	2.55	0.011563
4	0.7	1.26	355.6	13.78	2.55	0.008108
5	0.9	1.61	463.1	14.02	2.61	0.006391
6	1.1	1.96	571.7	14.60	2.74	0.005429
7	1.3	2.31	680.9	15.47	2.92	0.004856
8	1.5	2.66	790.6	16.54	3.13	0.004491
9	1.7	3.01	900.7	17.52	3.33	0.004191
10	1.9	3.36	1011.3	18.26	3.48	0.003902
11	2.1	3.71	1122.2	18.91	3.62	0.003652
12	2.3	4.05	1233.3	19.60	3.76	0.003452
13	2.5	4.40	1344.8	20.29	3.90	0.003284

Table 6. Calculated thermal parameters for the first row of PFTHE.

i	$w_{o,i}$, m/s	$w_{max,i}$, m/s	$Re^{IV}_{a,i}$	$\alpha^{IV}_{a,i}$	$Nu^{IV}_{a,i}$	$j^{IV}_{a,i}$
1	0.3	0.55	146.9	14.34	2.59	0.019199
2	0.4	0.73	196.5	13.92	2.52	0.013971
3	0.5	0.91	246.9	13.70	2.49	0.010986
4	0.7	1.27	349.7	13.71	2.51	0.007826
5	0.9	1.62	454.5	14.33	2.64	0.006350
6	1.1	1.98	560.4	15.88	2.94	0.005743
7	1.3	2.33	666.9	17.73	3.30	0.005419
8	1.5	2.68	773.9	19.11	3.57	0.005055
9	1.7	3.04	881.5	20.14	3.77	0.004694
10	1.9	3.39	989.8	21.06	3.96	0.004389
11	2.1	3.74	1098.5	22.11	4.17	0.004165
12	2.3	4.09	1207.4	23.45	4.43	0.004030
13	2.5	4.44	1316.6	24.93	4.72	0.003939

The following designations are used in Equation (5): $Re_{d_h,a}$ is the Reynolds number in case of air hydraulic diameter; d_h is the air hydraulic diameter; ν_a is the air kinematic viscosity.

$$w_{max} = \frac{(s \cdot p_l)}{\left(s - \delta_f\right) \cdot (p_l - d_{o,min})} \cdot \frac{\left(\overline{T}^i_a\right)}{\left(\overline{T}_{a,o}\right)} \cdot w_o \tag{6}$$

The symbol w_{max} designates the air velocity in the least cross-section of the air flow. The symbol s denotes fin pitch. Other symbols represent: p_l the longitudinal fin pitch δ_f the fin thickness, $d_{o,min}$ the minimal dimension between tubes, $\overline{T}^i_a$ the mass average air temperature in i-th row of PFTHE, and $\overline{T}_{a,o}$ the inlet mass average air temperature.

The approximation function was determined using the least-squares method. Using this function, the Nusselt number was approximated, depending on the Reynolds and Prandtl numbers, according to Equation (7). The range of Reynolds numbers considered is factored in the present study, and is also present in Equation (7).

$$Nu_a = x_1 \cdot Re^{x_2} \cdot Pr^{\frac{1}{3}} \qquad 140 < Re < 1500 \tag{7}$$

The heat transfer correlation for the entire heat exchanger is shown in Figure 6. The Nusselt numbers for the first and second rows of tubes as a function of the Reynolds number, calculated separately for each row, are illustrated in Figure 7a,b. The heat transfer correlations for the third and fourth rows of tubes are shown in Figure 8a,b. Figures 6–8 also show the confidence intervals for the Nusselt number calculated using Equation (7). The legend of each figure shows the corresponding correlation, either for the whole exchanger (Figure 6) or for a given row of tubes (Figures 7 and 8). Each figure also shows the limits of the 95% confidence intervals for the Nusselt number on a given tube row, where the coefficients x_1 and x_2 are determined by the least-squares method. For the 95% confidence interval, the values of the Nusselt number calculated for a given Reynolds number (Equation (7)) differ by $+/-2\,\sigma$, where the symbol σ denotes the mean standard deviation of the Nusselt numbers obtained by the CFD modelling. The process of determining 95% confidence intervals for correlations per Nusselt number obtained by least-squares is detailed in Chapter 11 of Taler's book [24].

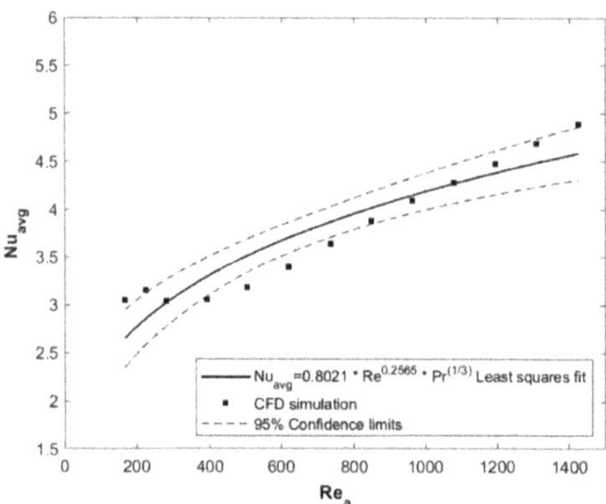

Figure 6. The average value of the Nusselt number as a function of the Reynolds number for the entire PFTHE.

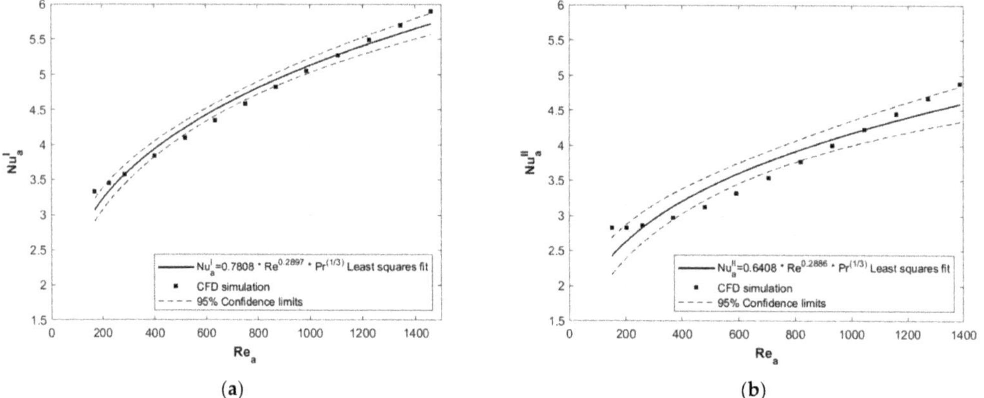

Figure 7. The Nusselt number as a function of the Reynolds number: (**a**) for the first row of PFTHE; (**b**) for the second row of PFTHE.

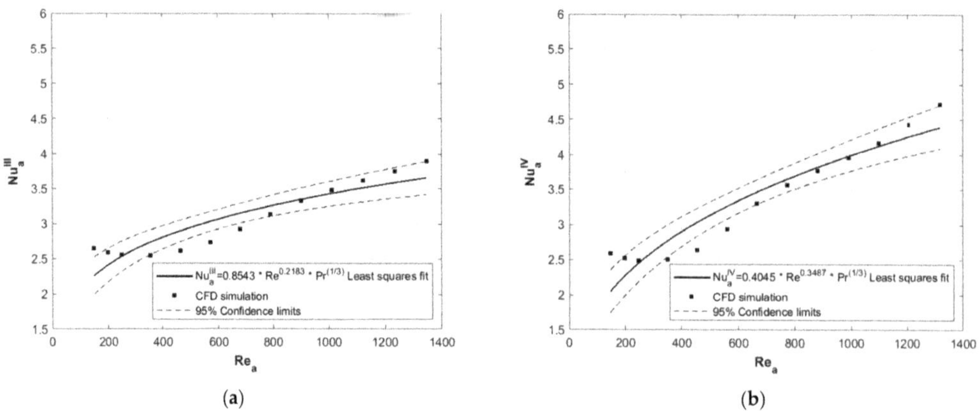

Figure 8. The Nusselt number as a function of the Reynolds number: (**a**) for the third row of PFTHE; (**b**) for the fourth row of PFTHE.

7. Discussion of the Obtained Results

Nusselt number correlations for considered PFTHE (Figure 1) have been determined using the CFD simulation data. New individual correlations (Figures 7 and 8) for particular rows are significantly different from the average Nusselt number correlation for the entire PFTHE (Figure 6), also determined by the CFD simulation. Parameter x_2 in Equation (7) determined by the least-squares method is compatible only for the first row with parameters known in the literature [2], while in the second row, there is a 46% discrepancy compared to the parameter x_2 found by [2] for the two-row car radiator. The heat transfer correlations for the last two rows, the third, and fourth, are new and there are no existing correlations in the literature for comparison.

Additionally, Nusselt numbers fall between Reynolds numbers from 140 to 500, although they also increase from 500 to 1400. This increase may occur when the air stream does not separate away from the tube for low air velocity (0.3 m/s). The Nusselt number value is larger for higher air velocity (0.5 m/s–0.7 m/s) where the detachment of the air stream from the tube surface occurs.

The determined heat transfer correlations showed that the third row is the least efficient. It has been noticed that, for the entire air velocity range, the Nusselt number falls to the third row and then increases in the fourth row. This could mean that airflow in the penultimate row is less turbulent or airflow has formed only a narrow flow channel, leaving a wide heat transfer dead zone in front of and behind the tubes in the third row.

Experimental verification of derived CFD-based correlations has not been carried out in this paper. Experimental verification is planned as an extension of this research. The future of this research topic also considers a wider Reynolds number range, different tube and fin pitches, and tube diameters of PFTHE.

8. Conclusions

New heat transfer correlations for the air-side Nusselt number have been proposed for four-row PFTHE. Consideration of individual correlations on individual rows of heat exchanger tubes can improve designing of the PFTHEs. In the analysed range of air velocities in front of the heat exchanger from 0.3 m/s to 2.5 m/s, the average HTC for the first tube row has the highest value. In the subsequent tube rows, the HTC is lower compared to the HTC on the first tube row. This paper also presents a simple method for the determination of Nusselt number correlation for individual rows, which can reduce or even eliminate experimental research. The adoption of a constant external surface temperature for the ribs and tubes simplifies the determination of HTC on individual tube rows. It is shown that, for constant fin and tube surface temperatures of 60 °C, 70 °C, and 80 °C, the resulting HTC values for the same airflow velocity differ slightly. However, CFD simulation must be precisely conducted with a proper finite volume mesh and turbulence model for turbulent flows. CFD modelling is an effective tool for determining individual correlations for calculating the air-side Nusselt number on the individual tube rows of the exchanger.

Author Contributions: Conceptualisation, D.T., J.T.; methodology, software, validation, formal analysis, investigation, writing—original draft preparation, M.M.; resources, data curation, M.M., K.W.; writing—review and editing; supervision, D.T., J.T. All authors have read and agreed to the published version of the manuscript.

Funding: This research received no external funding.

Institutional Review Board Statement: Not applicable.

Informed Consent Statement: Not applicable.

Data Availability Statement: The data presented in this study are available on request from the corresponding authors.

Conflicts of Interest: The authors declare no conflict of interest.

Nomenclature

A_{fin}	area of the fin sides, m^2
A_{tube}	tube outer surface area, m^2
d_h	hydraulic diameter, m
d_o	tube outer diameter, m
e	relative differences, %
j	Colburn factor
n	number of mesh elements
Nu	Nusselt number
p_l	longitudinal fin pitch, m
p_t	transversal fin pitch, m
Pr	Prandl number
r_{in}	outer radius of a plain tube, m
s	fin pitch, m
Re	Reynold number
$\overline{T}_a$	mass average air temperature, °C
$\overline{T}_{a,0}$	air inlet temperature, °C
$\Delta\overline{T}_{m,a}$	log mean temperature difference, °C
T_w	wall temperature (tube and fin), °C
w	air velocity, m/s
Q	total heat transfer rate, W
x_1	Nusselt number parameter
x_2	Nusselt number parameter
y^+	dimensionless wall distance
α	heat transfer coefficient, W/(m^2.K)
λ	thermal conductivity coefficient, W/(m.K)
δ_f	fin thickness, m
σ	standard deviation
ν	air kinematic viscosity, m^2/s
$0, I, II, III, IV$	number of PFTHE row
i	current row of PFTHE
$i+1$	next row of PFTHE
a	air
avg	average
i	next iteration
max	maximal

References

1. Thulukkanam, K. *Heat Exchanger Design Handbook*, 2nd ed.; CRC Press: Boca Raton, FL, USA, 2013.
2. Taler, D.; Taler, J.; Trojan, M. Thermal calculations of plate–fin–and-tube heat exchangers with different heat transfer coefficients on each tube row. *Energy* **2020**, *203*, 117806. [CrossRef]
3. Ali, H.M. Recent advancements in PV cooling and efficiency enhancement integrating phase change materials based systems—A comprehensive review. *Sol. Energy* **2020**, *197*, 163–198. [CrossRef]
4. Ali, H.M.; Zafar, M.A.; Bashir, M.A.; Nasir, M.A.; Ali, M.; Siddiqui, A.M. Effect of dust deposition on the performance of photovoltaic modules in city of taxila, Pakistan. *Therm. Sci.* **2017**, *21*, 915–923. [CrossRef]
5. Li, X.Y.; Li, Z.H.; Tao, W.Q. Experimental study on heat transfer and pressure drop characteristics of fin-and-tube surface with four convex-strips around each tube. *Int. J. Heat Mass Transf.* **2018**, *116*, 1085–1095. [CrossRef]
6. Matos, R.S.; Vargas, J.V.C.; Rossetim, M.A.; Pereira, M.V.A.; Pitz, D.B.; Ordonez, J.C. Performance comparison of tube and plate-fin circular and elliptic heat exchangers for HVAC-R systems. *Appl. Therm. Eng.* **2021**, *184*, 116288. [CrossRef]
7. Maarawi, A.; Anxionnaz-Minvielle, Z.; Coste, P.; Raimondi, N.D.M.; Cabassud, M. 1D/3D Numerical Approach for Modelling a Milli-Structured Heat Exchanger/Reactor. *Chem. Eng. Trans.* **2021**, *86*, 1441–1446.
8. Li, X.Y.; Mu, Y.T.; Li, Z.H.; Tao, W.Q. Numerical study on the heat transfer and pressure drop characteristics of fin-and-tube surface with four round-convex strips around each tube. *Int. J. Heat Mass Transf.* **2020**, *158*, 120034. [CrossRef]
9. Marcinkowski, M.; Taler, D. Calculating the Efficiency of Complex-Shaped Fins. *Energies* **2021**, *14*, 577. [CrossRef]
10. Khan, M.S.; Zou, R.; Yu, A. Computational simulation of air-side heat transfer and pressure drop performance in staggered mannered twisted oval tube bundle operating in crossflow. *Int. J. Therm. Sci.* **2021**, *161*, 106748. [CrossRef]

11. Lindqvist, K.; Skaugen, G.; Meyer, O.H.H. Plate fin-and-tube heat exchanger computational fluid dynamics model. *Appl. Therm. Eng.* **2021**, *189*, 116669. [CrossRef]
12. Łęcki, M.; Andrzejewski, D.; Gutkowski, A.N.; Górecki, G. Study of the Influence of the Lack of Contact in Plate and Fin and Tube Heat Exchanger on Heat Transfer Efficiency under Periodic Flow Conditions. *Energies* **2021**, *14*, 3779. [CrossRef]
13. Sadeghianjahromi, A.; Wang, C.C. Heat transfer enhancement in fin-and-tube heat exchangers—A review on different mechanisms. *Renew. Sustain. Energy Rev.* **2021**, *137*, 110470. [CrossRef]
14. Elmekawy, A.M.N.; Ibrahim, A.A.; Shahin, A.M.; Al-Ali, S.; Hassan, G.E. Performance enhancement for tube bank staggered configuration heat exchanger—CFD Study. *Chem. Eng. Process.-Process Intensif.* **2021**, *164*, 108392. [CrossRef]
15. Okbaz, A.; Pınarbaşı, A.; Olcay, A.B. Experimental investigation of effect of different tube row-numbers, fin pitches and operating conditions on thermal and hydraulic performances of louvered and wavy finned heat exchangers. *Int. J. Therm. Sci.* **2020**, *151*, 106256. [CrossRef]
16. Petrik, M.; Szepesi, G. Experimental and numerical investigation of the air side heat transfer of a finned tubes heat exchanger. *Processes* **2020**, *8*, 773. [CrossRef]
17. Petrik, M.; Szepesi, G.; Jármai, K. CFD analysis and heat transfer characteristics of finned tube heat exchangers. *Pollack Period.* **2019**, *14*, 165–176. [CrossRef]
18. González, A.M.; Vaz, M.; Zdanski, P.S.B. A hybrid numerical-experimental analysis of heat transfer by forced convection in plate-finned heat exchangers. *Appl. Therm. Eng.* **2019**, *148*, 363–370. [CrossRef]
19. Awais, M.; Bhuiyan, A.A. Heat and mass transfer for compact heat exchanger (CHXs) design: A state-of-the-art review. *Int. J. Heat Mass Transf.* **2018**, *127*, 359–380. [CrossRef]
20. Adam, A.Y.; Oumer, A.N.; Najafi, G.; Ishak, M.; Firdaus, M.; Aklilu, T.B. State of the art on flow and heat transfer performance of compact fin-and-tube heat exchangers. *J. Therm. Anal. Calorim.* **2020**, *139*, 2739–2768. [CrossRef]
21. Rich, D. The effect of the number of tube rows on heat transfer performance of smooth plate fin-and-tube heat exchangers. *ASHRAE Trans.* **1975**, *81*, 307–317.
22. Taler, D.; Taler, J.; Trojan, M. Experimental verification of an analytical mathematical model of a round or oval tube two-row car radiator. *Energies* **2020**, *13*, 3399. [CrossRef]
23. Kays, W.; London, A. *Compact Heat Exchangers*, 3rd ed.; Krieger Pub Co.: Malabar, FL, USA, 1998.
24. Taler, D. *Numerical Modelling and Experimental Testing of Heat Exchangers*; Springer Nature: Cham, Switzerland, 2019.

Article

Novel Method of the Seal Aerodynamic Design to Reduce Leakage by Matching the Seal Geometry to Flow Conditions

Damian Joachimiak

Institute of Thermal Engineering, Poznan University of Technology, 60-965 Poznan, Poland; damian.joachimiak@put.poznan.pl; Tel.: +48-61-665-22-09

Abstract: This paper presents a novel method of labyrinth seals design. This method is based on CFD calculations and consists in the analysis of the phenomenon of gas kinetic energy carry-over in the seal chambers between clearances. The design method is presented in two variants. The first variant is designed for seals for which it is impossible to change their external dimensions (length and height). The second variant enables designing the seal geometry without changing the seal length and with a slight change of the seal height. Apart from the optimal distribution of teeth, this variant provides for adjusting chambers geometry to flow conditions. As the result of using both variants such design of the seal geometry with respect to leakage is obtained which enables achieving kinetic energy dissipation as uniform as possible in each chamber of the seal. The method was developed based on numerical calculations and the analysis of the flow phenomena. Calculation examples included in this paper show that the obtained reduction of leakage for the first variant ranges from 3.4% to 15.5%, when compared with the initial geometry. The relation between the number of seal teeth and the leakage rate is also analyzed here. The second variant allows for reduction of leakage rate by 15.4%, when compared with the geometry with the same number of teeth. It is shown that the newly designed geometry reveals almost stable relative reduction of leakage rate irrespective of the pressure ratio upstream and downstream the seal. The efficiency of the used method is proved for various heights of the seal clearance.

Keywords: labyrinth seal; leakage; design method; kinetic energy; inverse problem; steam turbines; gas turbines; fluid-flow machines

Citation: Joachimiak, D. Novel Method of the Seal Aerodynamic Design to Reduce Leakage by Matching the Seal Geometry to Flow Conditions. *Energies* **2021**, *14*, 7880. https://doi.org/10.3390/en14237880

Academic Editor: Artur Bartosik

Received: 6 October 2021
Accepted: 18 November 2021
Published: 24 November 2021

Publisher's Note: MDPI stays neutral with regard to jurisdictional claims in published maps and institutional affiliations.

1. Introduction

Labyrinth seals are widely used in various types of fluid-flow machines, such as steam turbines, gas turbines, and compressors. Labyrinth seals enable reducing the leakage of working medium between two elements that are non-contacting. Concern for the protection of the natural environment imposes current trends to achieve greater efficiency of flow machines and flowing systems. High parameters of machines operation are required, which means that seals work in the environment characterized by high temperatures and large pressure drop. Labyrinth seals have particular impact on the efficiency of high power-generating machines. In the paper [1], the impact of the leakage in the internal gland of the steam turbine 13K215 on the power loss was analyzed. It was showed that the power loss resulting from the steam leakage through the internal seal of nominal geometry had achieved approx. 1 MW, and through the worn out seal it could achieve 2 MW. Therefore, the problem of leakage minimization is of great importance. The paper [2] presented the numerical analysis of fluid flow in the two-teeth straight through over-bandage seal in the axial turbine stage, with particular attention to the impact of the leakage on the main flow. Authors of this paper showed that the leakage had a considerable impact on the agitation loss and secondary flows generating serious energy losses in the turbine stage. It was proved in the paper [3] that 1% increase in the seal-tooth clearance height causes a significant decrease in performance and efficiency of the multistage axial compressor. Labyrinth seals also have a great influence on the efficiency of reciprocating machines [4–6].

185

Various calculation methods for estimating the leakage rate in the labyrinth seal are presented in the scientific literature. The first calculation model describing the gas flow in the straight through seal was developed by Martin [7]. Egli [8] and other researchers, Hodkinson [9], Zimmermann and Wolff [10], had included in their calculation models the phenomenon of kinetic energy carry over as well as the result of flow contraction. Those models were next developed by Neumann [11] and Scharrer [12], among others, who proposed the method for calculating the seal leakage tooth by tooth. There are new one-dimensional models appearing in the literature [13,14], enabling the determination of supercritical CO_2 leakage. The most advanced of these models provide for such features of the seal geometry as the clearance height, chamber length, and tooth thickness. These methods have different accuracy which depends on coefficients being determined experimentally. Applying the above-mentioned models to analyze the seal geometry in order to optimize it with respect to the leakage rate would be subjected to great error, which was proved in the paper [15].

Usage of the CFD flow analysis enables better observation of physical and flow phenomena, which in turn enables designing seals of higher leak tightness. In papers [16–20], the impact of the seal geometric parameters, such as the pitch or the chamber size, on the leakage rate was described. The degree of sealing wear and tear has a considerable impact on the failure of the seal integrity [21–23]. Slight changes in the dimensions of the clearance influence the leakage value and the flow coefficients, which is presented in the paper [24]. A seal with staggered helical teeth, which is characterized by reduced leakage, is presented in the paper [25]. In the paper [26], the reduction of leakage is achieved by slight changes in the chamber geometry, which were disturbing the gas jet flowing with high velocity. Surrogate models of optimization based on a set of input data, which features had been specified parametrically, obtained from CFD calculations, are discussed in the paper [27]. Radial clearance height and teeth spacing were variable parameters. The leakage minimization and decreasing total increase of enthalpy resulting from the friction were taken as the optimization criterion. The paper [28] presents the method for optimizing the straight through sealing with the use of CFD by changing the teeth inclination angle and spacing. The paper [29] presents an optimization model taking into account such design variables as the seal clearance, teeth width, teeth height, pitch, and teeth backward and forward expansion angle. The problem of optimizing the geometry of seal chambers is considered in the paper [30]. Semi-empirical model and parametric CFD analysis were used for this purpose. In the paper [15], the impact of the increasing number of teeth for constant length of the straight through seal on the leakage rate is investigated. Data presented therein indicate that there is a specific range of the seal pitch length in the straight through seal for which the minimum leakage is obtained. The increase of the number of seal teeth results in the decrease of pitch length and reduction of chambers size. It was proved in the paper [9] that the application of too many teeth leads to the effect of increasing leakage in the seal. This is the result of fading gas expansion in chambers of reducing size. Then the flow character becomes similar to that in the slot seal [31]. In this paper a new method for designing the geometry of the labyrinth seal is described. The inspiration for research work presented herein was to develop a relatively simple in-use method for designing such seal geometry which is based on the observation of flow phenomena intensifying leakage. So far, seals with equal chamber dimensions or with sequentially repeated chambers were designed.

The approach to designing seals by matching the seal geometry to flow conditions, and particularly based on the analysis of the gas kinetic energy distribution along the seal length, has not yet been presented in scientific literature. The method for labyrinth seal design, presented in this paper, enables quick obtaining of results. It enables determination of many dimensions of the geometry in one calculation step. Application of this method results in obtaining the geometry with a lower leakage than the geometry with the same external dimensions and teeth being spaced evenly (variant A) or the geometry with slightly increased chambers height (variant B). This problem belongs to the group of geometry-related inverse problems [32,33].

2. Method of Aerodynamic Design

The leakage rate in sealing is strongly affected by the gas expansion in clearances and by the gas kinetic energy dissipation in chambers. Gas flowing through subsequent clearances expands, which results in the density reduction. The relation between the clearance height and the pressure distribution in the straight-through seal was presented in the paper [34]. In each subsequent clearance the gas velocity is higher and higher (Figure 1). Widely used labyrinth seals are characterized by constant chamber size.

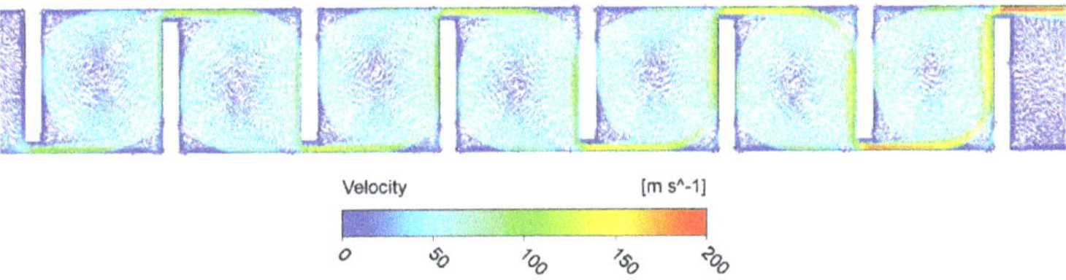

Figure 1. Example vectors of velocity fields in a representative segment for $p_{in}/p_{out} = 2$ and $T_{in} = 300$ K, $p_{out} = 10^5$ Pa.

In the seal chambers of constant dimensions one can observe the unequal degree of gas kinetic energy dissipation. The measure of the gas kinetic energy dissipation in chambers is the kinetic energy carry-over coefficient, which was analyzed in the paper [35]. The phenomenon of the gas kinetic energy carry-over has a great impact on the leakage rate [36–38]. The idea of a novel method of aerodynamic design is to adjust the seal geometry to flow conditions so that the gas kinetic energy dissipation as even as possible in the seal chambers is obtained.

The novel design method is discussed based on the geometry of a staggered labyrinth seal (Figure 2) of the outer diameter D, segment height H, radial clearance RC, segment length LS, pitch LP, and tooth thickness B.

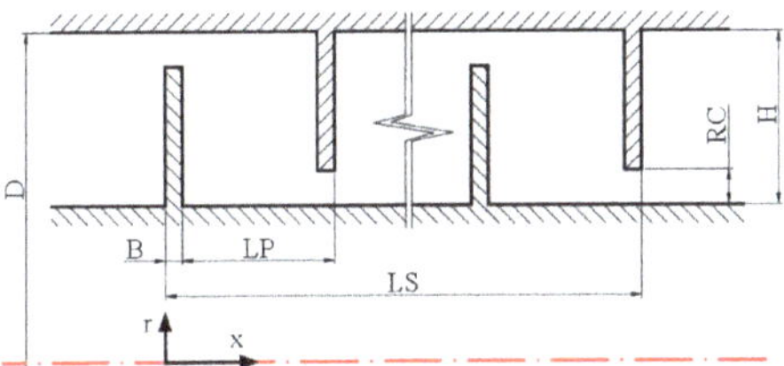

Figure 2. Illustrative geometry of a staggered labyrinth seal.

The design method presented in this paper is based on CFD calculations pertaining to gas flow in the labyrinth seal.

Labyrinth seals have axisymmetric geometries, therefore the flow can be described along the axis X. In the method presented herein, calculations were based on the local non-dimensional coordinate x of the seal length being parallel to the seal axis. The origin of the coordinate system is determined at the place being the beginning of the first seal tooth on the gas inflow side. The end of the non-dimensional coordinate system is located at the end of the segment (Figure 2). Hence,

$$x = \frac{X}{LS}.$$

(1)

The method consists in the analysis of the kinetic energy $E(x)$ distribution in the axial direction (the main flow) within the seal area. The gas kinetic energy in the axial direction was described by the dependency

$$E(x) = \frac{u^2}{2},\tag{2}$$

where u denotes the gas velocity in the axial direction. Non-dimensional kinetic energy was defined as follows:

$$e = \frac{E(x) - E_{min}}{E_{max} - E_{min}},\tag{3}$$

where E_{min} and E_{max} denote the minimal and maximal gas kinetic energy, respectively, in the axial direction in the seal. For calculations it was assumed that $E_{min} = 0$.

Illustrative fields of non-dimensional gas kinetic energy e are shown in Figure 3. To improve the readability of the Figure 3, the scale was set within the range from 0 to 0.6.

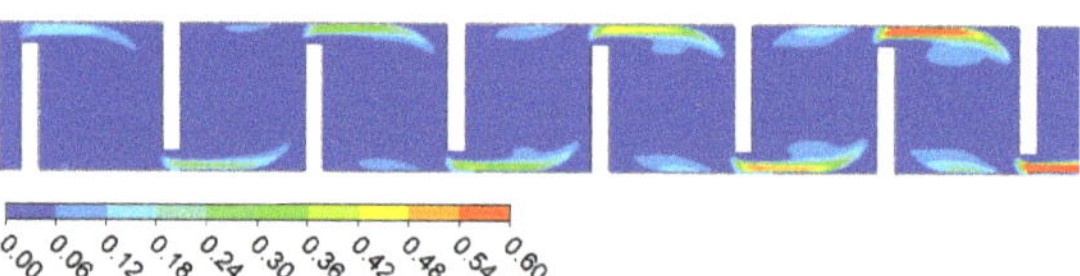

Figure 3. Example distribution of non-dimensional kinetic energy e in the axial direction in the staggered seal for $p_{in}/p_{out} = 2$, $T_{in} = 300$ K, $p_{out} = 10^5$ Pa.

The initial stage of the method of adjusting the seal geometry to flow conditions is searching for local maximal non-dimensional values of the gas kinetic energy in the axial direction. In staggered labyrinth seals local maxima of the kinetic energy occur in the area of clearances and in seal chambers just behind clearances (Figure 3).

Illustrative distribution of local maxima of non-dimensional kinetic energy $e_{max}(i)$ occurring in the staggered seal with the approximation function $e(x)$ is shown in Figure 4.

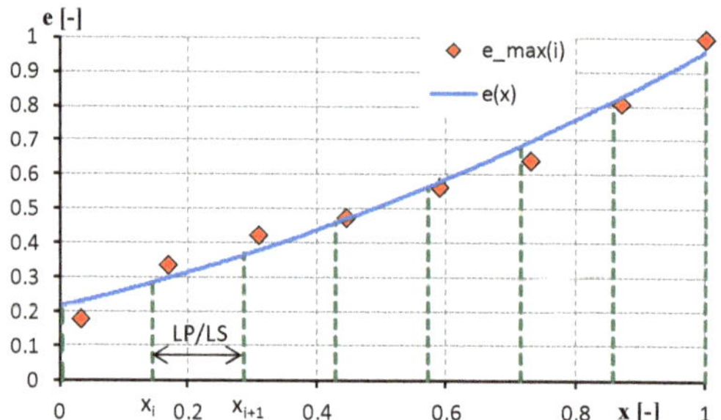

Figure 4. Local maxima of non-dimensional gas kinetic energy $e_{max}(i)$ in the staggered seal with the approximating line $e(x)$; discontinuous green line shows the location of tooth sidewalls on the inflow side (the distance between them is LP).

In non-dimensional coordinates, the field below the approximating line, hereinafter designated as $e(x)$, (Figure 4), is divided into n areas with beginnings and ends included between points (x_i, x_{i+1}) for $i = 1, 2, \ldots, n + 1$. The number n of areas is equal to the number of chambers in the seal segment, and their length corresponds with the length of pitch $x_{i+1} - x_i = LP$ of the initial geometry.

The area below the approximating curve can be described as follows:

$$A_e = \int_0^1 e(x)dx = \sum_{i=1}^{n} A_e(i),\tag{4}$$

where fields of respective areas are as follows:

$$A_e(i) = \int_{x_i}^{x_{i+1}} e(x)\,dx, \text{ where } i = 1, 2, \dots, n. \tag{5}$$

Fields below the curve $e(x)$ of respective areas are proportional by weight to the length of respective pitches of the new geometry.

The length of respective pitches of the improved geometry results from the following equation:

$$LP(i) = LS\frac{A_e(i)}{A_e}. \tag{6}$$

As a result of the applied method, a variable pitch length of the seal adapted to the flow conditions was obtained. Depending on the possibility of changing the inner diameter and the external geometry of the seal, two variants of the method were specified.

2.1. Variant A of the Design Method

Variant A of the design method consists in relevant arrangement of the seal teeth. Pitch lengths are determined according to the Equation (6) without changing the length LS and height H of the seal (Figure 2). External dimensions of the seal do not change in this variant. It can be applied for cases when there is no place for increasing the height of the seal in the geometry of the flow machine.

2.2. Variant B of the Design Method

Variant B of the design method can be applied when the change of the height of the seal chambers in the flow machine design is possible. This variant consists in changing the pitch lengths according to the Equation (6) without changing the length *LS* and with adapting the height of respective seal chambers to the flow conditions. The application of variant B results in obtaining variable chamber volumes in such a way that chambers are increasing in the direction of the gas flow (Figure 5). Illustrative diagram of the staggered seal geometry after the variant B of the design method was applied, is presented in Figure 5.

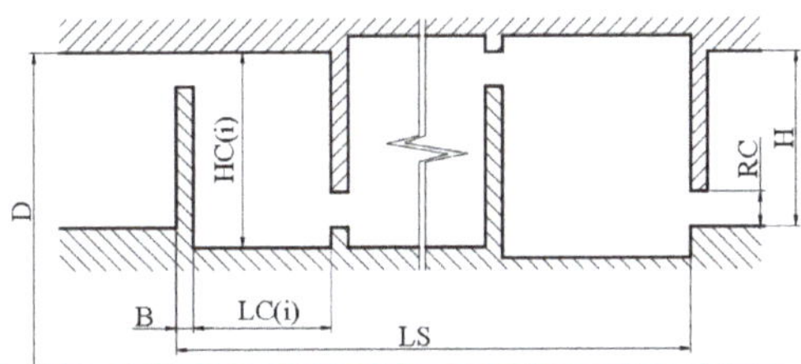

Figure 5. Diagram of geometry as per the comprehensive design method.

Variable chambers' height is selected so that the height of the last chamber is equal to its length $HC(n) = LC(n)$. It was assumed that the variability of chambers' height $HC(i)$ is linear, which can be described by the dependence:

$$HC(i) = H + (LC(n) - H)\frac{i}{n}, \tag{7}$$

where n is the number of the seal chambers, $i = 1, 2, \dots, n$.

Furthermore, it was assumed that the new geometry has clearances of the same flow field as the initial geometry. Height H and diameter D of the inlet and of the outlet channels are also the same as in the initial geometry.

2.3. Assumptions for CFD Calculations

CFD calculations were performed using Ansys Fluent 19. The investigated geometry has inflow section of the length of 6H (Figure 6), where H denotes the height of the inflow channel (Figure 2). The length of the outflow section was assumed as $11H$. At the inlet to the seal, the total pressure p_{in} and temperature T_{in} are known, while downstream the segment the static pressure p_{out} is known. Based on the above boundary conditions (Table 1.) CFD calculations are performed.

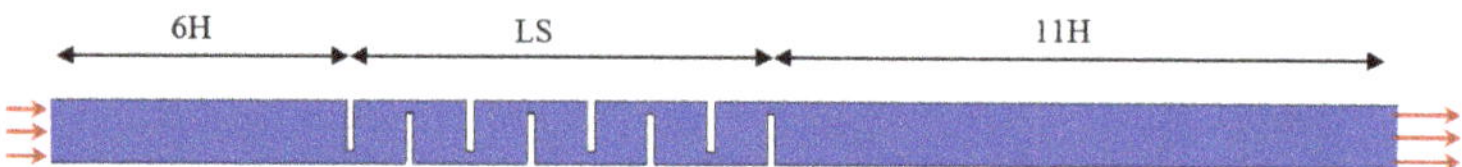

Figure 6. Computation domain.

Table 1. Boundary conditions.

Surface	Parameter	Value
inlet	total pressure	$2{\cdot}10^5 - 2.8{\cdot}10^5$ Pa
	total temperature	300 K
outlet	static pressure	10^5 Pa
walls	adiabatic, no slip (no rotation)	-

In these calculations air was treated as the compressible ideal gas. RANS type calculations were performed for 2D axisymmetric geometry. Equations (8)–(10) were taken into account in the CFD calculations. The continuity equation

$$\nabla\left(\rho \vec{v}\right) = 0,\tag{8}$$

and the momentum conservation equation

$$\nabla\left(\rho \vec{v}\,\vec{v}\right) = -\nabla p + \nabla{\cdot}\vec{\tau},\tag{9}$$

where $\vec{\tau}$ is the stress tensor, was analyzed. The energy conservation equation [39] was considered

$$\nabla\left(\vec{v}\left(\rho E_t + p\right)\right) = \nabla\left(k_{eff}\nabla T + \left(\vec{\tau}_{eff}{\cdot}\vec{v}\right)\right),\tag{10}$$

where E_t is the total energy, k_{eff} is the effective thermal conductivity, $\vec{\tau}_{eff}$ is the deviatoric stress tensor.

The next part of the paper deals with the method of selection of the grid for CFD calculations according to [40]. The grid convergence method was used, which is based on the Richardson extrapolation method. In order to select the mesh, a representative mesh size h was used

$$h = \left[\frac{1}{N}\sum_{i=1}^{N}(\Delta A_i)\right]^{1/2}.\tag{11}$$

where ΔA_i is area of the i-th cell, and N is total number of cells. Three significantly different mesh sets were selected for the analysis. In order to determine the mass flux change calculations were performed for the boundary conditions given in the Table 1 ($p_{in} = 2{\cdot}10^5$ *Pa*). Let $h_1 < h_2 < h_3$ and $r_{21} = h_2/h_1$, $r_{32} = h_3/h_2$ then calculate the apparent order p of the method using the expression

$$p = \frac{1}{ln(r_{21})}\left|ln|\varepsilon_{32}/\varepsilon_{21}| + q(p)\right|\tag{12a}$$

$$q(p) = ln\left(\frac{r_{21}^p - s}{r_{32}^p - s}\right)$$ (12b)

$$s = 1 \cdot sgn(\varepsilon_{32}/\varepsilon_{21})$$ (12c)

where $\varepsilon_{21} = \dot{m}_2 - \dot{m}_1$ and $\varepsilon_{32} = \dot{m}_3 - \dot{m}_2$.

Extrapolated mass flow was calculated from expression

$$\dot{m}_{ext}^{21} = \left(r_{21}^p \dot{m}_1 - \dot{m}_2\right)/\left(r_{21}^p - 1\right).$$ (13)

Then the following parameters were calculated and reported, along with the apparent order p:

- Approximate relative error,

$$e_a^{21} = \left(\dot{m}_1 - \dot{m}_2\right)/\dot{m}_1,$$ (14)

- Extrapolated relative error,

$$e_{ext}^{21} = \left(\dot{m}_{ext}^{21} - \dot{m}_1\right)/\dot{m}_{ext}^{21},$$ (15)

- The fine-grid convergence index,

$$GCI_{fine}^{21} = \frac{1.25 e_a^{21}}{\left(r_{21}^p - 1\right)}.$$ (16)

The analysis was performed for the initial geometry with dimensions given in point 3 and exact geometry presented in Figure 10a (t = 8, LP = 4 mm, RC = 0.315 mm). The results are summarized in Table 2.

Table 2. Parameters used in the grid convergence method [40].

Parameter	Mark	Value
Total number of cells	N_1	350,000
	N_2	290,378
	N_3	120,812
Mass flux difference	r_{21}	1.104 [-]
	r_{32}	1.55 [-]
Mass flow	$\dot{m}_1$	0.01981 [kg/s]
	$\dot{m}_2$	0.01986 [kg/s]
	$\dot{m}_3$	0.0201 [kg/s]
Apparent order	p	0.598 [-]
Extrapolated mass flow	$\dot{m}_{ext}^{21}$	0.01967 [kg/s]
	$\dot{m}_{ext}^{32}$	0.01968 [kg/s]
Approximate relative error	e_a^{21}	0.042 [%]
Extrapolated relative error	e_{ext}^{21}	0.69%
Fine-grid convergence index	GCI_{fine}^{21}	0.86%

According to Table 2, the obtained numerical uncertainty for the fine grid solution N_1 is 0.86%. This is a satisfactory value. The grid with the number of approximately N_1 = 354,000 elements was used for further analysis. In boundary layers, approx. 20 grid cells were assumed. Illustrative grid taken for calculations is shown in Figure 7. The k-ω SST turbulence model [41] was included in calculations. To obtain the grid of appropriate quality for the k-ω SST model, the condition that $y^+ < 2$ in the boundary layer was assumed.

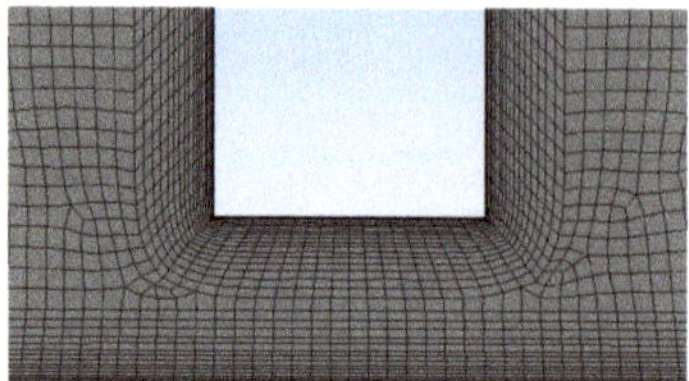

Figure 7. Illustrative grid used for calculations.

Stationary calculations were performed using the pressure-based coupled solver. Convergence tolerance of 4×10^{-6} was assumed for calculations. To obtain the required convergence criterion, for each case approx. 150 iterations of calculations were made.

For the model research presented in this paper, CFD simulations were performed for similar grid parameters and the same solver settings as those described in papers [15,24]. The same working medium and similar boundary conditions as in [15,24] were applied. Relative mass-flux error between the values obtained experimentally and from CFD calculations for the segment of the straight-through seal [15] was within the range from 0.3 to 0.7%.

This paper presents initial analysis of the new design method. It was decided to perform two-dimensional calculations due to reduced calculation time for many geometries and various boundary conditions. Data shown in the paper [42] indicate that the rotational speed has a little impact on the gas leakage value. Therefore, the impact of the rotational speed was not included in the general CFD analysis.

3. Application of the Method of Aerodynamic Design for the Staggered Seal of the Geometry Similar to the Element of the Front Sealing in Turbine 13CK60

The staggered seal consisting of eight teeth located alternately on the shaft and on the body of the sealing (Figure 2) was analyzed. The initial geometry is the staggered seal segment of the external diameter $D = 139.9$ mm, length $LS = 28.5$ mm, height $H = 4$ mm, equal pitch $LP = 4$ mm, the tooth thickness $B = 0.5$ mm, and the clearance height $RC = 0.315$ mm (Figure 2). The paper presents preliminary research. CFD calculations were performed for perfect air. Distribution of local maxima of non-dimensional kinetic energy dependent on the non-dimensional segment length, obtained from the CFD simulation, is shown in Figure 8.

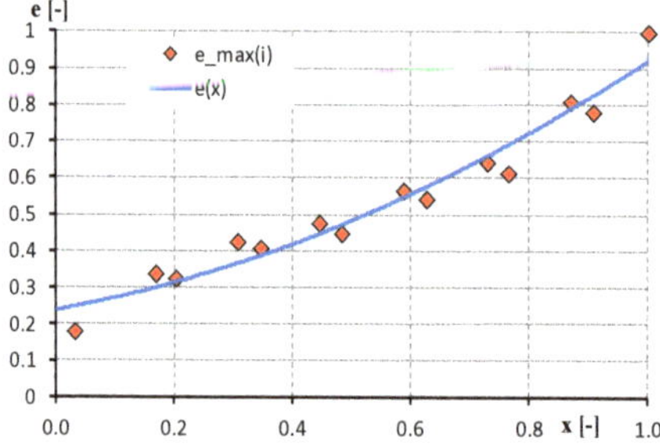

Figure 8. Local maxima of the non-dimensional gas kinetic energy $e_{max}(i)$ in the staggered seal with the approximating line $e(x)$ for boundary conditions $p_{in}/p_{out} = 2$, $T_{in} = 300$ K, $p_{out} = 10^5$ Pa.

Within the area of the first and last clearances, one local maximum of the kinetic energy was obtained (Figure 8). In other clearances two local maxima of the kinetic energy $e_{max}(i)$ were obtained. As a result of the design method, variable lengths of the seal pitches $LP(i)$ were obtained. The first seal pitch is 2.04-mm long and the last one is 6.58-mm long (Figure 9).

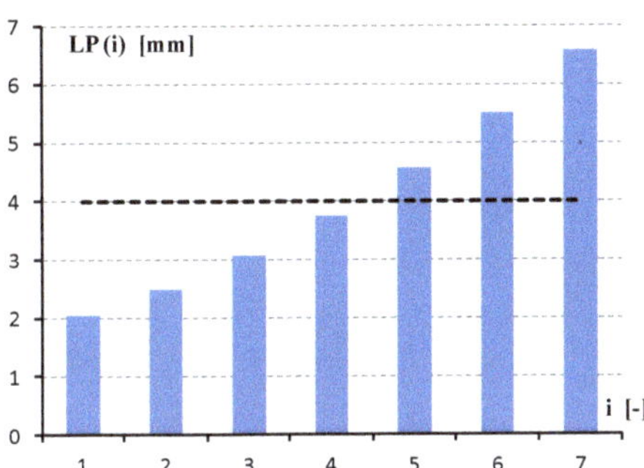

Figure 9. Resultant distribution of the staggered seal pitch lengths; discontinuous line marks the constant length of the pitch of the initial geometry.

The application of the design method results in the reduction of the leakage rate. To present results, a relative leakage change for the new geometry had been defined, which was in turn compared with the initial geometry comprising eight teeth, which was described by the following formula

$$\delta\dot{m}(t,8) = \frac{\dot{m}_{LPconst}(8) - \dot{m}_{LPdm}(t)}{\dot{m}_{LPconst}(8)} \times 100\% \tag{17}$$

The relative leakage change between the geometry of the seal with equal pitch and the improved one with the same number of teeth t was described by the relation

$$\delta\dot{m}(t) = \frac{\dot{m}_{LPconst}(t) - \dot{m}_{LPdm}(t)}{\dot{m}_{LPconst}(t)} \times 100\% \tag{18}$$

Further part of the paper presents results obtained with the use of the design method without changing external dimensions according to the variant A and results of the variant B with a slight change of the seal height.

3.1. Application of the Variant A of the Design Method

In this chapter, the impact of the variant A application on the leakage rate for various number of the seal teeth is analyzed. Figure 10 presents the gas velocity contours for the initial geometry and the improved one.

For the geometries under consideration (Figure 10) of uniformly spaced teeth, the change of gas velocity along the seal length is great. The gas velocity in the first clearance is approx. 70 m/s, and in the last one it exceeds 180 m/s. The effect of gas expansion in clearances impacts the intensity of gas vortices in subsequent chambers. The vortex in the last chambers is significantly greater than in the first few chambers. In the designed geometry, it was observed that chamber lengths were adapting to the increasing gas velocity in clearances. In result, the gas velocity close to the last tooth (upstream the clearance) is significantly smaller for the designed geometry, when compared with the initial geometry for cases 8t, 9t, and 10t, being considered. When the distribution of the gas velocity in the optimized geometries of 9 and 10 teeth is analyzed, it can be observed that gas vortices occur within the whole chambers' volume. Table 3 summarizes obtained values of mass flow for geometries shown in Figure 10.

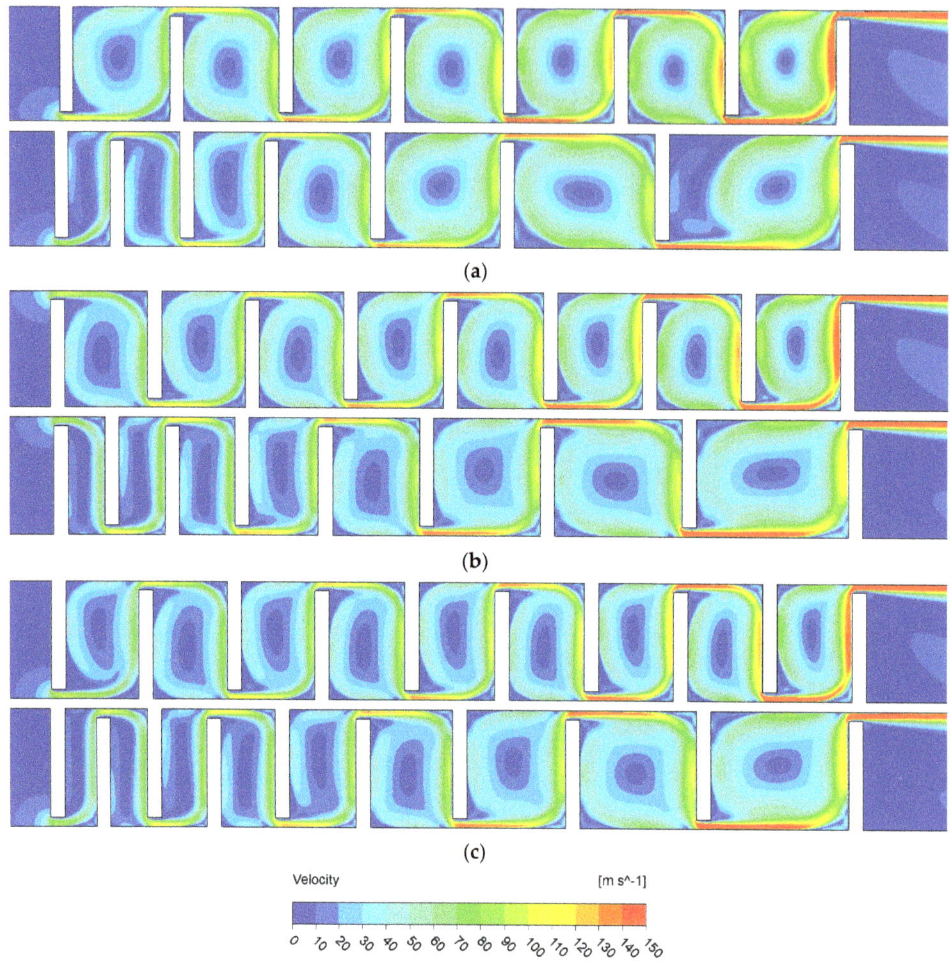

Figure 10. Distribution of air velocity in the seal of the initial geometry and of the designed one for the radial height RC = 0.315 mm comprising of (**a**) eight, (**b**) nine, and (**c**) ten teeth for boundary conditions p_{in}/p_{out} = 2.4, T_{in} = 300 K, $p_{out} = 10^5$ Pa.

Table 3. Leakage rate for the seal of the initial geometry and of the designed one and their relative changes according to Equations (17) and (18), RC = 0.315 mm and of different number of teeth for p_{in}/p_{out} = 2, T_{in} = 300 K, $p_{out} = 10^5$ Pa.

t	$\dot{m}_{LPconst}$ (t) [kg/s]	$\dot{m}_{LPdm}$ (t) [kg/s]	$\delta\dot{m}$ (t, 8) [%]	$\delta\dot{m}$ (t) [%]
8	0.0198	0.0191	3.4	3.4
9	0.0184	0.0179	9.7	2.8
10	0.0171	0.0167	15.5	2.2

The design method for geometries 8t, 9t, and 10t, when compared to the geometry 8t with evenly spaced teeth (constant pitch length) LPconst (Figure 10a), improves the leak-tightness of the relative value $\delta\dot{m}(t, 8)$ by 3.4%, 9.7%, and 15.5%, respectively. Comparing the relative difference of the mass flow (Table 3, Figure 11) obtained for the designed geometry $\dot{m}_{LPdm}(t)$ and the geometry with evenly spaced teeth $\dot{m}_{LPconst}(t)$ according to the Equation (12), the leak-tightness was improved by 3.4%, 2.8%, and 2.2%, respectively, for the geometry 8t, 9t, and 10t.

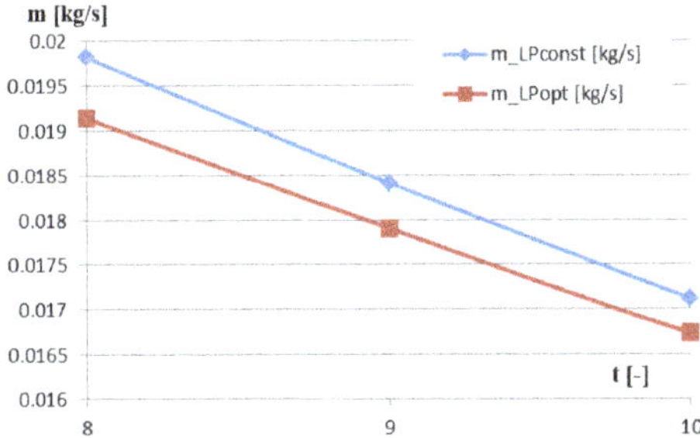

Figure 11. Mass flow obtained for the seal of equal pitch and for the designed one depending on the number of teeth t, for the clearance RC = 3.15 mm and boundary conditions p_{in}/p_{out} = 2, T_{in} = 300 K, p_{out} = 10^5 Pa.

Application of the design method and changing the number of teeth from eight to ten enable significant reduction of the leakage. There is a linear relationship between the number of teeth in the range from 8 and 10 and the relative reduction of the leakage rate for p_{in}/p_{out} = 2 (Figure 12). The limitation of the applied design method for the given seal length LS and many teeth is obtaining too small spaces between them. Therefore, the analysis has not been continued for a greater number of teeth.

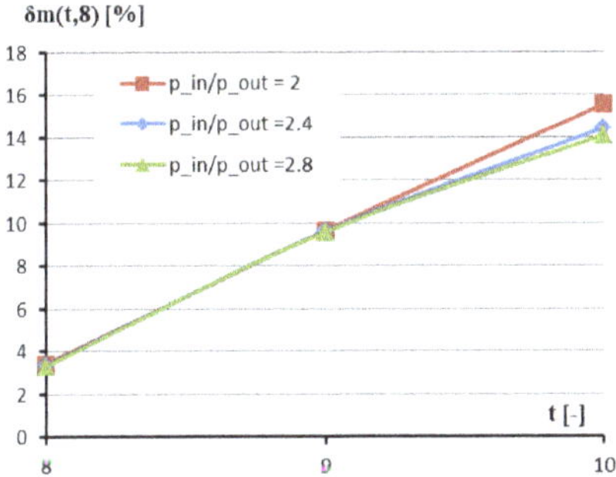

Figure 12. Relative reduction of the integrity of the designed geometry $\delta\dot{m}(t,8)$ [%] depending on the number of teeth t, for the clearance RC = 0.315 mm and the pressure ratio p_{in}/p_{out}, T_{in} = 300 K, p_{out} = 10^5 Pa.

Distribution of flow and thermodynamic parameters of the seal is affected by the ratio of pressure upstream and downstream the sealing. For the ratio p_{in}/p_{out} equal to 2.4 and 2.8, the increase of leakage rate for the designed geometry with 10 teeth was observed (Figure 12). Within the frames of research work a series of numerical calculations was performed for the geometry of the clearance height RC = 0.315; 0.542 and 0.77 mm for the pressure ratio p_{in}/p_{out} ranging from 2 to 2.8. Tables 4–6 summarize the obtained values of air mass flow for the initial geometry and the designed one for various heights of the radial clearance RC.

Table 4. Change of the leakage rate by Equation (17) depending on the pressure ratio p_{in}/p_{out} for the initial 8t and the designed geometry 9t of the staggered seal, RC = 0.315 mm, $T_{in} = 300$ K, $p_{out} = 10^5$ Pa.

p_{in}/p_{out} [-]	$\dot{m}_{LPconst}$ (t = 8) [kg/s]	$\dot{m}_{LPdm}$ (t = 9) [kg/s]	$\delta\dot{m}$ (t, 8) [%]	$\Delta\dot{m}$ [kg/s]
2	0.0198	0.0179	9.65	0.0019
2.2	0.0224	0.0202	9.62	0.0022
2.4	0.0249	0.0225	9.62	0.0024
2.6	0.0273	0.0247	9.62	0.0026
2.8	0.0297	0.0269	9.57	0.0028

Table 5. Change of the leakage rate by Equation (17) depending on the pressure ratio p_{in}/p_{out} for the initial 8t and the designed geometry 9t of the staggered seal, RC = 0. 542 mm, $T_{in} = 300$ K, $p_{out} = 10^5$ Pa.

p_{in}/p_{out} [-]	$\dot{m}_{LPconst}$ (t = 8) [kg/s]	$\dot{m}_{LPdm}$ (t = 9) [kg/s]	$\delta\dot{m}$ (t, 8) [%]	$\Delta\dot{m}$ [kg/s]
2	0.0302	0.0269	10.78	0.0033
2.2	0.0341	0.0304	10.87	0.0037
2.4	0.0379	0.0338	10.89	0.0041
2.6	0.0416	0.0371	10.90	0.0045
2.8	0.0453	0.0404	10.90	0.0049

Table 6. Change of the leakage rate by Equations (17) and (18) depending on the pressure ratio p_{in}/p_{out} for the initial 8t and the designed geometry 9t of the staggered seal, RC = 0. 77 mm, $T_{in} = 300$ K, $p_{out} = 10^5$ Pa.

p_{in}/p_{out} [-]	$\dot{m}_{LPconst}$ (t = 8) [kg/s]	$\dot{m}_{LPdm}$ (t = 9) [kg/s]	$\delta\dot{m}$ (t, 8) [%]	$\Delta\dot{m}$ [kg/s]
2	0.0385	0.0343	10.95	0.0042
2.2	0.0435	0.0388	10.95	0.0048
2.4	0.0484	0.0431	10.80	0.0052
2.6	0.0532	0.0474	10.76	0.0057
2.8	0.0580	0.0517	10.97	0.0064

The relative reduction of the leakage $\delta\dot{m}(t)$ for geometries of the staggered seal under consideration is slightly affected by the pressure ratio p_{in}/p_{out}. For geometries RC = 0.315, 0.542, 0.77 mm, it is included in the range 9.57–9.65%, 10.78–10.90%, and 10.76–10.97%, respectively. The leakage rate reduction depends on the clearance height. The greatest relative reduction of the leakage was obtained for the geometry RC = 0.542 and 0.77 mm, $t = 8$ (Tables 5 and 6).

For the increasing pressure ratio, the leakage reduction $\Delta\dot{m}$ is linear for the analyzed RC, (Figure 13).

3.2. Application of the Variant B of the Design Method

For the geometry as per the variant B (described in Section 2.2.), numerical simulations of air flow for various air pressure ratios p_{in}/p_{out} from the range 2–2.8 were performed.

In the geometry with chambers of variable height the gas velocity field is more complex than in the geometry from the variant A. During the analysis of the distribution of the gas velocity, a better adaptation of chambers' size to the gas vortex can be observed, when compared with the variant A. The higher is the gas velocity in the clearance, the greater are the chambers' volume and vortex.

The gas velocity before clearances is significantly lower than in the variant A (Figure 14). Furthermore, a significantly weaker gas stream of a high velocity is observed in chambers behind clearances. When compared with the initial geometry and data from the variant A, in chambers no. 2, 4, 5, and 6 additional gas vortices were obtained in the area of the outlet from the clearance behind back walls of shorter teeth, which made the gas flow be more dissipative.

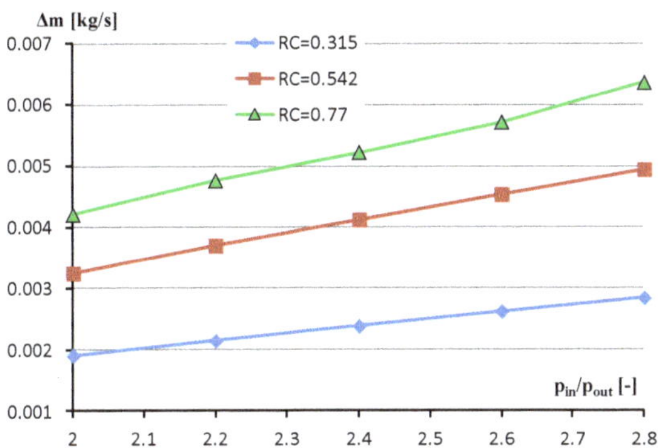

Figure 13. Leakage reduction for geometries 8t of the clearance height RC = 0.315, 0.542, and 0.77 mm depending on the pressure ratio for T_{in} = 300 K, p_{out} = 10^5 Pa.

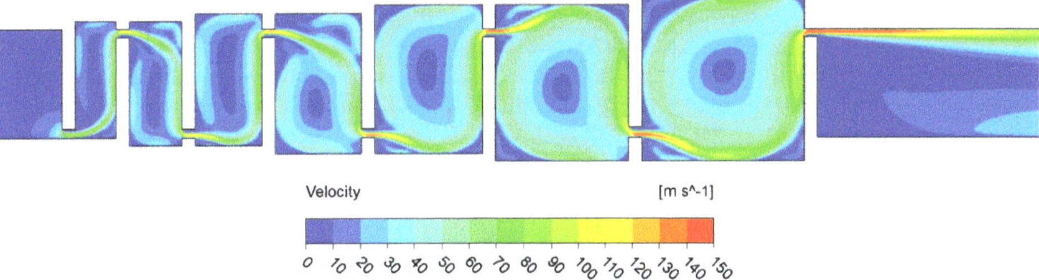

Figure 14. Velocity fields in the designed geometry as per variant B for p_{in}/p_{out} = 2.4, T_{in} = 300 K, p_{out} = 10^5 Pa.

To sum up, more complicated flow results in a significant reduction of the leakage rate. Obtained results concerning the gas mass flow for the geometry in the variant B are summarized in Table 7.

Table 7. Variant B—change of the leakage rate by Equation (18) depending on the pressure ratio p_{in}/p_{out} for the initial geometry 8t and the improved one of the staggered seal RC = 0.315 mm, T_{in} = 300 K, p_{out} = 10^5 Pa.

p_{in}/p_{out} [-]	$\dot{m}_{LPconst}$ (t = 8) [kg/s]	$\dot{m}_{LPdm}$ (t = 8) [kg/s]	$\delta\dot{m}$ (t) [%]	$\Lambda\dot{m}$[kg/s]
2	0.0198	0.0167	15.7	0.00311
2.2	0.0224	0.0189	15.5	0.00348
2.4	0.0249	0.0211	15.3	0.00381
2.6	0.0273	0.0232	15.1	0.00413
2.8	0.0297	0.0252	15.2	0.00452

Application of the method for the improved spacing of teeth and changing the chambers depth enabled a significant reduction of the leakage rate by approx. 15.4%. The relative reduction of the leakage rate reveals a slightly decreasing trend for growing pressure ratio p_{in}/p_{out}. It is a good result for the leakage rate reduction when a slight change of the seal height with keeping the length unchanged is taken into consideration. It should be noted that in the designed geometry the gas flow field in clearances has not changed.

Pressure drop has no significant impact on the relative leakage decrease, therefore representative results for Variant A and B are summarized in Table 8.

Table 8. Summary of change of the leakage rate by Equations (17) and (18) for application of the design method for Variant A and B, for the initial geometry (8t) and the improved one 8t and 9t, $p_{in}/p_{out} = 2$ $T_{in} = 300$ K, $p_{out} = 10^5$ Pa.

Variant	RC [mm]	$\delta\dot{m}$ (t = 8) [%]	$\delta\dot{m}$ (t = 9, 8) [%]
A	0.315	3.4	9.65
	0.542	3.5	10.78
	0.77	3.3	10.95
B	0.315	15.7	-

Variant A of the method for the geometry without change of teeth number ($t = 8$) brings improvement of leak-tightness $\delta\dot{m}(t)$ in the range from 3.3% to 3.5%. For geometry with the increased number of teeth (variant A), a significantly reduced leakage ($\delta\dot{m}(t, 8)$ was achieved in the range from 9.65% to 10.95%. Variant B enables a significant reduction of the leakage (even by 15.7%) without the increased number of teeth.

4. Conclusions

Minimization of the leakage rate in labyrinth seals is a problem of great importance since it enables improvement of high-power fluid-flow machines efficiency. This paper presented two variants of the method for aerodynamic design of the seal geometry. Variant A can be used when it is impossible to change external dimensions of the seal; it consists in changing the location of the seal teeth. In this variant, the design method anticipates the interference in the geometry of internal elements of the seal. Variant B can be used when there is a possibility to change slightly the height of seal chambers in the fluid-flow machine.

The design method in both variants, A and B, enables improvement of seals in different machines, newly designed as well as already installed but under modernization or repair. This method is based on the observation of thermodynamic and flow phenomena occurring in labyrinth seals, such as the gas flux of a high velocity formation, and the mechanism of kinetic energy dissipation. The above described method allows for determining the resultant geometry unequivocally.

The method in the variant A enables changing the pitch length for the staggered seals to obtain the reduced leakage. For the representative straight through seal geometry the leakage rate reduction ranging from 3.4% to 15.5% had been obtained, when compared with the initial geometry of a constant pitch.

The method in the variant B brings significant reduction of the leakage rate by 15.4% without changing the seal length and with a slight change of the seal height.

Results of calculations presented in this paper confirm the effectiveness of the design method for both variants irrespective of the value of the pressure ratio p_{in}/p_{out} in the range from 2 to 2.8. Results presented in the paper show that the use of the improved geometry enables the reduction of leakage of various degrees of wear of the seal, from radial clearance 0.315 to 0.77 mm.

Currently, there is no similar approach to the design of seals in the literature. The method discussed in this article opens up a new perspective on seal design. It is planned to use this design method to other geometries of sealing and to conduct experimental research.

Funding: This research received no external funding.

Institutional Review Board Statement: Not applicable.

Informed Consent Statement: Not applicable.

Data Availability Statement: Not applicable.

Conflicts of Interest: The authors declare no conflict of interest.

Nomenclature

A_e	area below the approximating curve [-]
B	tooth thickness [m]
D	seal diameter [m]
E	specific kinetic energy [J/kgK]
e	non-dimensional kinetic energy [-]
$e_{max}(i)$	i-th local maximum of non-dimensional kinetic energy [-]
e(x)	function approximating local maxima of non-dimensional kinetic energy $e_{max}(i)$ [-]
H	segment height [m]
HC(i)	height of the i-th chamber [m]
LC(i)	length of the i-th chamber [m]
LS	segment length [m]
LP	pitch length [m]
LP(i)	length of the i-th pitch [m]
$\dot{m}$	mass flow [kg/s]
$\delta \dot{m}$	relative change in mass flow [-]
n	number of chamber [-]
p	pressure [Pa]
RC	radial clearance [m]
t	number of teeth [-]
u	gas velocity in the axial direction [m/s]
$\vec{v}$	gas velocity vector [m/s]
X	coordinate along the seal axis [m]
x	non-dimensional coordinate along the seal axis [-]
indices:	
in	total parameter upstream the seal
LPconst	constant pitch
LPdm	method design
max	maximum value
min	minimum value
out	static parameter downstream the seal
Greek symbols:	
Δ	–absolute difference
δ	–relative difference

References

1. Krzyślak, P.; Winowiecki, M. A method of diagnosing labyrinth seals in fluid-flow machines. *Pol. Marit. Res.* **2008**, *15*, 38–41. [CrossRef]
2. Anker, J.E.; Mayer, J.F.; Stetter, H. Computational study of the flow in an axial turbine with emphasis on the interaction of labyrinth seal leakage flow and main flow. *High Perform. Comput. Sci. Eng.* **2002**, *1*, 363–371.
3. Liang, D.; Jin, D.; Gui, X. Investigation of seal cavity leakage flow effect on multistage axial compressor aerodynamic performance with a circumferentially averaged method. *Appl. Sci.* **2021**, *11*, 3937. [CrossRef]
4. Larjola, J.; Honkatukia, J.; Sallinen, P.; Backman, J. Fluid dynamic modeling of a free piston engine with labyrinth seals. *J. Therm. Sci.* **2010**, *19*, 141–147. [CrossRef]
5. Feng, J.; Wang, L.; Yang, H.; Peng, X. Numerical Investigation on the Effects of Structural Parameters of Labyrinth Cavity on Sealing Performance. *Math. Probl. Eng.* **2018**, *2018*, 5273582. [CrossRef]
6. Graunke, K. Labyrinth leakage flow in a labyrinth-piston compressor. *Sulzer Tech. Rev.* **1985**, *67*, 30–33.
7. Martin, H.M. Labyrinth packings. *Engineering* **1908**, *85*, 35–38.
8. Egli, A. The leakage of steam through labyrinth seals. *Trans. ASME* **1935**, *57*, 115–122.
9. Hodkinson, B. Estimation of the leakage through a labyrinth gland. *Proc. Inst. Mech. Eng.* **1939**, *141*, 283–288. [CrossRef]
10. Zimmermann, H.; Wolff, K.H. Comparison Between Empirical and Numerical Labyrinth Flow Correlations. *Am. Soc. Mech. Eng.* **1987**. [CrossRef]
11. Childs, D.W.; Scharrer, J.K. An Iwatsubo-Based Solution for Labyrinth Seals: Comparison to Experimental Results. *J. Eng. Gas Turbines Power* **1986**, *108*, 325–331. [CrossRef]
12. Scharrer, J.K. Theory versus Experiment for the Rotordynamic Coefficients of Labyrinth Gas Seals: Part I—A Two Control Volume Model. *J. Vib. Acoust. Stress Reliab.* **1988**, *110*, 270–280. [CrossRef]
13. Zhu, G.; Liu, W. Analysis of calculational methods on leakage for labyrinth seals. *Lubr. Eng.* **2006**, *4*.

14. Zhu, Y.; Jiang, Y.; Liang, S.; Guo, C.; Guo, Y.; Cai, H. One-dimensional computation method of supercritical CO_2 labyrinth seal. *Appl. Sci.* **2020**, *10*, 5771. [CrossRef]

15. Joachimmiak, D.; Krzyslak, P. Analysis of the Gas Flow in a Labyrinth Seal of Variable Pitch. *J. Appl. Fluid Mech.* **2019**, *12*, 921–930. [CrossRef]

16. Hur, M.S.; Lee, S.I.; Moon, S.W.; Kim, T.S.; Kwak, J.S.; Kim, D.H.; Jung, I.Y. Effect of clearance and cavity geometries on leakage performance of a stepped labyrinth seal. *Processes* **2020**, *8*, 1496. [CrossRef]

17. Baek, S.I.; Ahn, J. Optimizing the Geometric Parameters of a Straight-Through Labyrinth Seal to Minimize the Leakage Flow Rate and the Discharge Coefficient. *Energies* **2021**, *14*, 705. [CrossRef]

18. Hur, M.-S.; Moon, S.-W.; Kim, T.-S. A Study on the Leakage Characteristics of a Stepped Labyrinth Seal with a Ribbed Casing. *Energies* **2021**, *14*, 3719. [CrossRef]

19. Savvakis, S.; Mertzis, D.; Nassiopoulos, E.; Samaras, Z. A Design of the Compression Chamber and Optimization of the Sealing of a Novel Rotary Internal Combustion Engine Using CFD. *Energies* **2020**, *13*, 2362. [CrossRef]

20. Zou, Z.; Shao, F.; Li, Y.; Zhang, W.; Berglund, A. Dominant flow structure in the squealer tip gap and its impact on turbine aerodynamic performance. *Energy* **2017**, *138*, 167–184. [CrossRef]

21. Yang, J.; Zhao, F.; Zhang, M.; Liu, Y.; Wang, X. Numerical Analysis of Labyrinth Seal Performance for the Impeller Backface Cavity of a Supercritical CO_2 Radial Inflow Turbine. *Comput. Model. Eng. Sci.* **2021**, *126*, 935–953. [CrossRef]

22. Čížek, M.; Pátek, Z.; Vampola, T. Aircraft turbine engine labyrinth seal CFD sensitivity analysis. *Appl. Sci.* **2020**, *10*, 6830. [CrossRef]

23. Joachimiak, D.; Krzyślak, P. Experimental Research and CFD Calculations Based Investigations Into Gas Flow in a Short Segment of a Heavily Worn Straight Through Labyrinth Seal. *Pol. Marit. Res.* **2017**, *24*, 83–88. [CrossRef]

24. Joachimiak, D.; Frąckowiak, A. Experimental and Numerical Analysis of the Gas Flow in the Axisymmetric Radial Clearance. *Energies* **2020**, *13*, 5794. [CrossRef]

25. Zhou, W.; Zhao, Z.; Wang, Y.; Shi, J.; Gan, B.; Li, B.; Qiu, N. Research on leakage performance and dynamic characteristics of a novel labyrinth seal with staggered helical teeth structure. *Alex. Eng. J.* **2021**, *60*, 3177–3187. [CrossRef]

26. Zhirong, L.; Xudong, W.; Xin, Y.; Naoki, S.; Taro, N. Investigation and improvement of the staggered labyrinth seal. *Chin. J. Mech. Eng.* **2015**, *28*, 402–408. [CrossRef]

27. Cremanns, K.; Roos, D.; Hecker, S.; Dumstorff, P.; Almstedt, H.; Musch, C. Efficient multi-objective optimization of labyrinth seal leakage in steam turbines based on hybrid surrogate models. In Proceedings of the ASME Turbo Expo 2016: Turbomachinery Technical Conference and Exposition, Seoul, Korea, 13–17 June 2016; Volume 2C-2016, pp. 1–11.

28. Szymanski, A.; Wróblewski, W.; Fraczek, D.; Bochon, K.; Dykas, S.; Marugi, K. Optimization of the Straight-Through Labyrinth Seal With a Smooth Land. *J. Eng. Gas Turbines Power* **2018**, *140*, 122503. [CrossRef]

29. Zhao, Y.; Wang, C. Shape Optimization of Labyrinth Seals to Improve Sealing Performance. *Aerospace* **2021**, *8*, 92. [CrossRef]

30. Asok, S.P.; Sankaranarayanasamy, K.; Sundararajan, T.; Rajesh, K.; Sankar Ganeshan, G. Neural network and CFD-based optimisation of square cavity and curved cavity static labyrinth seals. *Tribol. Int.* **2007**, *40*, 1204–1216. [CrossRef]

31. Joachimiak, D.; Krzyślak, P. A model of gas flow with friction in a slotted seal. *Arch. Thermodyn.* **2016**, *37*, 95–108. [CrossRef]

32. Joachimiak, M. Analysis of thermodynamic parameter variability in a chamber of a furnace for thermo-chemical treatment. *Energies* **2021**, *14*, 2903. [CrossRef]

33. Frackowiak, A.; Olejnik, A.; Wróblewska, A.; Ciałkowski, M. Application of the protective coating for blade's thermal protection. *Energies* **2021**, *14*, 50. [CrossRef]

34. Joachimiak, D.; Krzyślak, P. Comparison of results of experimental research with numerical calculations of a model one-sided seal. *Arch. Thermodyn.* **2015**, *36*, 61–74. [CrossRef]

35. Joachimmiak, D. Universal Method for Determination of Leakage in Labyrinth Seal. *J. Appl. Fluid Mech.* **2020**, *13*, 935–943. [CrossRef]

36. Eser, D.; Dereli, Y. Comparisons of rotordynamic coefficients in stepped labyrinth seals by using Colebrook-White friction factor model. *Meccanica* **2007**, *42*, 177–186. [CrossRef]

37. Wang, W.; Liu, Y.; Jiang, P.; Chen, H. Numerical Analysis of Leakage Flow Through Two Labyrinth Seals. *J. Hydrodyn.* **2007**, *19*, 107–112. [CrossRef]

38. Morrison, G.L.; Al-Ghasem, A. Experimental and Computational Analysis of a Gas Compressor Windback Seal. In Proceedings of the ASME Turbo Expo 2007: Power for Land, Sea, and Air, Montreal, ON, Canada, 14–17 May 2007; Volume 4, pp. 1231–1247.

39. ANSYS, Inc. *ANSYS Fluent Theory Guide 15*; ANSYS, Inc.: Canonsburg, PA, USA, 2019.

40. Celik, I.B.; Ghia, U.; Roache, P.J.; Freitas, C.J.; Coleman, H.; Raad, P.E. Procedure for estimation and reporting of uncertainty due to discretization in CFD applications. *J. Fluids Eng. Trans. ASME* **2008**, *130*, 780011–780014. [CrossRef]

41. Menter, F.R. Two-equation eddy-viscosity turbulence models for engineering applications. *AIAA J.* **1994**, *32*, 1598–1605. [CrossRef]

42. Li, J.; Wen, K.; Wang, S.; Jiang, S.; Kong, X. Experimental and numerical investigations on the leakage flow characteristics of labyrinth seals. In Proceedings of the ASME Turbo Expo 2012: Turbine Technical Conference and Exposition, Copenhagen, Denmark, 11–15 June 2012; pp. 164–172.

Article

Low-Cost Air-Cooling System Optimization on Battery Pack of Electric Vehicle

Robby Dwianto Widyantara [1], Muhammad Adnan Naufal [1], Poetro Lebdo Sambegoro [1], Ignatius Pulung Nurprasetio [1], Farid Triawan [2], Djati Wibowo Djamari [2], Asep Bayu Dani Nandiyanto [3], Bentang Arief Budiman [1,*] and Muhammad Aziz [4,*]

[1] Faculty of Mechanical and Aerospace Engineering, Institut Teknologi Bandung, Jl. Ganesha No. 10, Bandung 40132, Indonesia; robbydwianto@gmail.com (R.D.W.); adnan.naufal22@hotmail.com (M.A.N.); poetro@ftmd.itb.ac.id (P.L.S.); ipn@ftmd.itb.ac.id (I.P.N.)
[2] Department of Mechanical Engineering, Faculty of Engineering and Technology, Sampoerna University, Jl. Raya Pasar Minggu No. 16, Jakarta 12780, Indonesia; farid.triawan@sampoernauniversity.ac.id (F.T.); djati.wibowo@sampoernauniversity.ac.id (D.W.D.)
[3] Departemen Kimia, Universitas Pendidikan Indonesia, Jl. Dr. Setiabudi No. 229, Bandung 40154, Indonesia; nandiyanto@upi.edu
[4] Institute of Industrial Science, The University of Tokyo, 4-6-1 Komaba, Meguro-ku, Tokyo 153-8505, Japan
* Correspondence: bentang@ftmd.itb.ac.id (B.A.B.); maziz@iis.u-tokyo.ac.jp (M.A.)

Citation: Widyantara, R.D.; Naufal, M.A.; Sambegoro, P.L.; Nurprasetio, I.P.; Triawan, F.; Djamari, D.W.; Nandiyanto, A.B.D.; Budiman, B.A.; Aziz, M. Low-Cost Air-Cooling System Optimization on Battery Pack of Electric Vehicle. *Energies* **2021**, *14*, 7954. https://doi.org/10.3390/en14237954

Academic Editor: Artur Bartosik

Received: 21 October 2021
Accepted: 22 November 2021
Published: 28 November 2021

Publisher's Note: MDPI stays neutral with regard to jurisdictional claims in published maps and institutional affiliations.

Abstract: Temperature management for battery packs installed in electric vehicles is crucial to ensure that the battery works properly. For lithium-ion battery cells, the optimal operating temperature is in the range of 25 to 40 °C with a maximum temperature difference among battery cells of 5 °C. This work aimed to optimize lithium-ion battery packing design for electric vehicles to meet the optimal operating temperature using an air-cooling system by modifying the number of cooling fans and the inlet air temperature. A numerical model of 74 V and 2.31 kWh battery packing was simulated using the lattice Boltzmann method. The results showed that the temperature difference between the battery cells decreased with the increasing number of cooling fans; likewise, the mean temperature inside the battery pack decreased with the decreasing inlet air temperature. The optimization showed that the configuration of three cooling fans with 25 °C inlet air temperature gave the best performance with low power required. Even though the maximum temperature difference was still 15 °C, the configuration kept all battery cells inside the optimum temperature range. This finding is helpful to develop a standardized battery packing module and for engineers in designing low-cost battery packing for electric vehicles.

Keywords: electric vehicle; battery thermal management system; optimization; lattice Boltzmann method

1. Introduction

In the last three decades, electric vehicles (EVs) have been developed rapidly and have brought a massive transformation in the automotive industry due to their low emission and energy-efficient advantages over internal combustion engine vehicles (ICEV) [1,2]. However, despite its rapid development, the current technology of EV still has several drawbacks compared to ICEV. One of the main disadvantages is the high battery cost, which is more than 40% of the total EV price [3]. This high cost is partly caused by the advanced technological implementation in the battery packing to assure high performance and high safety standards, such as complex cooling systems, massive structural packing protection, and advanced electrical system for energy management and control.

The batteries experience high charging and discharging rates during EV operation, increasing their surface temperature [4]. The frequency of charging and discharging also impacts battery degradation, decreasing the battery's lifetime [5]. To realize the best performance and longest lifetime, the batteries need to be operated under the optimal

201

temperature condition. This optimal temperature condition for a lithium-ion battery consists of two terms: (1) the optimal operating temperature range, which states the temperature range where a battery cell gives optimal charging and discharging rate while maintaining the longest life cycle; and (2) the maximum temperature difference, which states the maximum difference in temperature between each battery cells to one another to provide relatively uniform charging and discharging rates.

The optimal temperature for lithium-ion battery cells to operate is in the range of 25 to 40 °C, with a maximum temperature difference among battery cells of 5 °C [6]. Operating outside the optimal temperature range can decrease the battery's performance significantly [7]. Moreover, safety issues like thermal runaway may arise when the battery operates in high-temperature conditions [8]. Temperature differences among battery cells may occur due to an inappropriate cooling system, especially when air cooling is applied. These differences may cause each battery cell to have different charging–discharging rates. Therefore, they are not beneficial because the battery management system (BMS) must work hard in balancing the state of charge (*SOC*) of the batteries, which can degrade its lifespan faster.

A battery thermal management system (BTMS) plays an essential role in maintaining the temperature of batteries at the optimal operating temperature [9]. An optimum BTMS can also reduce the workload of the BMS by lowering the temperature differences among battery cells. Therefore, various kinds of BTMS are applied and installed in the battery packing, such as air cooling [10], liquid cooling [11], heat pipe [12], and phase change materials (PCMs) [13]. Being a novel medium for BTMS and having high efficiency and stable performance in extreme conditions, PCMs have gained popularity in recent times. However, PCMs have the disadvantages of having low conductivity and needing to be regenerated after completely melted [14].

On the other hand, heat pipe systems do not suffer from these drawbacks, and they do not require an external power supply. They have high conductivity and efficiency in reducing battery temperature rise, but the equipment is complicated and not conducive to the practical applications of EV [15,16]. Another thermal management system, liquid-cooled BTMS, is also complex with its many supporting devices, like pumps, fans, and pipes, making it costly and vulnerable to the risk of leakage that may lead to a short circuit [17].

Another cooling method in BTMS is an air-cooling system. This system has a simple configuration, low initial and maintenance cost, simple integration, and it possesses no risk of leakage, making it more favorable in the market compared to the methods mentioned above. Moreover, air-cooling systems can significantly lower battery manufacturing costs, which directly reduces the EV price. However, despite its advantages, low heat transfer coefficient, uneven temperature distribution, and low efficiency are the main drawbacks of air-cooled BTMS [18,19]. To overcome the low heat transfer coefficient of air as the cooling medium, a hybrid system of BTMS was developed either by combining air-cooling with PCMs [20] or by integrating air-cooling with mini-channel liquid cooling [21]. These studies successfully lower the battery's temperature, but on the other hand, increase the BTMS power consumption.

Adjusting the structure and flow configuration of the battery pack could also be beneficial to lower the temperature difference between battery cells and increase the efficiency of air-cooled BTMS. Several studies were conducted to achieve this result by optimizing the shape of battery pack [22,23]. Xu et al. [22] discovered that a horizontal battery pack with a double U-type duct could improve the heat dissipation performance of the air-cooling system in various conditions. Zhang et al. [23] minimized the temperature difference in battery packs for prismatic battery cells for Z-, U-, and I-types air-cooled BTMS by optimizing the widths of parallel cooling channels and divergence/convergence ducts. Other studies also have focused on improving the air-cooled BTMS by adding parts to the battery pack [24,25]. Mohammadian et al. [24] studied that thermal management of air-cooling systems of high-power lithium-ion batteries could be enhanced by implementing aluminum metal porous. Hong et al. [25] improved air-cooled BTMS performance in reducing temperature differences by adding a secondary vent. And some studies did both, optimizing the shape and adding parts to the battery pack. Jiaqiang et al. [26] improved the performance of the air-cooling strategy by locating the lateral inlet and outlet on different sides and utilizing the baffle plates. Other studies have tried to improve air-cooled BTMS performance by optimization. Liu et al. [27] performed manifold size optimization to improve J-type BTMS thermal performance under varying working conditions, resulting in the optimal configurations for each battery working condition. To achieve optimal performance for different battery working conditions, a valve control mechanism was needed to control the manifold size. Chen et al. [28] found that optimization of airflow parallel outlet position can improve the performance of J-type BTMS for prismatic battery cells; although, optimization for the airflow parallel inlet position is not as effective.

The studies mentioned above [22–28] may have successfully improved BTMS performance, but their practical implementation on EVs is still challenging, especially in reliability, energy density, and feasibility due to its complexity. Moreover, there has been no standard and design guideline regulating the battery packing until now, which leads to complex designs for battery packing. In addition, to improve the applicability, the optimization design should consider the air properties, number and position of cooling fans, spacing between cells, and other related factors [29]. These factors must be determined during the design process, together with maintaining lower manufacturing costs.

Despite the numerous studies conducted to improve the performance of air-cooling systems either by combining the air-cooling system with another type of cooling system or adjusting the structure and airflow configuration of the battery pack, not a lot have been done by optimizing the number of cooling fans and the inlet air temperature. By optimizing these parameters, this research aimed to develop a simple yet reliable air-cooling system that can simultaneously maintain high energy density by conditioning the battery temperature inside the optimal temperature condition while having low manufacturing cost.

Thermal analysis was necessary to investigate the effect of the number of cooling fans and inlet air temperature on the BTMS performance for the optimization. Some studies have performed thermal analysis for electric vehicle batteries using simulation. For example, Raharjo et al. [30] conducted a thermal analysis of modular battery by computational fluid dynamics (CFD) simulation to understand its thermal behavior, prevent overheating, and maintain battery life. Divakaran et al. [31] performed finite element simulation to analyze 18,650 lithium-ion batteries' thermal behavior under two conditions: with and without cooling systems.

To understand the cooling phenomenon, the battery packing was built in a 3D numerical model and analyzed using CFD simulation based on the lattice Boltzmann method (LBM). The effects of the number of cooling fans and inlets on air velocity and temperature distribution inside the battery pack were revealed, and the simulation results from the LBM analysis were compared. The effects of the number of cooling fans and inlets along with the inlet air temperature on the performance of BTMS were then discussed thoroughly. Furthermore, an optimized cooling strategy for air-cooled BTMS based on the consideration of temperature distribution and power consumption was developed. The results of this study are significant to develop a standardized battery packing module and enrich the literature on electric vehicle battery pack optimization.

2. Numerical Model and Simulation

2.1. Heat Generation in Battery Cell

In battery packing, the battery cells generate heat during the charging and discharging processes. The initial temperature was set according to the ambient temperature. The heat generation of the lithium-ion battery module consisted of two primary sources: the irreversible heat caused by the Joule's heating based on the internal resistance (Q_{irr}) and the reversible heat inside the battery (Q_{rev}). From these sources, the battery heat generation can be represented by Equation (1) [26].

$$Q = I^2 R_e - I\left[T\frac{dE}{dT}\right] \tag{1}$$

where I is the current flowing through the battery (A), R_e is the internal resistance (Ω), T is the temperature (K), dE/dT is the entropy coefficient (V/K), and Q is the battery heat generating rate (W). R_e and dE/dT are a function of battery temperature (T) and SOC as shown in Equations (2) and (3), respectively.

$$R_e = \left(\begin{array}{c} -112 \times SOC^3 - 0.203 \times SOC^2 \times T + 0.000737 \times SOC \times T^2 \\ +0.00000753 \times T^3 + 301 \times SOC^2 - 0.144 \times SOC \times T \\ -0.0061 \times T^2 - 188 \times SOC + 1.28 \times T + 23.6 \end{array}\right) \times 10^{-3} \tag{2}$$

$$\frac{dE}{dT} = \left(-0.342 + 0.979 \times SOC - 1.49 \times SOC^2 + 0.741 \times SOC^3\right) \times 10^{-3} \tag{3}$$

To model the battery in an extreme condition, we used a current of 2.6 A for 1 C discharge rate, battery temperature of 20 °C, SOC of 0.1, along with the parameters shown in Table 1. The parameters were under the assumption of operation in a tropical region with the already included ambient temperature. Based on Equations (1) to (3), the lower the SOC or the battery temperature, the higher the heat generated from the battery. SOC of 0.1 was the limit value used in the experiment by Jiaqiang et al. according to the recommendation of the battery manufacturer [26]. Meanwhile, the battery temperature of 20 °C was used as the low temperature in possible operating conditions in a tropical region with an average ambient temperature of 30 °C. The heat generation rate of the battery module was calculated with the result of 81.02 W. This value was used for the simulation input for battery cell heat generation.

Table 1. The specification of the lithium-ion battery investigated in this study.

Characteristics	Specifications
Nominal capacity [a]	2.6 Ah
Nominal voltage [a]	3.7 V
Cell mass [a]	0.0475 kg
Thermal conductivity in the axial direction [b]	37.6 W/m·K
Thermal conductivity in the radial direction [b]	0.2 W/m·K
Specific capacity [c]	1200 J/kg·K
Cell diameter	18.3 mm
Cell length	64.5 mm

[a] Ref. no [26]. [b] Ref. no [32]. [c] Ref. no [33].

2.2. System Design

The battery pack design in this study is shown in Figure 1. The interconnection of the battery cells developed within the battery pack created an electrical and mechanical connection, and the casing covered the mechanical requirements for the battery pack. The battery pack also consisted of stiffeners and acrylic to support the battery pack under static and dynamic loads transferred from an electric frame or chassis. The conductor plate was used to flow electricity from the battery to other electrical components and vice versa. The BMS, which could be an active or passive balancer, was placed inside the BMS casing. The inlet and outlet of the airflow were created on the lateral side of the battery pack casing to investigate the effect of forced convection by the cooling fan on the temperature distribution within the battery pack. The battery cells used in this battery pack were lithium nickel manganese cobalt oxide (NMC) 18,650 cylindrical batteries, and the specification is presented in Table 1. The 18,650 cylindrical battery type was employed due to its popularity in its application for electric vehicle battery. This module was suitable for electric trike and city cars. The electric trike, in particular, is currently being developed in our research facilities and has been presented in several studies, such as by Reksowardojo et al. [34].

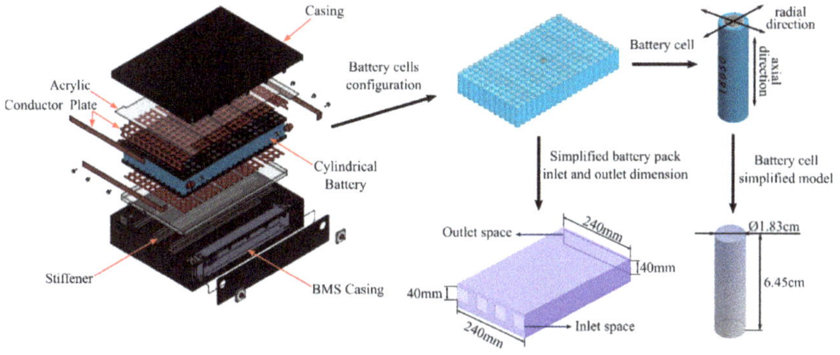

Figure 1. Battery pack design and its simplification model.

The original geometry of the battery pack was complex, so it was difficult to configure the airflow within the battery pack iteratively with the Ansys Discovery Live software. This situation could also lead to continuous errors and long simulation time. To achieve the optimum results using the software, the battery pack was modeled as a simplified form without removing the components sensitive to the heat generation within the battery pack. The battery cells themselves consisted of layers of cathode, anode, separator, and current collector. Although these different layers had their thermal properties, the detailed structure of the cylindrical cell presented in Figure 1 had an insignificant impact on the thermal performance of the battery, according to many references [35,36]. Therefore, the battery parameters listed in Table 1 were utilized as the equivalent values representing a whole battery cell.

The simplified model of the battery pack was created in the form of 240 cells enclosed by a box, as shown in Figure 1. Each battery cell is represented as a cylinder with a diameter of 18.3 mm and a height of 64.5 mm. The offset between the 240 cylinders and the box was 1 mm on each surface. This close proximity was applied to ensure the designed battery pack had a high density and compact geometry as space and weight in EV are the main constraints to increase energy density [37]. The distance between the center of the two adjacent cylinders was 20 mm, and the gap between them was 1.7 mm. The pack would act as the fluid, and the inlet came from one lateral side and the outlet on the other.

Forced convection utilized in this system was carried out using a cooling fan. The available area for the cooling fan was 240 mm × 40 mm; therefore, the maximum dimension of the cooling fan was 40 mm × 40 mm. The selected cooling fan was a chip cooler AP0405MX-J70 from Adda Corp. with a rated power of 0.7 W and a maximum airflow of 4.7 CFM or 7.99 m^3/h.

The effect of fluid flow within the battery pack on the temperature distribution was studied by conducting LBM-based transient CFD simulations. For the simulation, the airflow inlet and outlet were positioned at the lateral opposing end of the battery pack, while the other sides were set as the wall. Each inlet was accompanied by one cooling fan. For simplicity, from this point forth, the number of cooling fans and inlets will be called the number of inlets only, without mentioning the cooling fans. The number of inlets was varied into four configurations of 1-, 2-, 3-, and 4-inlets to investigate the effect of airflow configuration on temperature distribution inside the battery pack. Afterward, the inlet air temperature was varied by 20, 25, and 30 °C for each airflow configuration to investigate the effect of inlet air temperature on temperature distribution and to calculate the power required for each variation. Finally, the configuration with minimum temperature distribution and power consumption was chosen as the optimum cooling strategy.

2.3. Lattice Boltzmann Method

The simulation was conducted using LBM that discerned air at its mesoscopic scale. LBM has an advantage in its excellent numerical stability and can be solved efficiently on parallel computers [38]. Furthermore, due to its diffusive particle modeling, LBM could handle the simulation of complex topology [39]. In LBM, the motion of the air particles is modeled in a completely disordered direction and random manner whose velocities are distributed around the mean value. In this study, the structure is introduced as the Bhatnagar–Gross–Krook (BGK) (Equation (4)) model is discretized in the lattice unit of space x and time t [40].

$$f_i(x + e_i\Delta t, t + \Delta t) - f_i(x,t) = \Delta t \frac{1}{\tau}\left(f_i^{eq}(x,t) - f_i(x,t)\right) + \Delta t e_i F_k \qquad (4)$$

where f_i is the distribution function of air particles, Δt is the lattice time step, F_k is the external force, e_i is the discrete lattice velocity vector of a particle in a link, τ is the single relaxation time, f_i^{eq} is the local equilibrium distribution function, and i represents the air particle in specified lattice point.

The LBM simulation was conducted using ANSYS Discovery Live to generate instant results of the velocity and temperature distribution within the battery pack. The model used in the LBM simulation was created using ANSYS Space Claim that was integrated with ANSYS Discovery Live. The dimension of each inlet was 40 mm × 40 mm, and the outlet dimension was 240 mm × 40 mm. The volume of the fluid was then extracted from the inlet to the outlet that enclosed the cylinders. After this process, airflow from the inlet to the outlet could be seen in real-time or live. The flow velocity at the inlet was set to 1.4 m/s according to the cooling fan characteristics, and at the outlet, the relative pressure was set to 0 MPa. The ambient temperature was 30 °C according to the average ambient temperature for the tropical region where the BTMS was designed to be used. The air was assumed to be dry air having low thermal conductivity to represent the worst cooling condition. The thermal condition of the battery cells was set by adjusting the heat flow for each cylinder with the calculated heat generation rate of 0.338 W. The example of the model in ANSYS Discovery Live, consisting of battery cells and the volume of the fluid within the battery pack for four inlets, is shown in Figure 2.

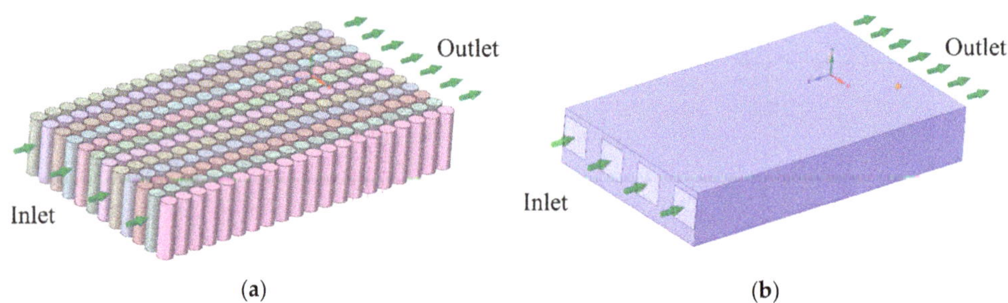

Figure 2. Example of LBM model (**a**) battery cells configuration (**b**) battery pack for 4 inlets.

The solution was created by assigning the inlet region and the outlet region enclosing the whole model. This region that enclosed the cylinders was extracted as the volume of fluid, which was the volume of air flowing. After the fluid volume was extracted, the velocity contour of the air flowing could be investigated from the maximum value or the small particles flowing within the battery packs. The number of particles flowing within the battery pack could be adjusted by setting the solution's speed or fidelity, where high-speed results reduced the fidelity, and the high fidelity reduced the speed results. This fidelity was based on the distance between the lattice points. The high speeds could generate the temperature contour within seconds, albeit not very accurately, because when the high-speed result was chosen, fewer particles of air were flowing within the cells. In this simulation for the 3D geometry model, the setting was adjusted to have higher fidelity than the speed, so the contour of the temperature distribution would resemble the real condition more accurately. The simulation in this work was performed in two configurations. The former was varying the number of inlets into 1, 2, 3, and 4 inlets with a constant inlet air temperature of 30 °C to study the effect of airflow configuration on temperature distribution. The latter was varying both the number of inlets (1, 2, 3, and 4 inlets) and the inlet air temperature (20, 25, and 30 °C) to find the optimum cooling strategy for the air-cooled BTMS.

3. Results and Discussion

3.1. LBM Simulation Results

The effect of airflow configuration on temperature distribution within the battery pack was studied by varying the number of inlets and simulated using LBM-based CFD. A constant air inlet temperature of 30 °C was used to represent the average air temperature of the tropical region where the BTMS was intended to be used. The performance of air-cooled BTMS could be represented by the air velocity and the temperature inside the battery pack. The air velocity affected the convective heat transfer of the air and indicated the amount of air flowing and taking the heat generated by the batteries. Meanwhile, the temperature was the main objective of BTMS and represented its performance in providing an optimal operating condition for the batteries. In this section, the distributions and the average values of air velocity and the temperature inside the battery pack were compared.

The velocity and temperature contour, along with the batteries whose temperatures were maximum and minimum of one, two, three, and four-inlets configuration, are presented in Figures 3–6 respectively. The velocity contour described the velocity distribution of the air traveling across the battery pack and indicated how the air could reach every spot inside the battery pack. The temperature contour showed the temperature distribution inside the battery pack after being cooled by the air.

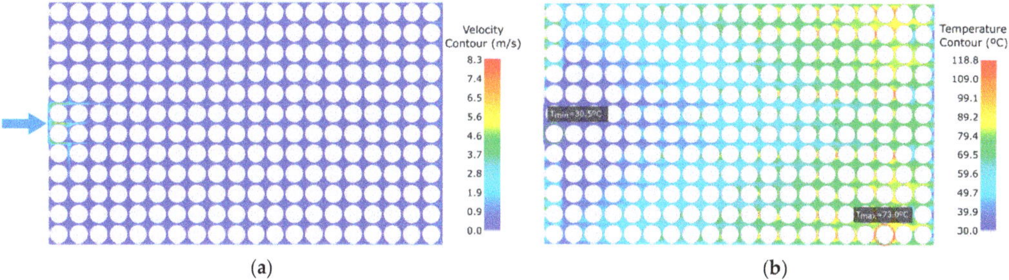

(a)   (b)

Figure 3. The contour of (**a**) velocity and (**b**) temperature of 1 inlet model.

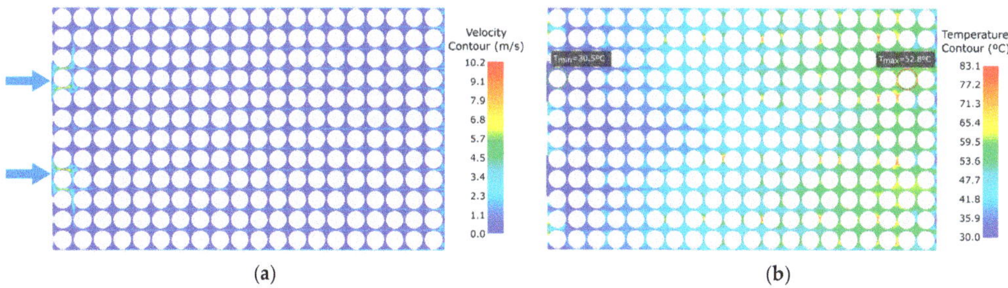

(a)   (b)

Figure 4. The contour of (**a**) velocity and (**b**) temperature of 2 inlet model.

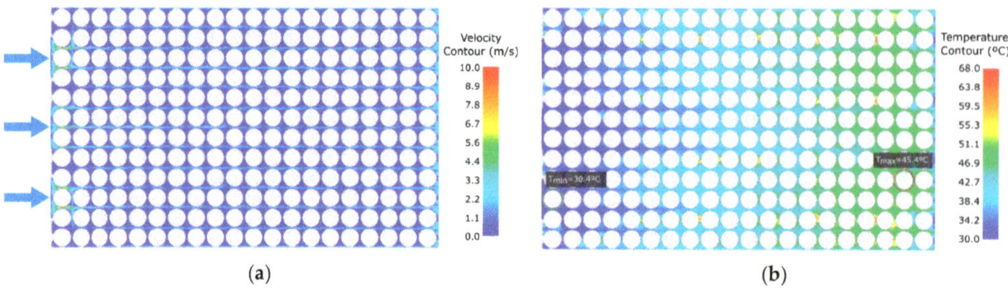

Figure 5. The contour of (**a**) velocity and (**b**) temperature of 3 inlet model.

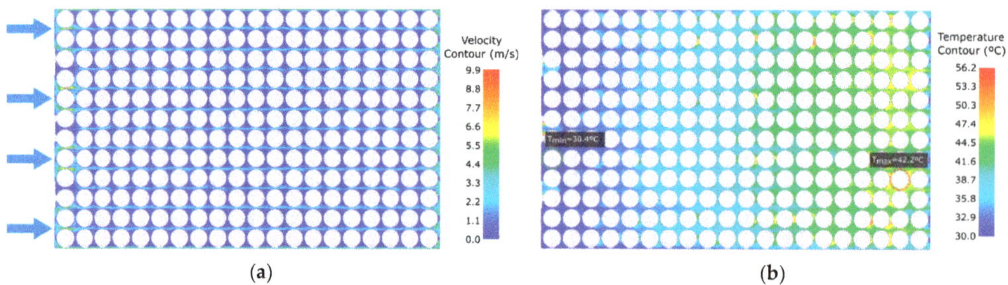

Figure 6. The contour of (**a**) velocity and (**b**) temperature of 4 inlet model.

The air entered the battery pack through the inlet on the left side and exited through the outlet on the right side while taking the heat generated by the batteries in the process. The arrows on the left side of the figure indicate the number and locations of the inlet and the airflow direction. The air velocity had its maximum value around the inlet as the nearest area to the cooling fan. It gradually decreased as it reached the outlet due to the decreasing kinetic energy. On the contrary, the battery temperature was at the minimum value around the inlet and gradually increased as it got further. This happened because the convective heat transfer decreased due to the rising temperature of the air as it traveled through the battery pack while taking the heat from the batteries. This phenomenon led to the temperature difference that resulted in different charging/discharging rates between battery cells, causing a heavy workload for BMS. Excessive workload caused inefficient equalization, especially for BMS with an active balancing method, leading to non-optimal power generation and reducing the batteries' lifespan faster. Ideally, the simulation result needs to be validated with an experiment, but obtaining velocity and temperature contour from an experiment might be challenging. However, the current simulation result was sufficient to show the temperature distribution, which represents the performance from each configuration.

The overall velocity and temperature results were represented by their average value. The average value of air velocity and batteries temperature from each configuration are plotted in Figure 7. The results generated showed that air velocity increased while the temperature decreased with the increasing number of inlets. The increase in the number of inlets meant an increase in the number of cooling fans. This led to more power pushing the air from the inlet to the outlet, resulting in higher air velocity and more air particles transferring the heat from the batteries. This condition, in turn, caused a lower temperature as the number of inlets increased. The lowest average temperature of 36.4 °C was achieved by the four inlets configuration; however, the maximum battery temperature was still 42.2 °C, which is higher than the maximum optimal temperature range of 40 °C. Therefore, an additional optimization strategy was needed for the air-cooled BTMS to meet the optimal temperature requirement. In the next section, we varied the inlet air temperature and the number of inlets to obtain the optimum cooling strategy.

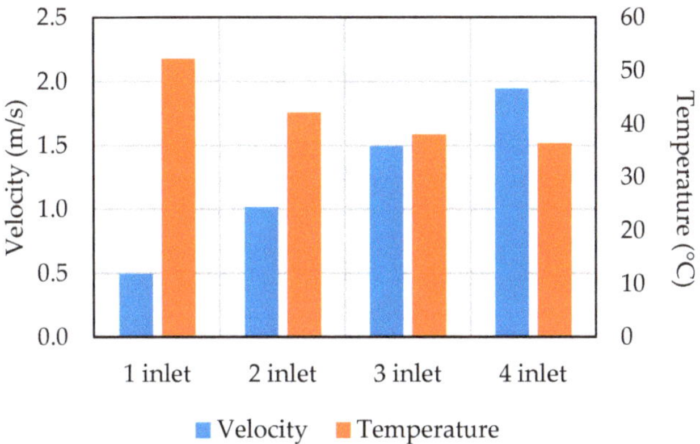

Figure 7. The average values of velocity and temperature vs. the number of inlets.

3.2. Optimization of Battery Packing Design

In addition to the number of inlets, the inlet air temperature was varied to 20, 25, and 30 °C to find the optimum temperature distribution inside the battery pack. Before entering the battery pack, the inlet air temperature was the conditioning of air temperature from its ambient temperature. The simulation is meant to model the battery operation in a tropical region, with no winter and summer, and where a temperature drop is unlikely to happen. Therefore, the ambient temperature does not change drastically throughout the year and can be considered a constant. We set the ambient temperature at 30 °C as it was the average temperature for the tropical region. Therefore, the inlet air temperature of 20 °C and 25 °C was obtained by cooling the ambient temperature. The number of inlet and inlet air temperature were fundamental factors that affected the BTMS performance, which was also simple and easy to manufacture once the design was optimized.

The simulation results of varying the inlet air temperature to each inlet configuration are presented in Figure 8 as a plot of temperature difference to the average temperature. It was fair to use average temperature as the representative parameter because one of the main objectives of the optimization was to obtain the lowest temperature difference that was still located within the optimal temperature range. The horizontal shaded area represents the optimal operating temperature, while the vertical one represents the optimal temperature difference for the lithium-ion battery, and the intersection between the two areas represents the optimal criteria for an optimized cooling strategy. From Figure 8, it can also be seen that for constant inlet air temperature, the temperature difference decreases as the number of inlets increases. The temperature difference was significantly affected by the number of inlets because higher inlet numbers had a wider inlet area that made it easier for air to reach every spot inside the battery pack.

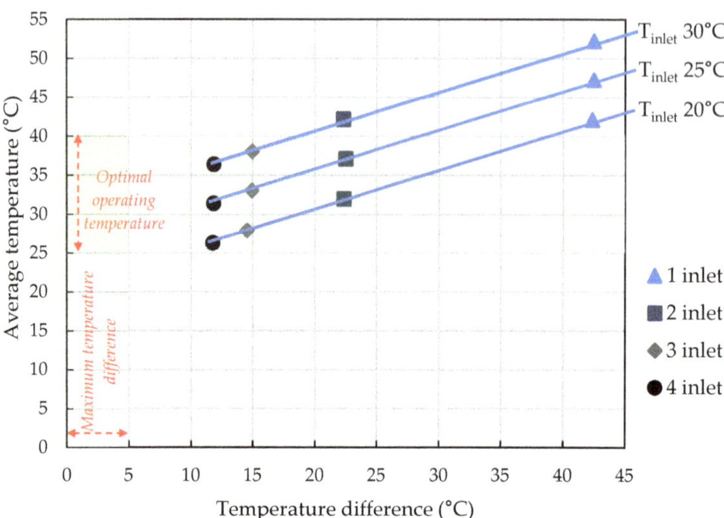

Figure 8. Temperature distribution of LBM model.

Meanwhile, the average temperature was affected by the inlet air temperature more than by the inlet number. This gave the idea of adjusting the inlet number and air temperature parameters to achieve any desired performance. The best airflow configuration was the four inlets configuration with 25 °C air inlet temperature. This configuration gave the lowest average temperature and the temperature differences that were still in the optimum range compared to all other configurations. However, using four cooling fans meant that the configuration cost more power to operate while potentially being over capacity. It is unjustified to have a high-performance BTMS that consumes a large amount of power since the objective of BTMS is to optimize the power generation for the vehicle operation. Therefore, the power required for each cooling fan and the inlet air temperature should also be considered to obtain the most energy-efficient cooling strategy giving the best performance.

The required power was calculated as the sum of the power required for operating the cooling fans and generating the sensible heat. The sensible heat represented the power needed to obtain the inlet air temperature of 20 °C and 25 °C by cooling the air from its ambient temperature of 30 °C before entering the battery pack. The sensible heat and the power required were calculated from Equations (5) and (6), respectively.

$$H_s = C_p \times \rho \times \forall \times dT \tag{5}$$

$$P = n \times P_f + H_s \tag{6}$$

where H_s is the sensible heat (kW), C_p is the specific heat of air (1.006 kJ/kg·°C), ρ is the density of air (1.202 kg/m^3), $\forall$ is air volume flow (m^3/s), dT is the temperature difference (°C), P is the power required (W), n is the number of cooling fans, and P_f is the cooling fan rated power (W).

The power required and the resulting temperature of all cooling strategies are plotted in Figure 9. Each strategy is represented by a marker that indicates its required power in x axis, and average temperature in y axis. The marker shape represents the inlet number of one, two, three, and four inlets by triangle, square, diamond, and circle, respectively. The color represents the inlet air temperature of 20 °C, 25 °C, and 30 °C by blue, green, and red, respectively. The dashed vertical line extending from the top and bottom of the markers represent the temperature range of the battery cells resulting in each corresponding configuration. The green shaded area represents the area of optimal operational temperature for the lithium-ion battery.

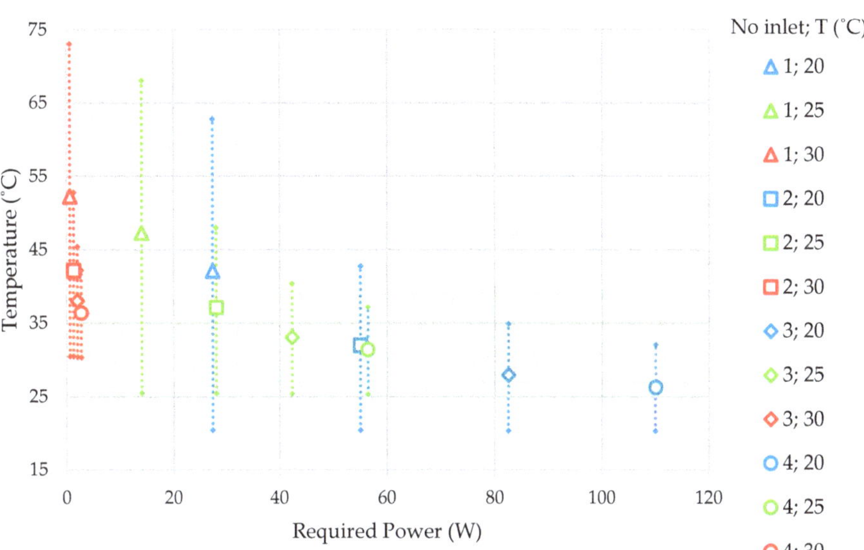

Figure 9. Temperature versus required power.

The optimal cooling strategy has the lowest median temperature and minimum temperature difference that meets the optimal temperature condition while having the least power required. Therefore, the optimal cooling strategy in Figure 9 is represented by the closest marker to the objective point (0 W, 25 °C), whose marker and dashed line are located inside the green shaded area. The optimization utilized the Qhull algorithm to find the nearest marker to the objective point. As a result, it was found that the three inlets configuration with the inlet air temperature of 25 °C was the best cooling strategy in terms of performance and power consumption and was chosen as the optimized cooling strategy for air-cooled BTMS. The configuration gave a mean temperature of 33.1 °C, which met the optimal temperature condition for a lithium-ion battery. However, the maximum temperature difference in this configuration was 14.9 °C, which did not meet the requirement. In fact, the temperature differences in all simulated configurations did not meet the requirement. This is the challenge of air-cooling applications for a battery pack.

Compared to the available BTMSs, the optimized design from this study has several advantages. Firstly, the optimized cooling strategy is low cost and easy to manufacture since there are no complex parts needed and it only consists of a few parts. In addition, the optimized design can meet the optimal temperature condition for a lithium-ion battery with low power consumption compared to other cooling methods such as water cooling, heat pipe, and PCMs. Therefore, the three inlets configuration with the inlet air temperature of 25 °C can be proposed as a standard design for a battery pack air-cooling system. For future work, when the simplicity is not the main objective, employing variable cooling fan speed might improve the performance while increase the system complexity for the additional control system. This study only discusses the optimization from the temperature distribution perspective. Therefore, further research is required to evaluate other important aspects, such as structural strength and water protection, to complement the design optimization of the battery pack air-cooling system.

4. Conclusions

The effect of the number of inlets and the inlet air temperature on air-cooled BTMS performance was studied using a LBM-based CFD simulation. To find the optimum cooling strategy, the power required to operate the configuration was also considered. In the end, an optimized cooling strategy for an air-cooled BTMS consisting of three inlets configuration with an inlet air temperature of 25 °C is recommended to achieve optimum performance in terms of temperature distribution and power consumption. The configuration gave the average temperature of 33.1 °C and the maximum temperature difference of 14.9 °C. Although the temperature difference is wider than the requirement of 5 °C, the chosen cooling strategy kept all the batteries inside the range of optimum operating temperature. Therefore, the optimized design can be proposed as a standard for a battery packing cooling system. For future work, an investigation on the effect of employing variable speed for cooling fans and a study for structural strength and water protection for the air-cooling systems are recommended.

Author Contributions: Conceptualization, P.L.S., F.T., B.A.B. and M.A.; methodology, P.L.S., I.P.N., B.A.B. and M.A.; modelling and design, M.A.N., F.T. and D.W.D.; validation, P.L.S. and I.P.N.; formal analysis, R.D.W., M.A.N. and D.W.D.; writing—original draft preparation, R.D.W., M.A.N., A.B.D.N. and B.A.B.; writing—review and editing, R.D.W., P.L.S., F.T., D.W.D., A.B.D.N., B.A.B. and M.A.; illustration, R.D.W. and I.P.N.; supervision, P.L.S., I.P.N., A.B.D.N. and B.A.B.; funding acquisition, B.A.B. All authors have read and agreed to the published version of the manuscript.

Funding: This research was funded by the Indonesia Endowment Fund for Education (LPDP) under Research and Innovation Program (RISPRO) for electric vehicle development with contract no. PRJ-85/LPDP/2020.

Institutional Review Board Statement: Not applicable.

Informed Consent Statement: Not applicable.

Conflicts of Interest: The authors declare that they have no known competing financial interests or personal relationships that could have appeared to influence the work reported in this paper. The funders had no role in the design of the study; in the collection, analyses, or interpretation of data; in the writing of the manuscript, or in the decision to publish the results.

References

1. Yuan, X.; Li, X. Mapping the technology diffusion of battery electric vehicle based on patent analysis: A perspective of global innovation systems. *Energy* **2021**, *222*, 119897. [CrossRef]
2. Aziz, M.; Oda, T.; Mitani, T.; Watanabe, Y.; Kashiwagi, T. Utilization of Electric Vehicles and Their Used Batteries for Peak-Load Shifting. *Energies* **2015**, *8*, 3720–3738. [CrossRef]
3. Tao, Y.; Huang, M.; Chen, Y.; Yang, L. Orderly charging strategy of battery electric vehicle driven by real-world driving data. *Energy* **2020**, *193*, 116806. [CrossRef]
4. Peng, X.; Cui, X.; Liao, X.; Garg, A. A Thermal Investigation and Optimization of an Air-Cooled Lithium-Ion Battery Pack. *Energies* **2020**, *13*, 2956. [CrossRef]
5. Huda, M.; Koji, T.; Aziz, M. Techno Economic Analysis of Vehicle to Grid (V2G) Integration as Distributed Energy Resources in Indonesia Power System. *Energies* **2020**, *13*, 1162. [CrossRef]
6. Malik, M.; Dincer, I.; Rosen, M.; Fowler, M. Experimental Investigation of a New Passive Thermal Management System for a Li-Ion Battery Pack Using Phase Change Composite Material. *Electrochim. Acta* **2017**, *257*, 345–355. [CrossRef]
7. Gandoman, F.; Jaguemont, J.; Goutam, S.; Gopalakrishman, R.; Firouz, Y.; Kalogiannis, T.; Omar, N.; Mierlo, J.V. Concept of reliability and safety assessment of lithium-ion batteries in electric vehicles: Basics, progress, and challenges. *Appl. Energy* **2019**, *251*, 113343. [CrossRef]
8. Sambegoro, P.L.; Budiman, B.A.; Philander, E.; Aziz, M. Dimensional and Parametric Study on Thermal Behaviour of Li-ion Batteries. In Proceedings of the 2018 5th International Conference on Electric Vehicular Technology (ICEVT), Surakarta, Indonesia, 30 October 2018; pp. 123–127. [CrossRef]
9. Al-Zareer, M.; Dincer, I.; Rosen, M. Heat transfer modeling of a novel battery thermal management system. *Numer. Heat Transf. Part A Appl.* **2018**, *73*, 277–290. [CrossRef]
10. Liu, Y.; Zhang, J. Self-adapting J-type air-based battery thermal management system via model predictive control. *Appl. Energy* **2020**, *263*, 114640. [CrossRef]

11. Dong, F.; Song, D.; Ni, J. Investigation of the effect of U-shaped mini-channel structure on the thermal performance of liquid-cooled prismatic batteries. *Numer. Heat Transf. Part A Appl.* **2019**, *77*, 105–120. [CrossRef]
12. Smith, J.; Singh, R.; Hinterberger, M.; Mochizuki, M. Battery thermal management system for electric vehicle using heat pipes. *Int. J. Therm. Sci.* **2018**, *134*, 517–529. [CrossRef]
13. Talluri, T.; Kim, T.H.; Shin, K.J. Analysis of a Battery Pack with a Phase Change Material for the Extreme Temperature Conditions of an Electrical Vehicle. *Energies* **2020**, *13*, 507. [CrossRef]
14. Lazrak, A.; Fourmigué, J.; Robin, J. An innovative practical battery thermal management system based on phase change materials: Numerical and experimental investigations. *Appl. Therm. Eng.* **2018**, *128*, 20–32. [CrossRef]
15. Chi, R.G.; Rhi, S.H. Oscillating Heat Pipe Cooling System of Electric Vehicle's Li-Ion Batteries with Direct Contact Bottom Cooling Mode. *Energies* **2019**, *12*, 1698. [CrossRef]
16. Putra, N.; Ariantara, B.; Pamungkas, R. Experimental investigation on performance of lithium-ion battery thermal management system using flat plate loop heat pipe for electric vehicle application. *Appl. Therm. Eng.* **2016**, *99*, 784–789. [CrossRef]
17. Fang, G.; Huang, Y.; Yuan, W.; Yang, Y.; Tang, Y.; Ju, W.; Chu, F.; Zhao, Z. Thermal management for a tube–shell Li-ion battery pack using water evaporation coupled with forced air cooling. *RSC Adv.* **2019**, *9*, 9951–9961. [CrossRef]
18. Li, X.; Zhao, J.; Yuan, J.; Duan, J.; Liang, C. Simulation and analysis of air cooling configurations for a lithium-ion battery pack. *J. Energy Storage* **2021**, *35*, 102270. [CrossRef]
19. Qin, P.; Sun, J.; Yang, X.; Wang, Q. Battery thermal management system based on the forced-air convection: A review. *eTransportation* **2021**, *7*, 100097. [CrossRef]
20. Ling, Z.; Wang, F.; Fang, X.; Gao, X.; Zhang, Z. A hybrid thermal management system for lithium ion batteries combining phase change materials with forced-air cooling. *Appl. Energy* **2015**, *148*, 403–409. [CrossRef]
21. Yang, W.; Zhou, F.; Zhou, H.; Wang, Q.; Kong, J. Thermal performance of cylindrical lithium-ion battery thermal management system integrated with mini-channel liquid cooling and air cooling. *Appl. Therm. Eng.* **2020**, *175*, 115331. [CrossRef]
22. Xu, X.; He, R. Research on the heat dissipation performance of battery pack based on forced air cooling. *J. Power Source* **2013**, *240*, 33–41. [CrossRef]
23. Zhang, J.; Wu, X.; Chen, K.; Zhou, D.; Song, M. Experimental and numerical studies on an efficient transient heat transfer model for air-cooled battery thermal management systems. *J. Power Source* **2021**, *490*, 229539. [CrossRef]
24. Mohammadian, S.; Rassoulinejad-Mousavi, S.; Zhang, Y. Thermal management improvement of an air-cooled high-power lithium-ion battery by embedding metal foam. *J. Power Source* **2015**, *296*, 305–313. [CrossRef]
25. Hong, S.; Zhang, X.; Chen, K.; Wang, S. Design of flow configuration for parallel air-cooled battery thermal management system with secondary vent. *Int. J. Heat Mass Transf.* **2018**, *116*, 1204–1212. [CrossRef]
26. Jiaqiang, E.; Yue, M.; Chen, J.; Zhu, H.; Deng, Y.; Zhu, Y.; Zhang, F.; Wen, M.; Zhang, B.; Kang, S. Effects of the different air cooling strategies on cooling performance of a lithium-ion battery module with baffle. *Appl. Therm. Eng.* **2018**, *144*, 231–241. [CrossRef]
27. Liu, Y.; Zhang, J. Design a J-type air-based battery thermal management system through surrogate-based optimization. *Appl. Energy* **2019**, *252*, 113426. [CrossRef]
28. Chen, K.; Wu, W.; Yuan, F.; Chen, L.; Wang, S. Cooling efficiency improvement of air-cooled battery thermal management system through designing the flow pattern. *Energy* **2019**, *167*, 781–790. [CrossRef]
29. Tete, P.; Gupta, M.; Joshi, S. Developments in battery thermal management systems for electric vehicles: A technical review. *J. Energy Storage* **2021**, *35*, 102255. [CrossRef]
30. Raharjo, J.; Wikarta, A.; Sidharta, I.; Yuniarto, M.N.; Rusli, M.R. Thermal analysis simulation of parallel cell in modular battery pack for electric vehicle application. *J. Phys. Conf. Ser.* **2020**, *1517*, 012023. [CrossRef]
31. Divakaran, A.M.; Hamilton, D.; Manjunatha, K.; Minakshi, M. Design, Development and Thermal Analysis of Reusable Li-Ion Battery Module for Future Mobile and Stationary Applications. *Energies* **2020**, *13*, 1477. [CrossRef]
32. Saw, L.H.; Ye, Y.; Tay, A.A.O. Electrochemical–thermal analysis of 18650 Lithium Iron Phosphate cell. *Energy Convers. Manag.* **2013**, *75*, 162–174. [CrossRef]
33. Rao, Z.; Qian, Z.; Kuang, Y.; Li, Y. Thermal performance of liquid cooling based thermal management system for cylindrical lithium-ion battery module with variable contact surface. *Appl. Thermal Eng.* **2017**, *123*, 1514–1522. [CrossRef]
34. Reksowardojo, I.K.; Arya, R.R.; Budiman, B.A.; Islameka, M.; Santosa, S.P.; Sambegoro, P.L.; Aziz, A.R.A.; Abidin, E. Energy Management System Design for Good Delivery Electric Trike Equipped with Different Powertrain Configurations. *World Electr. Veh. J.* **2020**, *11*, 76. [CrossRef]
35. Lu, Z.; Yu, X.; Wei, L.; Qiu, Y.; Zhang, L.; Meng, X.; Jin, L. Parametric study of forced air cooling strategy for lithium-ion battery pack with staggered arrangement. *Appl. Therm. Eng.* **2018**, *136*, 28–40. [CrossRef]
36. Wu, Y.; Li, K.; Wang, J.; Ji, S.; Wang, S. Experimental study and numerical modeling on cylindrical lithium-ion power battery thermal inertia. *Energy Procedia* **2019**, *158*, 4396–4401. [CrossRef]
37. Wahid, M.R.; Budiman, B.A.; Joelianto, E.; Aziz, M. A Review on Drive Train Technologies for Passenger Electric Vehicles. *Energies* **2021**, *14*, 6742. [CrossRef]
38. Arumuga Perumal, D.; Dass, A.K. A Review on the development of lattice Boltzmann computation of macro fluid flows and heat transfer. *Alex. Eng. J.* **2015**, *54*, 955–971. [CrossRef]

39. Zhang, C.; Fakhari, A.; Li, J.; Luo, L.; Qian, T. A comparative study of interface-conforming ALE-FE scheme and diffuse interface AMR-LB scheme for interfacial dynamics. *J. Comput. Phys.* **2019**, *395*, 602–619. [CrossRef]
40. Sheikholeslami, M.; Hayat, T.; Alsaedi, A. Numerical simulation for forced convection flow of MHD CuO-H_2O nanofluid inside a cavity by means of LBM. *J. Mol. Liq.* **2018**, *249*, 941–948. [CrossRef]

Article

A Case Study of Open- and Closed-Loop Control of Hydrostatic Transmission with Proportional Valve Start-Up Process

Paweł Bury [1], Michał Stosiak [1,*], Kamil Urbanowicz [2], Apoloniusz Kodura [3], Michał Kubrak [3] and Agnieszka Malesińska [3]

[1] Department of Technical Systems Operation and Maintenance, Faculty of Mechanical Engineering, Wrocław University of Science and Technology, 50-371 Wrocław, Poland; pawel.bury@pwr.edu.pl

[2] Faculty of Mechanical Engineering and Mechatronics, West Pomeranian University of Technology in Szczecin, 70-310 Szczecin, Poland; kamil.urbanowicz@zut.edu.pl

[3] Faculty of Building Services, Hydro and Environmental Engineering, Warsaw University of Technology, 00-653 Warsaw, Poland; apoloniusz.kodura@pw.edu.pl (A.K.); michal.kubrak@pw.edu.pl (M.K.); agnieszka.malesinska@pw.edu.pl (A.M.)

* Correspondence: michal.stosiak@pwr.edu.pl

Abstract: This paper concerns the start-up process of a hydrostatic transmission with a fixed displacement pump, with particular emphasis on dynamic surplus pressure. A numerically controlled transmission using a proportional directional valve was analysed by simulation and experimental verification. The transmission is controlled by the throttle method, and the variable resistance is the throttling gap of the proportional spool valve. A mathematical description of the gear start-up process was obtained using a lumped-parameters model based on ordinary differential equations. The proportional spool valve was described using a modified model, which significantly improved the performance of the model in the closed-loop control process. After assuming the initial conditions and parameterization of the equation coefficients, a simulation of the transition start-up was performed in the MATLAB–Simulink environment. Simulations and experimental studies were carried out for control signals of various shapes and for various feedback from the hydraulic system. The pressure at the pump discharge port and the inlet port of the hydraulic motor, as well as the rotational speed of the hydraulic motor, were analysed in detail as functions of time. In the experimental verification, complete measuring lines for pressure, speed of the hydraulic motor, flow rate, and temperature of the working liquid were used.

Keywords: hydrostatic transmission; hydrostatic transmission start up; hydraulic drive

Citation: Bury, P.; Stosiak, M.; Urbanowicz, K.; Kodura, A.; Kubrak, M.; Malesińska, A. A Case Study of Open- and Closed-Loop Control of Hydrostatic Transmission with Proportional Valve Start-Up Process. *Energies* **2022**, *15*, 1860. https://doi.org/10.3390/en15051860

Academic Editors: Artur Bartosik and Helena M. Ramos

Received: 4 February 2022
Accepted: 1 March 2022
Published: 3 March 2022

Publisher's Note: MDPI stays neutral with regard to jurisdictional claims in published maps and institutional affiliations.

1. Introduction

In the case of heavy working machinery, some actuators often require low rotational speed values, ranging from a few to tens of rotations per minute. A crane rotation mechanism is an example of such equipment.

Hydrostatic drive units for low rotational speed movement can be constructed using two methods: with a high-speed hydraulic motor combined with an additional mechanical transmission, or with a low-speed motor coupled directly to the driven mechanism. In practice, however, a solution based on a hydrostatic transmission with a high-speed motor and mechanical transmission is used, as this is the only solution that can be applied in the case of a crane rotation mechanism [1]. This results from the need to comply with industry-specific legal regulations, which require using a mechanical brake to securely block the rotation mechanism in the event of external forces (wind, sloping ground within admissible limits, etc.). When a direct drive with a low-speed motor is used, the dimensions of the mechanical brake are quite large, particularly in the case of high load values.

The designer of the drive unit, apart from basic parameters, such as output power, speed range of the driven element, efficiency [2], etc., has to ensure specific dynamic characteristics corresponding to the nature of the designed machinery [3]. In some cases, dynamic surplus is not allowed; for example, in the case of CNC machining tools, where the tool must be positioned right next to the machined surface without overshoot of the tool position.

Furthermore, the requirements applicable to machinery in dynamic states are constantly increasing. Modern machine and equipment evaluation criteria have recently been expanded to include vibration- and noise-level criteria, especially for hydrostatic drive machines [4]. Apart from their well-known advantages, hydrostatic drive units have a significant drawback—they generate relatively high noise emission levels, a factor which can disqualify this type of drive by causing them to exceed the standard noise emission levels (which are being gradually reduced), determined by ergonomic considerations [5–8]. It also means the necessity to reduce the risk of cavitation in hydrostatic systems [9] and the need for its proper modelling. However, limiting the maximum pressure during start-up will result in a reduction in the global noise emission level generated by the transmission in a transient state [10]. On the other hand, reducing the time during which the maximum pressure is generated will result in shortening of the time of the maximum noise levels of the unit during the start-up of the transmission. For a number of years, there has been an increasing tendency to reduce energy losses in mechanical systems, including hydraulic ones [11]. An increase in the efficiency of hydraulic systems can be achieved through control and adjustment of the hydraulic elements (valves, pumps, receivers), the use of systems featuring energy recuperation or hybrid systems, and reduction in dynamic loads [12]. Specific solutions make it possible to increase the efficiency of hydrostatic systems by several tens of percent. In the case of systems offering energy recuperation, the consumption of energy can be reduced by approx. 30% [13]. In hydrostatic systems, this can be achieved in a number of ways, depending on whether the system features fixed or variable displacement pumps. In systems with fixed displacement pumps controlled by throttle methods, the predominant approach is to limit the operation of the safety valve (where serial throttle control is used). The introduction of proportionally controlled valves (a proportional relief valve or a proportional spool valve) can also result in a reduction in energy consumption by the hydrostatic system, particularly in a transient state [14,15].

Additionally, dynamic surpluses (pressure and speed) occurring within the system contribute to excessive wear of the system elements and reduce the uptime [16].

Hydrostatic Transmission Control Methods

There are two basic methods for controlling a hydrostatic transmission in a hydrostatic drive unit: a throttle method and a volumetric method [17]. The throttle method involves an intentional modification of the flow resistance value (e.g., via an adjustable throttle valve) to regulate the value of the usable flow supply to the receiver. There are two types of control methods depending on the positioning of the adjustable throttle valve relative to the receiver—serial or parallel. In practice, the variable throttling gap is often obtained on the gaps of the spool–sleeve couple of the proportional spool valve. In the serial throttle control method, the use of a proportional valve for the purpose of throttling the usable flow supply to the receiver is justified by the fact that this solution enables changing the direction of the movement of the hydraulic receiver, among other factors. The continuous volumetric control method involves using a variable displacement pump and/or receiver.

This paper analyses a drive unit based on an M2C 1613 high-speed, hydraulic gear motor and an OH-500 three-stage planetary gear with a total gear ratio of $i = 69.7$. The paper investigates the impact of the parameters of the control signal for the proportional spool valve on the waveforms of the pressure and speed of the transmission, with a particular emphasis on the transient state during start-up. The transmission start-up process was examined according to the serial throttle control method. Additionally, a system featuring feedback from the speed of the hydrostatic motor and a PI numerical control system was analysed.

In the research on the control of hydraulic systems using proportional spool valve, many approaches can be found on how to model the opening characteristics of a hydraulic distributor. Very often, a linear or quadratic relationship between the spool displacement and the flow rate at a given differential pressure is adopted [18,19]. These methods often give satisfactory results for basic analysis of control systems that rarely operate at small valve openings. Unfortunately, when analysing closed-loop control systems, this approach can produce simulation results that differ significantly from experiment, which makes a proper machine control design process difficult.

Another method found in the literature is a very detailed modelling of the spool shape, which allows one to determine the cross-section of the orifice responsible for the flow [20]. This approach gives significantly better results than those of assuming a linear relationship; however, it does not take into account the phenomena associated with flow through a variable orifice. Additionally, such a spool modelling process requires one to obtain detailed documentation from the valve manufacturer, or the disassembly and detailed measurement of the spool and sleeve.

In this paper, a different method of spool modelling is presented, which does not require any additional documentation of the valve beyond the opening characteristic curve typically provided by the manufacturer.

It was assumed that the characteristic curve can be approximated with sufficient accuracy by a polynomial equation. A similar approach was presented by the authors of paper [21], but they did not present the implementation of this solution in the control system.

Figure 1 shows a diagram of a serial throttle control system for a hydrostatic transmission based on the use of a proportional valve, while Figure 2 shows schematics of the analysed control systems.

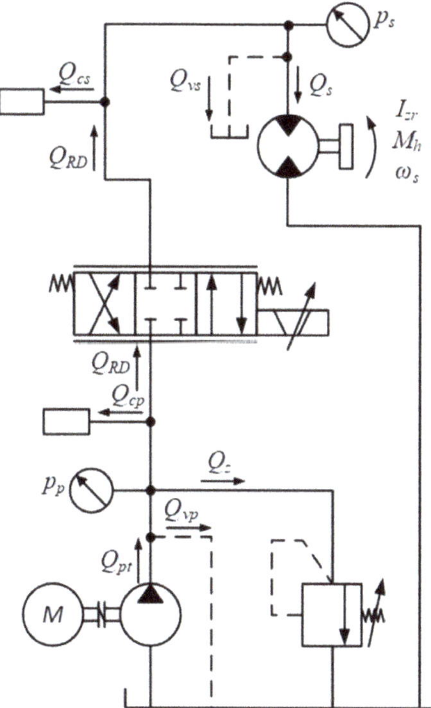

Figure 1. Diagram of the hydraulic system of a hydrostatic transmission with serial throttle control.

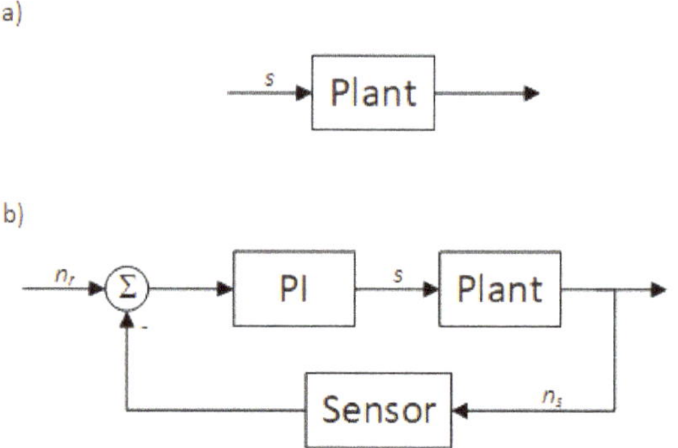

Figure 2. Diagrams of investigated control systems: (**a**) open-loop system; (**b**) closed-loop system.

2. Mathematical Model for the Serial Throttle Control Method

A mathematical description of the start-up of a hydrostatic transmission with serial throttle control using a proportional valve was obtained based on a set of ordinary differential equations (a model with focused parameters). One of the equations for the model is the flow continuity equation at particular points of the hydraulic circuit, and the other is the equation of the equilibrium of torque values on the shaft of the hydrostatic motor [22].

In order to solve this set of equations, it is also necessary to formulate the initial conditions.

In the presented mathematical model, the following simplifying conditions were adopted (among others):

- The temperature of the medium, and consequently its viscosity, remains constant throughout the simulation;
- The pressure has no effect on the viscosity of the medium;
- The compressibility of the medium and the deformability of the hydraulic elements was reduced to the concentrated capacitance at particular points of the system;
- There is no air in the system;
- There is no backlash in the mechanical system;
- The load of the motor is focused on its shaft (inertia);
- There are no wave phenomena;
- There are no external leaks in the system;
- The speed of the electric motor driving the pump is constant and is independent of the pump load;

The flow continuity equation can be formulated as follows:

$$Q_{pt} = Q_{vp} + Q_{cp} + Q_{RD} + Q_z \tag{1}$$

The throttling valve flow rate is determined as follows:

$$Q_{RD} = Q_s + Q_{vs} + Q_{cs} \tag{2}$$

$$Q_{RD} = G_{RD}\sqrt{p_p - p_s - p_d} \tag{3}$$

In a system with a proportional flow valve, it is usually assumed that there is a linear dependency between the area of the surface through which the liquid flows and the displacement of the spool. Unfortunately, this simplification does not work in simulations of systems where even the slightest displacement of the spool is a significant factor.

In the model, it was assumed that the aforementioned relation is a polynomial function (4), whose degree and values of coefficients were selected based on the characteristics listed in the catalogue specification provided by the manufacturer of the spool valve. The result of model matching has been presented in Figure 3.

$$s_m = A_3 s^3 + A_2 s^2 + A_1 s + A_0 \tag{4}$$

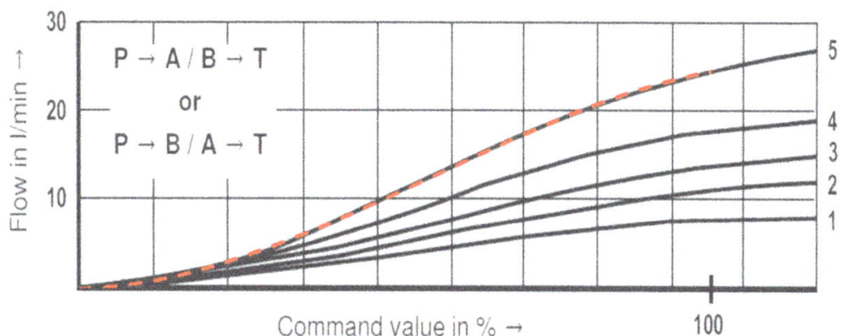

Figure 3. Opening characteristic curve of the spool valve from the data sheet with a fitted curve of model based on Equation (4). Curves 1–5 correspond to different valve pressure differential, the red dashed line is the fitted curve.

In the system with a proportional flow valve, it was assumed that the dynamics of the proportional flow valve are characterised by the following first-order differential equation, identical to the first-order inertial element:

$$s_m \cdot G_{RDmax} = G_{RD} + T_{RD} \frac{dG_{RD}}{dt} \tag{5}$$

The flow through the hydraulic motor is described by the following equation:

$$Q_s = q_s \omega_s \tag{6}$$

The flow caused by the compressibility of the working medium [23] and the deformation of the elements of the system was assumed to be as follows:

- On the section from the pump to the proportional spool valve:

$$Q_{cp} = c_p \frac{dp_p}{dt} \tag{7}$$

- On the section from the proportional spool valve to the motor:

$$Q_{cs} = c_s \frac{dp_s}{dt} \tag{8}$$

Losses caused by leakages in the pump and in the motor can be described linearly with the following equations:

$$Q_{vp} = a_{vp} p_p \tag{9}$$

$$Q_{vs} = a_{vs} p_s \tag{10}$$

The pressure drop caused by the total loss of pressure resulting from viscous drag values (Hagen–Poiseuille equation) as well as the turbulent flow (Bernoulli's equation) were modelled using the following relation:

$$p_d = \frac{8\mu L Q_{RD}}{\pi R^4} + \sum_j \zeta_j \frac{\rho}{2} \left(\frac{Q_{RD}}{\pi R^2} \right)^2 \tag{11}$$

The equation of the flow through the safety valve can be presented in the following form [24]:

$$Q_z(t) = \begin{cases} h_z(p_p - p_0) - T_z\frac{dQ_z}{dt} & \text{for} \quad p_p > p_0 \\ 0 - T_z\frac{dQ_z}{dt} & \text{for} \quad p_p \leq p_0 \end{cases} \tag{12}$$

In the analysed case, the motor load torque consists of three components: the constant one coming from static friction, the one coming from viscous friction in the motor and gearbox, and the moment of inertia. The torque value equilibrium condition on the hydrostatic motor shaft is described by the following relation:

$$q_s p_s = M_b + f\omega_s + I_{zr}\frac{d\omega_s}{dt} \tag{13}$$

To solve the above equations, the following initial conditions were assumed (slightly different from those found in the literature):

$$p_p(0) = p_0 + \frac{Q_{pt}}{h_z} \tag{14}$$

$$p_s(0) = 0 \tag{15}$$

$$Q_z(0) = Q_{pt} - a_{vp}p_0 \tag{16}$$

$$\omega_s(0) = 0 \tag{17}$$

The limit condition for the hydrostatic motor was defined as follows:

$$\text{if } q_s p_s \leq M_b, \text{ then } \omega_s = 0, \frac{d\omega_s}{dt} = 0 \tag{18}$$

Solving the above equations numerically requires their parametrization; this was carried out based on the catalogue data and the information found in the literature. However, the available literature does not specify the value of some of the coefficients for the equations; therefore, experiments were conducted in order to determine the friction coefficient of the hydraulic motor and of the coupled planetary gear.

In hydrostatic drive units, damping is caused predominantly by internal leakage, resistance related to the flow of the working medium, and friction forces caused by the movement of the hydraulic motor and the driven mechanism. In the dynamic model of hydrostatic transmission, leakages are taken into account in the flow balance equation, while the resistance of the movement of the hydraulic motor and of the coupled mechanism (independent of the speed in the case of Coulomb friction and linearly dependent on the speed in the case of viscous friction) are described via the equation of the equilibrium of the torque values acting on the shaft of the motor. The viscous friction coefficients of the hydraulic motor and of the coupled planetary gear were determined by measuring the resistance to idle running motion as a function of the angular velocity of the shaft of the motor. The pressure differences, p_s, in the connection pipes of the hydraulic motor were taken as the measure of the aforesaid resistance values. Figure 4 shows the relation p_s as a function of the angular velocity, ω_s, of idle running. As the graph of this function shows, it is a linear relation, which—with accuracy sufficient for practical purposes—can be approximated with a straight line, which confirms the assumption of viscous friction.

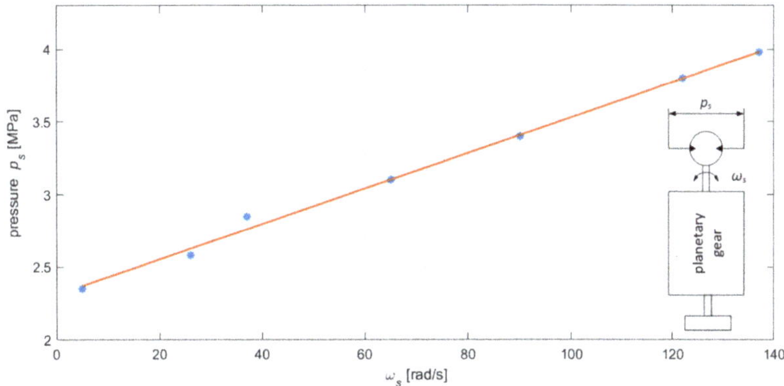

Figure 4. Resistance to motion of the gear motor M2C1613 and of the coupled planetary gear OH-500 as a function of angular velocity, ω_s, of idle running.

The findings presented here pertain to an M2C1613 gear motor coupled with an OH-500 planetary gear. The planetary gear was filled with SAE 85W90 transmission oil, whose temperature was kept within the range $t_1 = 25 - 28\,^{\circ}\mathrm{C}$ during the measurements, and the hydraulic motor was supplied with ISO-VG 46 hydraulic oil at a temperature of $t_2 = 40\,^{\circ}\mathrm{C} \pm 2\,^{\circ}\mathrm{C}$.

Based on the presented measurement results, one can determine the value of the viscous friction torque, and, in consequence, also the coefficient f of that friction, according to the following equation:

$$f = \frac{(p_s - p_c)q_s}{\omega_s} \tag{19}$$

Considering the specific absorptivity of the motor ($q_s = 5.03 \times 10^{-6}\,[\mathrm{m}^3/\mathrm{rad}]$), the viscous friction coefficient—determined based on the data presented in Figure 4, in accordance with Equation (19)—was $f = 6.3 \times 10^{-2}\,\left[\mathrm{N} \cdot \mathrm{m} \cdot \mathrm{s}/\mathrm{rad}^2\right]$.

Due to the specific nature of the rotational speed measurement, it was assumed in the model that the speed measurement system can be described using a first-order inertial element with the following transmittance:

$$G_n(s) = \frac{1}{T_n s + 1} \tag{20}$$

Once the equations of the mathematical model had been parametrized and the initial conditions adopted, it was possible to solve the model numerically, and subsequently to present in graphical form the pumping pressure of the pump, p_p, the pressure on the motor, p_s, and the angular velocity of the motor's shaft, n_s, over time, for various waveforms of the control signal, s, supplied to the coils of the proportional electromagnet, as described by Equation (21) and presented in graphical form in Figure 5:

$$s = \begin{cases} s_0 + \frac{s_{\max} - s_0}{t_0}t & \text{for } 0 < t < t_0 \\ s_{\max} & \text{for } t > t_0, \end{cases} \tag{21}$$

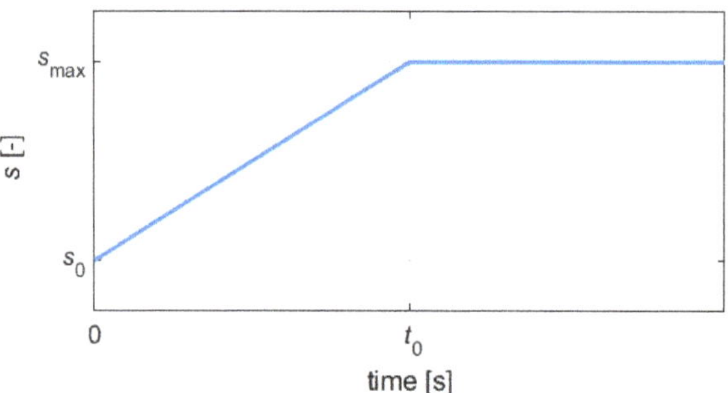

Figure 5. Waveform of the control signal, s, for the proportional valve.

The coefficient s_0 was taken in such a form as to compensate for the idle stroke of the spool of the proportional spool valve resulting from the stationary overlap, s_{max} was assumed to be the maximum value of the signal as specified in the specification sheet of the spool valve, and t_0 is the signal rise time from s_0 to s_{max}.

Figures 6–8 show the waveforms obtained by solving the mathematical models for, respectively, the pressure in the discharge flange of the pump p_p, the pressure at the inlet port of the motor p_s, and the rotational speed of the motor shaft n_s.

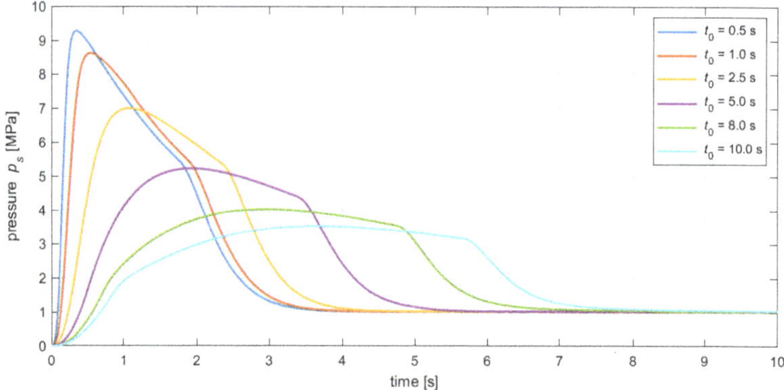

Figure 6. Result of simulation of the waveform of pressure on the motor p_s for various signal rise times, t_0.

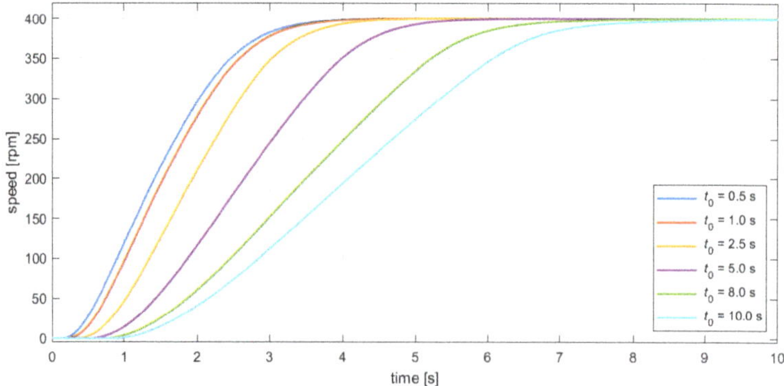

Figure 7. Result of simulation of the waveform of rotational speed n_s for various signal rise times, t_0.

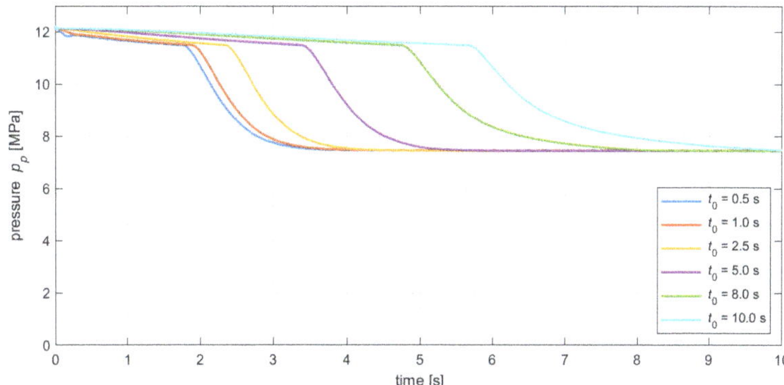

Figure 8. Result of simulation of the waveform of pressure on the pump p_p for various signal rise times, t_0.

The simulations were carried out for various values of the ramp time (edge rise time) $t_0 = \{\, 0.5\,\text{s}, 1\,\text{s}, 2.5\,\text{s}, 5\,\text{s}, 8\,\text{s}, 10\,\text{s} \,\}$.

For the obtained results, the following parameters enabling assessment of the transitional state of the transmission were determined:

- Dynamic surplus pressure $p_{s\,\text{max}}$ as the maximum pressure recorded during the start-up phase;
- Start-up time t_s as the time after which the rotational speed reaches 95% of the set value from the moment of supplying the control signal;
- Reaction time t_r as the time after which the speed of the motor reaches 5% of the set value;
- Energy E_s generated by the pump during the start-up process (first 10 s).

Table 1 presents a comparison of the parameters of the control signal along with the above-described values of the parameters for assessment of the transition state.

Table 1. Comparison of the control signal parameters and dynamic surplus pressure values at the inlet port of the motor, and the transmission start-up duration and reaction time. Transmission controlled with the serial throttle method without feedback.

t_0 [s]	$p_{s\,\text{max}}$ [MPa]	t_s [s]	t_r [s]	E_s [Wh]
0.5	9.29	3.00	0.48	5.58
1.0	8.63	3.12	0.58	5.62
2.5	6.99	3.61	0.81	5.76
5.0	5.23	4.60	1.13	6.08
8.0	4.03	5.91	1.44	6.54
10	3.53	6.87	1.65	6.87

Analysing the simulation results obtained for the open control system (Figures 6–8 and Table 1), we can observe that for a signal rise time below $t_0 < 5\,\text{s}$ the start-up time, t_s, varies in a small range. This is directly due to the performance of the system, specifically the large moment of inertia and the maximum pressure set on the relief valve. Additionally, Table 1 shows the energy generated by the pump during start-up process. It can be observed that increasing the rise time of the signal increases the energy used for start-up. This is due to the fact that the pressure on the pump remains high for a longer period of time when the signal rise time is increased.

As the next step, the control system was equipped with a feedback loop from the angular velocity on the shaft of the hydraulic motor of the tested transmission. The following relation describes the transmittance of the PI controller used:

$$G_{PI}(s) = K_P\left(1 + \frac{1}{T_I s}\right) \tag{22}$$

The linear rise from $n_0 = 0$ rpm to the set value of $n_{max} = 200$ rpm was taken to be the signal of the set value to the controller, and $t_0 = 10$ s was taken to be the signal rise time. The following is the mathematical description of the signal of the set value:

$$n_r = \begin{cases} n_0 + \frac{n_{max} - n_0}{t_0}t & \text{for} & 0 < t < t_0 \\ n_{max} & \text{for} & t > t_0, \end{cases} \tag{23}$$

A series of simulations for various values of parameters of the PI controller was conducted for the assumed control signal. To assess the state of adjustment, apart from the above-mentioned parameters, two additional parameters were introduced:

- Overshoot parameter described by the following relation:

$$\kappa_n = \frac{\max\{n\}}{n_{max}} \tag{24}$$

- Steady-state error for the ramp input measured at the end of the ramp signal:

$$e_r = n_r - n_s \tag{25}$$

Table 2 presents a comparison of the parameters of the PI controller for which the simulation results are presented, as well as the determined values of the assessment parameters.

Table 2. Comparison of the parameters of the PI controller and the obtained values of the dynamic surplus pressure on the motor, the start-up time, the reaction time, and the overshoot.

K_P	T_I	$p_{s\,max}$ [MPa]	t_s [s]	t_r [s]	κ_n [%]	e_r [rpm]
0.020	0.50	2.70	9.72	1.49	102.7	4.4
0.020	0.75	2.34	9.84	1.63	101.8	6.4
0.020	1.00	2.15	9.94	1.73	100.9	8.8
0.015	0.50	2.51	9.79	1.70	102.7	5.7
0.015	0.75	2.19	9.94	1.87	101.6	8.6
0.015	1.00	2.03	10.09	2.00	100.5	11.5
0.010	0.50	2.28	9.93	2.05	102.6	8.5
0.010	0.75	2.04	10.16	2.28	101.2	12.9
0.010	1.00	1.91	10.39	2.45	-	17.4

The diagrams in Figures 9 and 10 illustrate the significant impact of the control and adjustment parameters on the waveforms of the pressure on the pump and on the motor, as well as the speed on the hydrostatic motor shaft.

Figures 11 and 12 show simulation results for different moments of inertia of the system in the range $I = 0.5 - 2.5 I_{zr}$. The presented model corresponds to other rotating systems of machines in which the moment of inertia changes significantly by changing the value and position of the load. In addition, the changes of the static load are insignificant, so it is assumed in the simulation that they are constant.

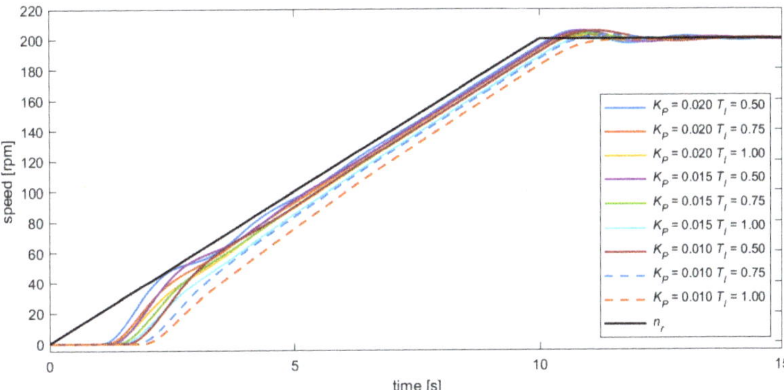

Figure 9. Result of simulation of the waveform of rotational speed, n_s, for various parameters of the PI controller.

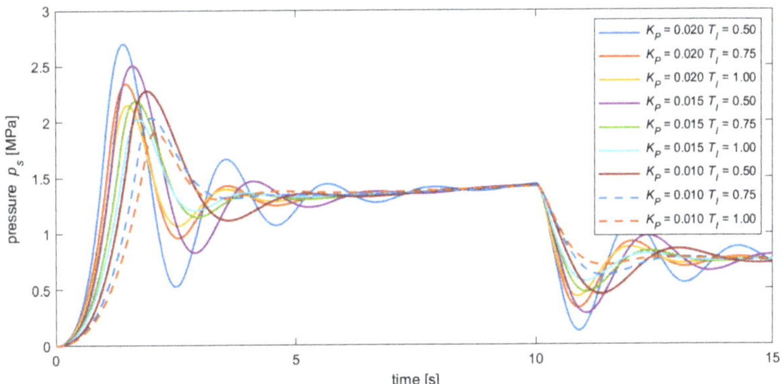

Figure 10. Result of simulation of the waveform of pressure on the motor p_s for various parameters of the PI controller.

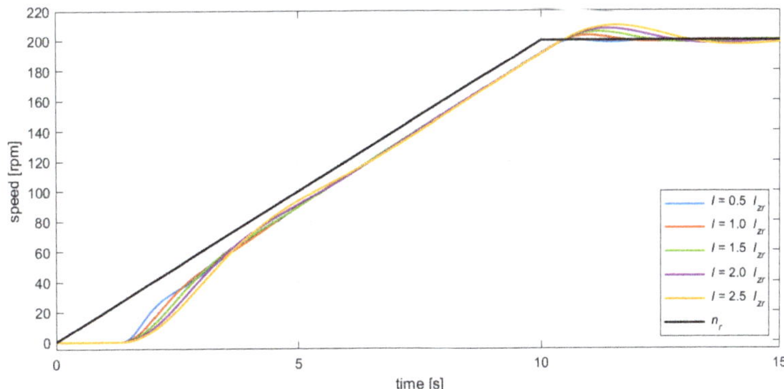

Figure 11. Result of simulation of the waveform of rotational speed, n_s, for various moments of inertia of the system ($K_P = 0.015$, $T_I = 0.75$ s).

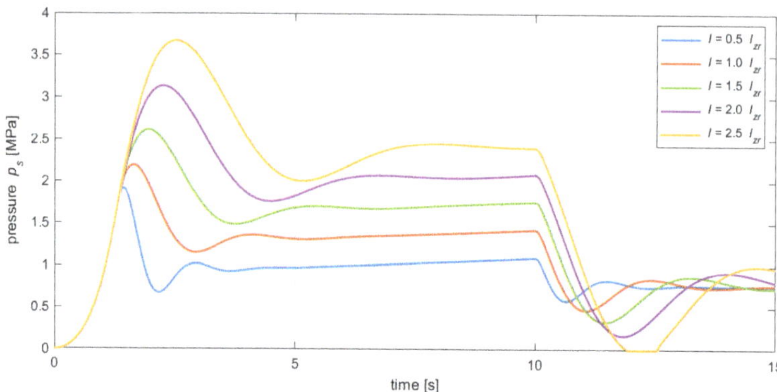

Figure 12. Result of simulation of the waveform of pressure on the motor p_s for various moments of inertia of the system ($K_P = 0.015$, $T_I = 0.75$ s).

3. Experimental Verification of the Mathematical Model

Experimental studies were conducted to verify the mathematical model, the assumed values of the parameters, and the assumptions. The experimental studies also made it possible to determine the relation between the shape (parameters) of the control signal for the proportional spool valve and the dynamic surplus pressure (at the inlet port of the motor) and the speed on the shaft of the motor. Additionally, the duration of the start-up process was analysed for various waveforms of the control signal for the proportional spool valve.

Figure 13 shows the test stand that was used to verify the model. The hydraulic power source is based on an axial variable displacement piston pump (flow was set to 14 L/min). The system is protected by a direct operating relief valve (nominal flow 25 L/min). A direct proportional spool valve (nominal flow 16 L/min) with spool position feedback was used to control the gear motor (size 32 cm^3/rev). The control signal was generated using a multifunction DAQ device with analogue I/O (NI USB-6001) connected to a PC with dedicated LabVIEW-based software. The measurement system was based on a 16-bit recorder (Hydrotechnik Multi System 8050) and piezoresistive pressure transducers (Hydrotechnik HySense PR100). The speed was measured using a tachometer (PZO E2/CPPB4).

Figure 13. The hydrostatic transmission test stand: 1—planetary gear; 2—coupling housing; 3—hydraulic gear motor; 4—hydraulic power source.

Figures 14–16 show the waveforms obtained during the start-up of the transmission controlled by the serial throttle method with a proportional spool valve.

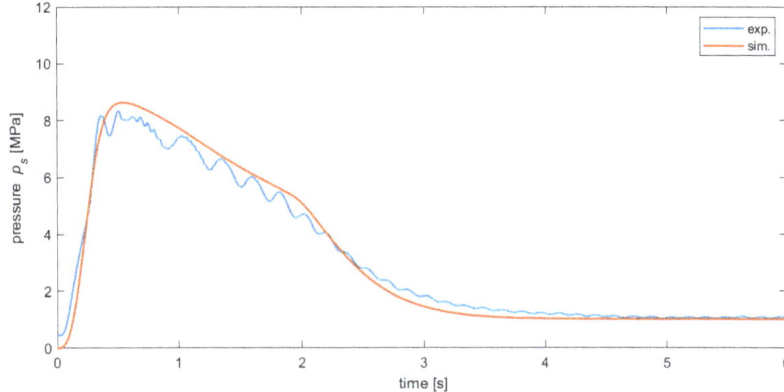

Figure 14. Measurement and result of simulation of the waveform of pressure on the motor p_s for the control signal rise time $t_0 = 1$ s.

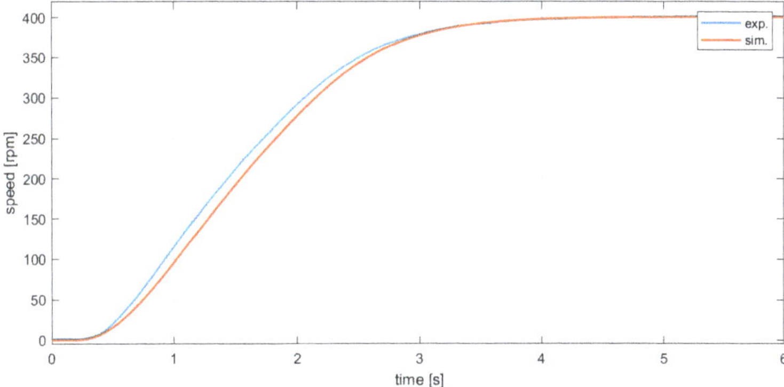

Figure 15. Measurement and result of simulation of the waveform of rotational speed on the motor, n_s, for the control signal rise time $t_0 = 1$ s.

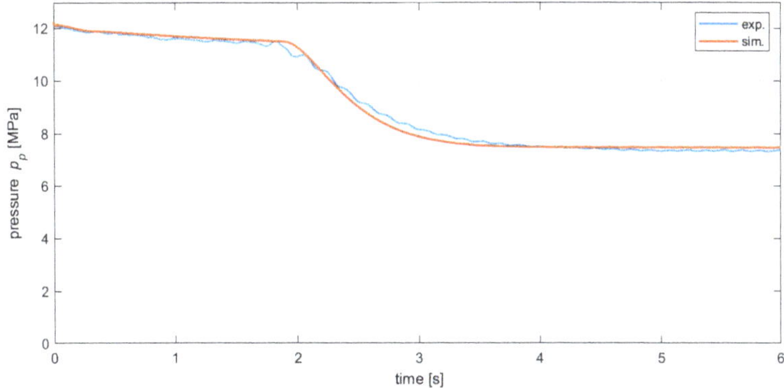

Figure 16. Measurement and result of simulation of the waveform of pressure on the pump p_p for the control signal rise time $t_0 = 1$ s.

In Figure 14, for the simulation time $t = 0$, the difference in pressure between the experiment and the result of simulation can be noticed. This difference is caused by the residual pressure remaining in the line from directional spool valve to the motor.

Figure 17 presents the result of two simulations compared with the experimental run. The simulation labelled "sim." is the result of the simulation including the opening characteristic fit (Equation (4)). In addition, the simulation labelled "sim.*" is shown, which assumes a linear dependence of flow on slider displacement ($s = s_m$), as in [18]. In the first control phase, a significant difference between the responses of different modelling approaches can be observed, which has a significant impact on the evaluation of the control dynamics.

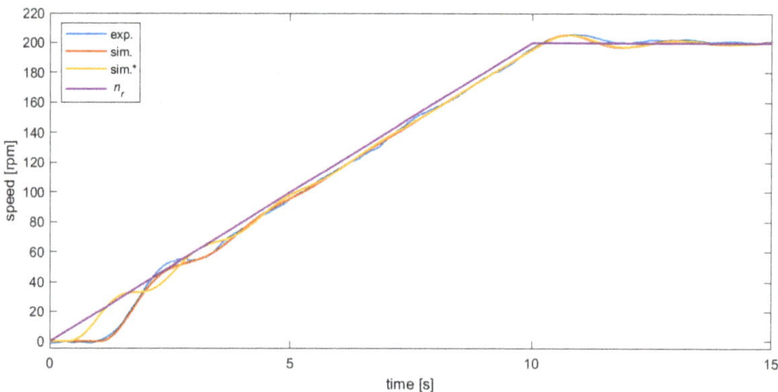

Figure 17. Measurement and result of simulation of the waveform of speed on the motor n_s for the control system ($K_P = 0.02$, $T_I = 0.5$ s). Waveforms labelled "sim." and "sim.*" shows the difference between simulation results with and without the modified spool valve model (Equation (4)).

The experimental verification studies demonstrate that the simulations satisfactorily model the actual transmission. This enables further work intended to optimise the parameters and the waveform of the control signal, which will result in a reduction in the dynamic surplus of the selected parameter (e.g., pressure at the inlet port of the motor) while maintaining control of the duration of the transitional process.

4. Conclusions

The paper describes the start-up process of a hydrostatic transmission controlled with a serial throttle. Additionally, it includes analyses of a transmission installed in a system equipped with a PI regulator controller. Based on the resulting waveforms over time, it can be concluded that during the start-up phase of an open-loop-controlled transmission with the serial throttle method, the positive-displacement pump operates under high load regardless of the ramp time of the proportional spool valve control signal. The pump high load time is correlated with the start-up time and the pressure waveform on the motor. However, the high load time of the pump can be reduced by suitably adjusting the parameters of the control signal supplied to the coils of the proportional spool valve. It was demonstrated that the maximum pressure value at the inlet port of the hydraulic motor during its start-up can be modified by adjusting the shape of the proportional spool valve control signal. In this way, it is also possible to reduce the noise generated by the transmission during start-up [25,26]. Therefore, reduction in the maximum pressure is an effective method for reducing the noise emission levels of a hydrostatic transmission in both transient and steady states. By adjusting start-up parameters (the control signal waveform), it is possible to regulate the maximum pressure value, the start-up time, and the reaction time. The start-up process can be also modified by selecting the shape of the spool of the proportional spool valve (e.g., symmetrical, asymmetrical) or by adjusting the stationary overlap and operational overlap, but these methods are not analysed in this paper. These methods also result in reduction in the load of the elements of the transmission, which extends its operational life.

The method of modelling the hydraulic system presented in this paper, in particular the method of modelling the opening characteristics of the proportional spool valve, allows improving the simulation results obtained. The detailed construction of the model and conducting simulations enable initial adjustment and analysis of the control or regulation parameters prior to their application on the target machine. This is extremely important because it reduces the risk of errors during the prototype start-up phase, and therefore it reduces the time and costs of prototyping.

The presented detailed model for the start-up of the hydrostatic transmission equipped with a proportional valve can be used for the purpose of optimising the start-up process in terms of the execution of the objective function for the following parameters: dynamic surplus of a selected value, start-up time, reaction time, or the energy efficiency of the system.

Author Contributions: Conceptualization, M.S., P.B. and K.U.; methodology, M.S. and P.B; software, M.S. and P.B.; validation, M.S. and P.B.; formal analysis, M.S., P.B. and K.U.; investigation, M.S., P.B., A.K, M.K. and K.U.; resources, M.S. and P.B.; data curation, P.B.; writing—original draft preparation, M.S. and P.B; writing—review and editing, M.S., P.B., A.K., M.K. and A.M.; visualization, P.B., M.S. and A.M.; supervision, M.S. and P.B.; project administration, M.S.; funding acquisition, M.S. All authors have read and agreed to the published version of the manuscript.

Funding: This research received no external funding.

Institutional Review Board Statement: Not applicable.

Informed Consent Statement: Not applicable.

Data Availability Statement: All coded generated during the study and experimental data are available from the corresponding author by request.

Conflicts of Interest: The authors declare no conflict of interest.

Nomenclature

a_{vp}	pump leakage coefficient $[\mathrm{m}^3/\mathrm{Pa\cdot s}]$
a_{vs}	motor leakage coefficient $[\mathrm{m}^3/\mathrm{Pa\cdot s}]$
A_0, A_1, A_2, A_3	coefficients of the polynomial function describing the opening of the spool valve $[-]$
c_p	capacitance of the liquid and of the pipes on the section from the pump to the proportional spool valve $[\mathrm{m}^3/\mathrm{Pa}]$
c_s	capacitance of the liquid and of the pipes on the section from the proportional spool valve to the motor $[\mathrm{m}^3/\mathrm{Pa}]$
e_r	steady-state error to ramp input $[\mathrm{rpm}]$
E_s	energy generated by the pump during the first 10 s $[\mathrm{Wh}]$
f	viscous friction coefficient $[\mathrm{N\cdot m\cdot s/rad}^2]$
G_n	transmittance of the rotational speed measuring system
G_{PI}	PI controller transmittance
G_{RD}	conductivity of the spool valve $\left[\mathrm{m}^3/\mathrm{s}\sqrt{\mathrm{Pa}}\right]$
$G_{RD\max}$	maximum conductivity of the spool valve $\left[\mathrm{m}^3/\mathrm{s}\sqrt{\mathrm{Pa}}\right]$
h_z	amplification factor of the relief valve $[\mathrm{m}^3/\mathrm{Pa\cdot s}]$
i	gear ratio of the planetary gear $[-]$
j	summation index $[-]$
s_0	minimum value of the control signal $[-]$
$s_{\max}$	maximum value of the control signal $[-]$
I_{zr}	reduced mass moment of inertia of rotational masses $[\mathrm{kg\cdot m}^2]$
K_P	proportional gain of the PI controller $[-]$
M_b	braking torque on the shaft of the hydrostatic motor $[\mathrm{N\cdot m}]$
L	total length of the pipes between the pump and the motor $[\mathrm{m}]$

n	rotational speed of the hydrostatic motor [rpm]
n_0	initial rotational speed in the controlled system [rpm]
n_{max}	target rotational speed in the controlled system [rpm]
n_r	current rotational speed set point in the controlled system [rpm]
n_s	rotational speed on the shaft of the hydrostatic motor [rpm]
p_0	opening pressure of the relief valve [Pa]
p_d	total pressure drop resulting from flow rate losses [Pa]
p_p	pressure on the pump [Pa]
p_s	pressure on the hydraulic motor [Pa]
$p_{s\,max}$	dynamic surplus pressure on the motor [Pa]
p_c	pressure at the hydraulic motor corresponding to static resistance values (for $\omega \approx 0$) [Pa]
R	inner radius of the hydraulic pipes [m]
s	control signal [−]
s_m	modified control signal [−]
t	simulation time (step) [s]
t_0	control signal edge rise time [s]
t_1	oil temperature in the planetary gear [°C]
t_2	oil temperature in the hydraulic system [°C]
t_r	transmission reaction time [s]
t_s	transmission start-up time [s]
T_I	time constant of the integrating element of the PI controller [s]
T_n	time constant of the system for measuring the rotational speed [s]
T_{RD}	time constant of the spool valve [s]
T_Z	time constant of the relief valve [s]
q_s	displacement of the hydraulic motor [m^3/rad]
Q_{pt}	theoretical pump output flow [m^3/s]
Q_{vp}	flow value resulting from losses in the pump [m^3/s]
Q_Z	flow through the relief valve [m^3/s]
Q_{RD}	flow through the proportional valve [m^3/s]
Q_{cp}	flow caused by compressibility in volume between the pump and the spool valve [m^3/s]
Q_{cs}	flow caused by the compressibility in volume between the proportional spool valve and the hydraulic motor [m^3/s]
Q_s	flow towards the hydraulic motor [m^3/s]
Q_{vs}	flow value resulting from losses in the hydraulic motor [m^3/s]
ζ	coefficient of pressure losses caused by turbulent flow [−]
κ_n	speed overshoot in a PI controlled system [%]
μ	dynamic viscosity of the working medium [N · s/m^2]
ρ	density of the working medium [kg/m^3]
ω_s	angular velocity of the hydrostatic motor [rad/s]

References

1. Pullen, K.R.; Dhand, A. Mechanical and Electrical Flywheel Hybrid Technology to Store Energy in Vehicles. In *Alternative Fuels and Advanced Vehicle Technologies for Improved Environmental Performance*; Elsevier: Amsterdam, The Netherlands, 2014; pp. 476–504. ISBN 978-0-85709-522-0.
2. Comellas, M.; Pijuan, J.; Potau, X.; Nogués, M.; Roca, J. Efficiency sensitivity analysis of a hydrostatic transmission for an off-road multiple axle vehicle. *Int. J. Automot. Technol.* **2013**, *14*, 151–161. [CrossRef]
3. Xiong, S.; Wilfong, G.; Lumkes, J.J. Components Sizing and Performance Analysis of Hydro-Mechanical Power Split Transmission Applied to a Wheel Loader. *Energies* **2019**, *12*, 1613. [CrossRef]
4. Hubballi, B.; Sondur, V. Noise Control in Oil Hydraulic System. In Proceedings of the 2017 International Conference on Hydraulics and Pneumatics-HERVEX, Băile Govora, Romania, 8–10 November 2017.
5. Liu, J.; Yang, Y.; Suh, S. Hydraulic Fluid-Borne Noise Measurement and Simulation for Off-Highway Equipment. In Proceedings of the 2017 SAE/NOISE-CON Joint Conference (NoiseCon 2017), Grand Rapids, MI, USA, 11 June 2017.
6. Fiebig, W.; Wrobel, J. System Approach in Noise Reduction in Fluid Power Units. In Proceedings of the BATH/ASME 2018 Symposium on Fluid Power and Motion Control, Bath, UK, 12–14 September 2018; American Society of Mechanical Engineers: Bath, UK, 12 September 2018; p. V001T01A025.

7.	Wróbel, J.; Blaut, J. Influence of Pressure Inside a Hydraulic Line on Its Natural Frequencies and Mode Shapes. In *Advances in Hydraulic and Pneumatic Drives and Control 2020*; Stryczek, J., Warzyńska, U., Eds.; Lecture Notes in Mechanical Engineering; Springer International Publishing: Cham, Switzerland, 2021; pp. 333–343. ISBN 978-3-030-59508-1.
8.	Fiebig, W.; Wrobel, J.; Cependa, P. Transmission of Fluid Borne Noise in the Reservoir. In Proceedings of the BATH/ASME 2018 Symposium on Fluid Power and Motion Control, Bath, UK, 12–14 September 2018; American Society of Mechanical Engineers: Bath, UK, 12 September 2018; p. V001T01A026.
9.	Zarzycki, Z.; Urbanowicz, K. Modelling of transient flow during water hammer considering cavitation in pressure pipes. *Inżynieria Chem. I Proces.* **2006**, *27*, 915–933.
10.	Makaryants, G.M.; Prokofiev, A.B.; Shakhmatov, E. Vibroacoustics Analysis of Punching Machine Hydraulic Piping. *Procedia Eng.* **2015**, *106*, 17–26. [CrossRef]
11.	Skorek, G. Study of Losses and Energy Efficiency of Hydrostatic Drives with Hydraulic Cylinder. *Pol. Marit. Res.* **2018**, *25*, 114–128. [CrossRef]
12.	Karpenko, M.; Bogdevičius, M. Review of Energy-saving Technologies in Modern Hydraulic Drives. *Moksl.-Liet. Ateitis* **2017**, *9*, 553–558. [CrossRef]
13.	Karpenko, M.; Bogdevičius, M. The Hydraulic Energy-Saving System Based on the Hydraulic Impact Effect. In Proceedings of the 10th National Conferences "Jūros ir krantų Tyrimai", Palanga, Lithuanian, 26–28 April 2017.
14.	Lin, T.; Chen, Q.; Ren, H.; Miao, C.; Chen, Q.; Fu, S. Influence of the energy regeneration unit on pressure characteristics for a proportional relief valve. *Proc. Inst. Mech. Eng. Part I J. Syst. Control. Eng.* **2017**, *231*, 189–198. [CrossRef]
15.	Domagała, Z.; Kędzia, K.; Stosiak, M. The use of innovative solutions improving selected energy or environmental indices of hydrostatic drives. *IOP Conf. Ser. Mater. Sci. Eng.* **2019**, *679*, 012016. [CrossRef]
16.	Sobczyk, A.; Pobędza, J. Hydraulic Systems Safety by Reducing Operation and Maintenance Mistakes. *Syst. Saf. Hum.-Tech. Facil.-Environ.* **2019**, *1*, 700–707. [CrossRef]
17.	Larsson, L.V. *Control of Hybrid Hydromechanical Transmissions*; Linköping Studies in Science and Technology; Division of Fluid and Mechatronic Systems, Department of Management and Engineering Linköping University: Linköping, Sweden, 2019.
18.	Yao, Z.; Liang, X.; Zhao, Q.; Yao, J. Adaptive Disturbance Observer-Based Control of Hydraulic Systems with Asymptotic Stability. *Appl. Math. Model.* **2022**, *105*, 226–242. [CrossRef]
19.	Acuna, W.; Canuto, E.; Malan, S. Embedded Model Control Applied to Mobile Hydraulic Systems. In Proceedings of the 18th Mediterranean Conference on Control and Automation, MED'10, Marrakech, Morocco, 23–25 June 2010; IEEE: Marrakech, Morocco, 2010; pp. 715–720.
20.	Saleem, A.M.; Alyas, B.H.; Mahmood, A.G. Mathematical Model for a Proportional Control Valve of a Hydraulic System. *Int. J. Mech. Prod. Eng. Res. Dev. (IJMPERD)* **2020**, *10*, 8433–8444.
21.	Tenesaca, J.; Carpio, M.; Saltarén, R.; Portilla, G. Experimental determination of the nonlinear model for a proportional hydraulic valve. In Proceedings of the 2017 IEEE International Autumn Meeting on Power, Electronics and Computing (ROPEC), Ixtapa, Mexico, 8–10 November 2017; pp. 1–4. [CrossRef]
22.	Singh, R.B.; Kumar, R.; Das, J. Hydrostatic Transmission Systems in Heavy Machinery: Overview. *Int. J. Mech. Prod. Eng.* **2013**, *1*, 47–51.
23.	Dasgupta, K.; Mandal, S.; Pan, S. Dynamic analysis of a low speed high torque hydrostatic drive using steady-state characteristics. *Mech. Mach. Theory* **2012**, *52*, 1–17. [CrossRef]
24.	Morselli, S.; Gessi, S., Marani, P.; Martelli, M.; De Hieronymis, C.M.R. Dynamics of pilot operated pressure relief valves subjected to fast hydraulic transient. *AIP Conf. Proc.* **2019**, *2191*, 020116. [CrossRef]
25.	Chenxiao, N.; Xushe, Z. Study on Vibration and Noise for the Hydraulic System of Hydraulic Hoist. In Proceedings of the 1st International Conference on Mechanical Engineering and Material Science, Shanghai, China, 28–30 December 2012; Atlantis Press: Zhengzhou, China, 2012.
26.	Stosiak, M. The impact of hydraulic systems on the human being and the environment. *J. Theor. Appl. Mech.* **2015**, *53*, 409–420. [CrossRef]

MDPI

St. Alban-Anlage 66

4052 Basel

Switzerland

Tel. +41 61 683 77 34

Fax +41 61 302 89 18

www.mdpi.com

Energies Editorial Office

E-mail: energies@mdpi.com

www.mdpi.com/journal/energies